A Story of Us

A Story of Us

A New Look at Human Evolution

Lesley Newson and Peter J. Richerson

OXFORD

UNIVERSITY PRESS

OXFORD
UNIVERSITY PRESS

Oxford University Press is a department of the University of Oxford. It furthers
the University's objective of excellence in research, scholarship, and education
by publishing worldwide. Oxford is a registered trade mark of Oxford University
Press in the UK and certain other countries.

Published in the United States of America by Oxford University Press
198 Madison Avenue, New York, NY 10016, United States of America.

Library of Congress Cataloging-in-Publication Data
Names: Newson, Lesley, author. | Richerson, Peter J., author.
Title: A story of us : a new look at human evolution /
by Lesley Newson & Peter J. Richerson.
Description: New York, NY : Oxford University Press, [2021] |
Includes bibliographical references and index.
Identifiers: LCCN 2020032769 (print) | LCCN 2020032770 (ebook) |
ISBN 9780190883201 (hardback) | ISBN 9780190883225 (epub)
Subjects: LCSH: Human evolution.
Classification: LCC GN281.4 N49 2021 (print) |
LCC GN281.4 (ebook) |DDC 599.93/8-dc23
LC record available at https://lccn.loc.gov/2020032769
LC ebook record available at https://lccn.loc.gov/2020032770

DOI: 10.1093/ oso/ 9780190883201.001.0001

1 3 5 7 9 8 6 4 2

Printed by Sheridan Books, Inc., United States of America

*To our descendants: Scott, Kate, Emily, Sophie, Alex, Peter, Jonah,
and those who haven't been born yet.*

Contents

Acknowledgments

The ideas and information we drew upon to write this book have been thought of, discovered, and shaped by thousands of people. A good portion of those people are mentioned in the notes so that readers can find out more about their work.

But we owe a special debt to the many friends and colleagues who helped shape our ideas of human evolution. First and foremost is Pete's long-time co-conspirator and friend Rob Boyd. The others we want to especially thank are Billy Baum, Bob Bettinger, Don Campbell, Tim Caro, Dwight Collins, Bill Davis, Jerry Edelman, Russ Genet, Alex Haslam, Joe Henrich, Katie Hinde, Sarah Hrdy, Kevin Laland, Stephen Lea, Bill Mason, Richard McElreath, Monique Borgerhoff Mulder, Robert Murphey, Tom Postmes, Joan Silk, John Odling-Smee, Mark Thomas, Colin Tudge, Paul Webley, Bruce Winterhalder, and Andy Whiten. And then there are the graduate students who taught us so much. These include Bret Beheim, Adrian Bell, Lien-Siang Chou, Viken Hillis, Nicole Naar, Brian Paciotti, Lore Ruttan, Bryan Vila, Tim Waring, and Matt Zefferman. And thank you to the Department of Environmental Science and Policy at University of California, Davis, which has been such a congenial home.

There were also people who provided ideas and advice on the actual writing of the book, listening to us, giving us feedback and reading chapters (including some appalling early drafts): Marion Blute, Barry Bogin, Joseph Carroll, Marilu Carter, Howard Cornell, Andreas De Block, Michael Fitzgerald, James Gaasch, Mark Grote, Susan Harrison, Sue Hodgson, Beth Jaffe, Graham Jelfs, Don Lotter, Mary Brooke McElreath, Cristina Moya, Peter Pascoe, Susan Pitcher, Dorothy Place, Sarit Richerson, Peter Thompson, Roman Wittig, Sydney Wood, Vedder Wright, and Devon Zagory.

Five people put a great deal of work into helping us, reading each of the chapters, suggesting improvements and keeping us company. They are Jeff Alexander, Sarah Brearly, Kristin Rauch, Nancy Redpath, and Debbie Worland.

We are very grateful to our illustrator, Jan Nerding, who also made many useful comments on the manuscript, and to Maurice Simmons who imagined two of our early ancestors.

And finally, thanks so much to Luba Ostashevsky, our agent, friend, muse, and the midwife of this book.

1
Getting Beyond the Apemen

What is it to be human? With your first breath, you began your lifelong exploration of this question. You slowly tried to work out how to be comfortable in your body and the world around you. You needed others to care for you, and you automatically behaved in ways that encouraged them to care. Your eyes were attracted to their faces and you stared back into their eyes. Your lips stretched themselves into little smiles. You cried when you felt uncomfortable. Your life has been intertwined with the lives of other people from the moment you were conceived.

Scientists want to understand how our species evolved because they believe it will shed light on the question of what it is to be human. Speculation about human evolution has been going on for over 150 years, but it wasn't until the middle of the 20th century that stories of human evolution began to be told to the general public. The scientists telling these stories had very little evidence to go on—just a few fragments of human-like bones and teeth that archaeologists had found. But they thought they understood enough about people to put together a good story. They confidently described apemen fighting over women and territory on the African savanna. It was a story that made sense to people who had recently experienced a devastating war and felt a little uncomfortable talking about sex in "polite company."[1] Talking about our animal ancestors' sex lives was acceptable but still a little titillating.

The world has moved on since these days, but the apemen stories didn't change much. By the 1990s, scientists were less interested in how the apemen fought and more interested in how they *thought*. The stories told by these "evolutionary psychologists" were still about apemen in the savanna, but these apemen were processing information and calculating how to defend territory and get the most females.

In recent years, scientists have released a torrent of new information about what our ancestors were like and how they lived their lives.[2] It's now possible to tell a bigger, bolder, and richer story of human evolution, and our purpose in writing this book is to tell such a story. When we were in the final stages of writing it, world events provided a new demonstration of why a new story is needed. In a matter of days, people found their lives dramatically changed as population after population attempted to slow the spread of a new virus. Many people were compelled to stay at home to reduce their risk of catching the virus, while others were asked to risk infection and go to work, caring for the sick and supporting those sheltering at home. Scientists with relevant expertise worked at trying to better understand the virus and its spread, and to find ways of treating and/or preventing infections. There was talk of "national emergencies" and being "at war," but this was a different kind of war. It wasn't a war of one group of humans against another group. Instead, *all humans* were urged to unite to confront a common enemy—the virus.

Efforts to fight this invisible enemy were often shambolic, and any "unity" achieved was far from perfect. But the global response to this virus in 2020 was very different from what would have happened if a virus like this had emerged in the middle of the 20th century. Most 21st-century people expected (and hoped) that humans would work together, not only to defeat the virus, but also to protect the complex social and economic bonds of our shared world.

Our story of human evolution explains how the descendants of apes living in an African forest could have evolved into the kind of animal that behaves like this (see Figure 1.1). It's not a complete story—far from it. And, without doubt, future investigations will turn up evidence that reveals new details. These may support or challenge parts of this story. That's how science works. But our story is certainly "truer" and more complete than the old apemen stories.[3] Also, it's long past time to tell a story that weaves in the new evidence and talks about all the interesting things that women and children were doing as human beings evolved.[4]

New ways of thinking about thinking

When humans puzzle over something abstract like "what is thinking?," we can't help but draw on our experiences with the concrete everyday world. Once scientists started using computers, they began to conceive of "thinking" as being like computing. They started to see a brain as a data processing and storage device like a computer. Since then, ideas about brain evolution have closely tracked the development of computers. In the 1990s scientists shared their offices with a desktop computer that had a floppy disk drive and a chunky monitor. They were excited about the performance of their new PC clone with a Pentium chip and a Windows 95 operating system. So when scientists in the 1990s thought about how the brain evolved, they were inclined to think about

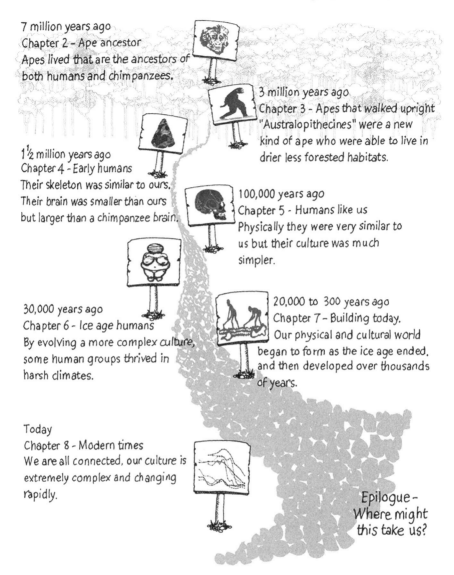

7 million years ago
Chapter 2 - Ape ancestor
Apes lived that are the ancestors of
both humans and chimpanzees.

3 million years ago
Chapter 3 - Apes that walked upright
"Australopithecines" were a new
kind of ape who were able to live in
drier less forested habitats.

1½ million years ago
Chapter 4 - Early humans
Their skeleton was similar to ours.
Their brain was smaller than ours
but larger than a chimpanzee brain.

100,000 years ago
Chapter 5 - Humans like us
Physically they were very similar to
us but their culture was much
simpler.

30,000 years ago
Chapter 6 - Ice age humans
By evolving a more complex culture,
some human groups thrived in
harsh climates.

20,000 to 300 years ago
Chapter 7 - Building today.
Our physical and cultural world
began to form as the ice age ended,
and then developed over thousands
of years.

Today
Chapter 8 - Modern times
We are all connected, our culture is
extremely complex and changing
rapidly.

Epilogue -
Where might
this take us?

Figure 1.1   The path from the forest.

hardware upgrades, imagining improvements in things like processing speed
and storage capacity. They saw genes as being like programs, and natural selection as being like a software designer.

Computers have changed a lot in the past 25 years. Computer networks
were rudimentary in the 1990s. The explosive growth of the Internet and
the proliferation of connected devices changed everything. It also gave us a
much better metaphor for how the human brain works. Networked, mobile

computers can do much more than the desktop computer of the 1990s. This isn't just because they're better at storing and handling data. What's really important is their ability to share data, programs, and processing power. The connectedness of today's computers has helped scientists to realize that the computer inside the human skull also owes much of its power to its ability to be in a network—to be part of an "us." We humans learn from others. We put our heads together and think things over. We share our feelings and feel each other's pain. Connectedness is vital for our mental health. Children who have been neglected and not allowed social contact don't just grow up ignorant of the world; their brain development is often delayed or impaired.[5]

Our ability and our need to be connected evolved because being connected made our ancestors successful. The stories of human evolution that we share and tell our children need to change to reflect this new understanding. But even more important, they need to reflect all the new evidence about our ancestors that doesn't fit into the violent apemen on the savanna story. For example, we now know much more about what our female ancestors were doing and what children's lives were like.[6] This is in no small part due to the many brilliant women scientists who have risen through the academic ranks to positions that allow them to influence research topics.[7] Knowing more about children's lives is vitally important for understanding human evolution, because it's during childhood that the competition for survival is most acute.[8] Even a hundred years ago it was common in some populations for half the babies to die before reaching adulthood. We must never forget that, over the millennia, our ancestors were the ones who not only made it to adulthood, but also had children who made it to adulthood and went on to have children themselves.

You might think that there is nothing complicated or special about humans raising their young. Producing offspring is, after all, something all animals do. But this view is deceptive. In evolutionary terms, producing offspring that survive is the most important thing animals do. In most species of mammal, the female does all the work of raising the young. In humans, mothers get help and support from other women and from the men in their lives. Whether they were hunters, farmers, soldiers, or something else, men's work was important for our evolution because it contributed to the survival of children. A key and recurring theme in the story of human evolution is how men's efforts have been harnessed and applied to the raising of offspring.

Today's evolutionary scientists are also less keen to try to learn about a gene-defined "human nature." In fact, the whole "human nature" idea is looking more and more bogus.[9] Awkward family gatherings may be difficult to endure, but they provide ideal opportunities to observe that people who are raised together and share a lot of the same genes don't necessarily behave in the same way. Neither genes nor environment do a good job of predicting how our children will turn out. Neither factor (alone or in combination) can explain why a challenge that makes one child resilient can scar another child for life. There is other stuff going on to make us what we are, including random factors. We may have to accept, like our ancestors did, that some stuff just happens—call it luck or "fate." Not all the influences in our lives can be pinned down and explained. Humans haven't evolved a mind programmed to respond in certain ways. We evolved a mind that could evolve, and it has made us what we are today—a very different animal.

Stories are tools for thinking, and so is evidence

Seven million years ago, our ancestors were apes that lived in forests in Africa. These apes are also the ancestors of the chimpanzees and bonobos that still live in Africa today. While their side of the family stayed put, our ancestors left the forest and produced descendants that eventually became scattered all over the world. The paths our ancestors took over the last seven million years are being investigated by legions of scientists, and their work is revealing more and more details.

Evidence is always better at showing what could *not* have happened than revealing for certain what *did* happen. That's why stories are valuable. By pulling together threads of evidence from many disciplines, it's possible to weave together ideas about the dramas that our ancestors might have starred in. We can even make educated guesses about what it felt like to be them. It was the challenges they faced, the solutions they found, and the trade-offs they made that shaped what we have become. To understand our ancestors, it's not enough to simply imagine how *we* might have faced those challenges or how *we* might have felt. We must think ourselves outside of the time and place we live in now. Our ancestors didn't just face different problems; they had lifetimes of different experiences and were steeped in different knowledge and beliefs. Our more ancient ancestors were physically different, and their minds had different capacities. They weren't capable of thinking and feeling like we do. And even quite recent ancestors, who may have looked just like people today, would have thought and felt differently.

Humans have long used stories to launch their imagination into situations they could never experience in their own lives. That's why this book is illustrated with stories about events that could have taken place in our ances- tors' lives. All scientists are storytellers, constantly using their imagination to try to make sense of the evidence they're gathering. The stories others have told are the inspiration for the ones that we have written for this book. The foundations of our ancestors' evolutionary success were laid during their childhood, when their lives were most at risk, so our stories are mostly set in this part of their lives. Our ancestors didn't just survive the dangers; they were able to make the most of the situations they found themselves in.

The stories we tell get longer as the evolutionary process brings increasing complexity to our ancestors' lives. And each one is followed by a summary of the evidence and theories that we drew on to write the story. Readers ac- customed to reading scientific literature will probably want to know more about the evidence we relied on. They may also want to know the names of the academics whose research and scholarship provided this evidence. For the benefit of these readers, the chapters are peppered with numbers that link to a note in a section at the back of the book. These notes give more detail and/or suggest further reading. We have written the book in this "multi-level" way because we believe that stories of our origins belong to all humans, not just to those who are trained in science or fascinated by it. If you're just interested in the story and find the labors of academics boring, don't bother with the notes. Or try them out when you have insomnia, in lieu of some other sleep aid!

Many people would rather not be told a story about humans being animals and a product of the evolutionary processes described by Charles Darwin. People's reactions to Darwin's ideas provide a good illustration of the diver- sity of our species. Some of us see it as obviously true that humans evolved by natural selection—"survival of the fittest"—and believe this view to be well supported by evidence. Others reject the idea, see it as obviously false, and quote plenty of evidence to show why it can't be true. Both sides have a point. The close connectedness of humans and our reliance on one another is con- vincing evidence that human evolution wasn't just a matter of survival of the fittest. Darwin saw that too. Humans may be animals, but we aren't *just* an- other animal. The evolution of an animal like us is so unlikely that scholars have puzzled and argued for a long time over how it could have happened. An animal like us could have only evolved in certain very unusual circumstances.

This means that, if humans are a product of evolution, we can rule out a lot of scenarios about our past that can't possibly be true. In developing the story of our evolution, we must imagine the kinds of things that might have hap- pened to make the evolution of an animal like us possible. The story of our evolution is amazing and unique. We humans have a history we can be proud of—well, maybe not all of it.

The special species

Nowadays, many people believe there's nothing special about humans—that we are "just" another animal. This wasn't the case in 1858 when Charles Darwin presented his idea of evolution by natural selection. His book *On the Origin of Species*, published the following year, was a bestseller (for a nonfiction book), and readers were impressed by the large body of evidence he had pulled together. Even so, a lot of them struggled with the basic idea. It seemed to strip life of its meaning and purpose except for some all-consuming need to compete, survive, and leave offspring. They didn't feel that this was what *their* lives were all about. This feeling (or need for a feeling) that our lives are *about* something is one of the things that is special about humans. Darwin went on to write *The Descent of Man* in 1871, another great book, but it left most of his critics unsatisfied.[10]

Alfred Russel Wallace[11] believed that understanding human specialness required looking beyond the physical world. Like Darwin, Wallace was a 19th-century British naturalist and explorer with a beard. He spent much of the early part of his life exploring in the tropics, observing the animals, plants, and people who lived there. Like Darwin, he collected many specimens and sent them back to England. It was in 1858, when he was 34 years old and exploring the islands that are now part of Indonesia, that he worked out the theory of evolution by natural selection.

Like most well-read English men and women of his time, Wallace was familiar with the evidence that the Earth and its life had changed (or "evolved") over time. There were fossilized remains of strange plants and animals that no longer existed. More and more of these remains were being found all the time. What's more, remains of sea creatures were often found embedded in rock on the tops of mountains, hundreds of miles from the coast. What force or forces were driving these changes? Wallace enjoyed puzzling over such questions with friends and colleagues. It was when he was recovering from malaria (and perhaps a little feverish) that he realized that there didn't need to be a *supernatural* force driving the change in living organisms.

The struggle for survival could be that force. This struggle was something Wallace witnessed all the time as he explored the tropical islands. When living things reproduced, they created far more offspring than could possibly survive, so there was constant competition. Animals competed to find food, trees competed to get their leaves into the sunlight, and everything tried to avoid being food for something else. Individual organisms were not all the same, and some had characteristics that made them better suited to the environment. They were the ones most likely to survive long enough to reproduce. If the offspring of these survivors inherit their parents' characteristics, then the population of organisms is bound to change with each generation. The new generation will be slightly better suited to the environment than the previous one.

As soon as he was well enough, Wallace wrote to Charles Darwin explaining his idea. He and Darwin had been exchanging letters for two years, but Wallace didn't know that Darwin had already had the same idea. Back in England, Darwin had been quietly developing his theory about how competition to survive and leave offspring could be a natural mechanism for evolution. He had talked about it with close friends but didn't want to discuss it more widely until he had built up a good body of evidence and examples to show that it was more than just a notion. Once he realized that someone else was thinking along the same lines, he arranged for Wallace's letter, along with a paper of his own, to be read at a meeting of the Linnean Society of London in 1858. At that time, the Linnean Society was one of the primary organizations discussing new discoveries in natural history, and by having the idea presented in this way, Darwin ensured that both he and Wallace would be credited with thinking of it (see Figure 1.2).

Still in Indonesia, Wallace knew nothing of this, but he eventually received letters telling him what a stir the idea was causing among naturalists. He continued his exploration of the islands for another four years, returning to England in 1862 to find that everyone interested in evolution was discussing Darwin's *Origin of Species*, which had been out for over two years. There's no evidence that Wallace had any grievance with Darwin about how he and his idea had been treated. In fact, he began to give public lectures explaining his and Darwin's shared view of how natural selection could have "created" the vast diversity of life.

It soon became clear, however, that Wallace's view of *human* evolution differed from Darwin's in an important way. Wallace believed that the evolution of the *physical* characteristics of living things could have come about through millions of generations of organisms competing to survive. But he couldn't see how the consciousness and the conscience of humans could have come about in this way. In his travels, he had seen countless examples of animals and plants competing. He had observed many times the suffering and death of creatures with an injury or illness that had made them less able to survive. A small wound or slight malaise was often fatal.

DARWIN-WALLACE MEDAL
1st July, 1908.

Figure 1.2 The two sides of the medal issued by the Linnean Society of London to commemorate the 50th anniversary of the reading of Darwin and Wallace's papers on evolution by natural selection.

This was far less likely in the humans that Wallace had observed. During his travels in South America and Indonesia, he had met many peoples whom his friends back in Europe would call "savages." The various tribes looked and behaved very differently from Europeans, and also very differently from each other. But he found that all the humans he met were the same in an important way. In an essay he wrote for the *Journal of the Anthropological Society of London*, he described it like this:

In the rudest tribes the sick are assisted at least with food; less robust health and vigor than the average does not entail death. Neither does the want of perfect limbs or other organs produce the same effects as among animals. Some division of labor takes place; the swiftest hunt, the less active fish, or gather fruits; food is to some extent exchanged or divided. The action of natural selection is therefore checked; the weaker, the dwarfish, those of less active limbs, or less piercing eyesight, do not suffer the extreme penalty which falls upon animals so defective.[12]

Wallace concluded that humans don't compete like other animals, and that this meant natural selection would not work in the same way as it did with other living things. He also wondered what could account for the generosity that seemed to be universal in humans and yet, he thought, completely absent in other animals. He speculated that human evolution might be different and that some other evolutionary process might be necessary to explain the development of the moral beliefs and mental life of humans. He wondered if this evolution could be occurring outside of the immediately observable physical world.

And why not? Up in Scotland, the physicist James Clerk Maxwell was demonstrating that electric and magnetic fields travel through space as invisible waves, moving at the speed of light. So much of what had been thought of as "supernatural" was becoming understood to be part of nature. There was a sense that many amazing new discoveries were just around the corner. In the 1860s there was much discussion of the possibility that an invisible "spirit world" might also be part of nature. Wallace, like a number of scientists at the time, believed that if such a world did exist, the possibility of communicating with spirits should be systematically investigated. He thought that, in spirit form, the human mind might evolve separately from a physical body.

Wallace attended some séances and heard knocking alleged to be made by disembodied spirits. He spoke to his dead relatives through mediums, and he even had his photograph taken with an apparition of his dead mother. Some of the people who produced such "supernatural phenomena" later admitted that they were frauds. Even so, Wallace remained convinced that some of the communications he'd had with the spirit world were genuine.

Nowadays, Darwin is given most of the credit for developing the theory of evolution by natural selection because of the huge amount of work he did to present evidence to support the theory. Today's scientists admire Darwin for his determination to be guided by that evidence rather than the fashionable ideas of his time. Darwin didn't think much of the spirit world idea, but he completely agreed with Wallace that explaining human evolution presented a problem. They both believed that all living things are related and that it's possible to compile a great family tree of life. Gaps in this family tree were constantly being filled in during Darwin's lifetime as his colleagues explored the world finding more living specimens and fossils of organisms that had died out. Many more gaps have been filled in since Darwin's time. Our new ability to sequence the DNA of living organisms and, in some cases, DNA recovered from the remains of long dead organisms is making a big contribution as well. Biologists are now able to more precisely position the many limbs and branches of the tree of life.

Darwin saw that humans had a place on the tree, even if our species had to be put way out on a limb. It must have irritated him that popular writers of the time misquoted him as claiming that humans are descended from monkeys. Darwin put humans on the same branch of the tree as the great apes of Africa (chimpanzees and gorillas) and believed us to be more distantly related to another kind of great ape, the orangutan, found on the Indonesian islands of Borneo and Sumatra. Monkeys are more distant relatives. Their faces may look human-like, but their bodies are very different. Most of them have a tail, for example. Darwin lived long before anyone knew anything of DNA or genes or how characteristics are passed from parents to offspring. His judgments were based on painstaking observations—both his own and those of his naturalist colleagues. DNA analyses have shown these judgments to be mostly correct.

Darwin's determination to base his judgments on evidence made his views about *human* evolution, which he published in 1871,[13] less popular than his earlier book on the origin of species. At the time, his views were less influential than those of some other scholars. Herbert Spencer was the celebrity pundit for human evolution. Spencer wasn't just very eloquent; he also seemed to be very sensitive to what the public wanted to hear.[14] Darwin disagreed with Spencer and many others who justified their countrymen's treatment of non-Europeans with the claim that other "races" are inferior or somehow less human. Darwin argued that the physical differences between peoples are so superficial that a biologist must conclude that not only do all humans belong to the same species, but that we are all closely related. He agreed that people raised in different environments behave differently, but he argued that experiences influence behavior. In his 1871 book *The Descent of Man*, he presented evidence that when young people from different parts of the world spend time in Europe and are encouraged to adopt European ways, they behave just like Europeans.

Darwin gained firsthand experience of this thanks to his five-year voyage around the world on the British naval vessel called HMS *Beagle*. He spent quite a bit of time with a young man whom the English called Jemmy Button.[15] Darwin and Button were on the *Beagle* together when the 90-foot-long, 242-ton bark set sail from England in 1829, crammed with 73 people and the provisions and equipment they would need for the first part of their journey. The main purpose of the journey was to survey South America to make more detailed maps of its coast.

Darwin, aged 22, was going along to be the ship's naturalist. His job was to disembark and explore the land as the *Beagle* sailed along the coasts. He planned to collect specimens of plants, animals, and rocks and send them back to England. Jemmy Button was on the *Beagle* to be taken back to his homeland. On the ship's previous voyage to South America two years earlier, the captain had picked up (kidnapped) four natives and decided to take them back to England. Button was one of the three that survived. Living with English people had transformed him. When the *Beagle* crew first encountered him, he was a scrawny child swimming naked with his friends in the near-freezing seawater that surrounded the ship. By the time Darwin met him, he was a rather plump teenager, dressed in the latest London fashion, with perfect table manners and the Christian values of the English people he had lodged with. The captain hoped that, once back with his people, he would be able to teach them Christianity and the "civilized" ways he had learned in England.

After many adventures, including nearly sinking in a storm, the *Beagle* was able to moor off Button's homeland on Tierra del Fuego on the southern tip of South America. Contact was made with his family and, after building him a small house and planting some crops for him to harvest, the officers and crew of the *Beagle* reluctantly left their well-dressed adoptee surrounded by people whom they all, including Button, regarded as murderous and thieving savages.

A year later, after completing a survey of the east coast of South America, the *Beagle* sailed back to Button's homeland to see how he was getting along. The crew feared they would find him ill, half-starved, or worse. As soon as Button heard that the *Beagle* had returned, he came out in a canoe to where the ship was anchored. The lean young savage who climbed aboard had dirty matted hair and, despite the cold, was naked apart from a small cloth tied around his private parts (see Figures 1.3 and 1.4). Darwin wrote in a letter home that "[i]t was quite painful to behold him . . . when he left us, he was very fat and so particular about his clothes that he was always afraid of even dirtying his shoes; scarcely ever without gloves and his hair neatly cut. I never saw so complete and grievous a change."

After Button boarded the *Beagle*, he immediately went below to wash and borrow some clothes. He emerged much more like the man they had known and was clearly thrilled to be with his English friends again. But when they urged him to stay on board and sail back to England with them, he looked surprised. He insisted that life with his people was good, telling the captain, "I am hearty, sir, never better." In his letter home, Darwin wrote that when they'd left him with his fellow Fuegians in 1833, Button had been appalled at the ignorance of his people and declared them "damn fools." But he now saw them, according to Darwin, as "very good people with too much to eat and all the luxuries of life." Button and some of his family members stayed on board for the rest of the day, and the crew continued trying to persuade him to come

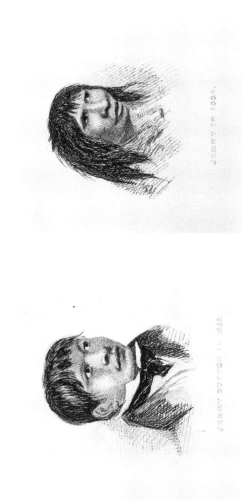

JEMMY BUTTON 1833.

JEMMY IN 1834.

Figures 1.3 and 1.4 Sketches of Jemmy Button by Conrad Martens, the artist who was part of the crew of HMS *Beagle*. The one on the left was made before they left Button among his people on the island of Tierra del Fuego at the southern tip of South America. The one on the right is of Button a year later when the ship returned and offered to take him back to England.

with them back to England. Then a canoe arrived at the ship carrying a young woman who was clearly upset. She was revealed to be Button's wife, worried that her husband would be spirited away. Darwin and the *Beagle* crew then agreed that Button couldn't possibly leave and gave him their enthusiastic congratulations on his marriage.

Button stayed in South America but had several more contacts with Europeans as they tried to claim and "tame" his land. He used his knowledge of his own culture and that of the English to protect his people as best he could, drawing on his understanding of the Europeans' Christian values. He died in his late forties when an epidemic struck his people. It was probably one of the infectious diseases, like plague or measles, that Europeans brought to the Americas.

The transformations in Button's behavior had a big impact on Darwin. He saw that no other animal, including apes, behaves as we do,[16] and that there are no examples of primitive humans that can serve as "missing links" between humans and the animals that we are most closely related to. What's more, no other animal shows more than a fraction of the diversity of behavior that can be seen in a small village of humans. A biologist observing human behavior might think to divide the human population into hundreds of different species. But humans are closely related and very similar physically. Darwin's time with Jemmy Button showed him how flexible the behavior of a single human can be, and that we don't inherit our behavior from our parents in the same way as we inherit the shape of our nose or the color of our eyes.

There was what Darwin called a "great gap" on the ape branch of the family tree between humans and chimpanzees, gorillas, and orangutans. In the years since Darwin's death, scientists have struggled to define the nature of that "gap." It's taken for granted that humans are much more intelligent than other animals, but what does that mean, exactly? The precise *kind of* intelligence that humans possess may be unique, but other animals show great ingenuity in solving the problems that are important in their own lives. Once scientists developed tests of intellect that were better suited to the animals they were studying, or when they simply started watching animals in the wild more closely, they found them to be cleverer than they had imagined.[17] Animals with brains much smaller than ours are good learners with amazing memories. They're also creative. Caledonian crows readily make tools out of twigs and leaves to pull edible insect larvae out of holes. Humpback whales work together to capture schools of tiny fish by blowing out a curtain of bubbles that acts like a net of air the fish can't swim through. Chimpanzees have the ability to work out what others know and are ignorant of[18]—a skill psychologists sometimes refer to as "mind-reading." It was once thought that only humans had "theory of mind." And even the human generosity that Wallace described turns out not to be unique to our species. Life for other animals isn't constant competition. Many species of animals help to look after each other's young and share food.[19] Some mammals happily suckle infants that are not their own.[20]

And a special evolution

In the years that have passed since the theory of evolution by natural selection was first proposed, scientists have solved the mystery of genetic inheritance and now understand many of its mechanisms in astonishing detail. But while some mysteries have been solved, quite a few things that once seemed straightforward have been revealed to be mysterious. For example, in the 19th century it seemed obvious how children "inherit" their parent's language. They listen to people around them and copy the way they talk. The mystery was how a child inherited her curly hair from her mother and her shyness from her father. Today we know how information coded in the DNA directs the formation of our hair follicles and psychologists are testing theories about how certain genes influence personality. But the details of how we "inherit" information by learning is now seen to be much more of a mystery. For example, children don't inherit their parent's way of talking, but instead tend to pick up the dialect of their friends. Sometimes when siblings attend different schools and have different friends, they end up speaking with different accents. What goes on to make this happen?

Over the last few decades, more and more evolutionists have come to believe that Alfred Russel Wallace's idea of a second evolutionary process might not be so far-fetched. They don't think that this evolution takes place in a spirit world. (If a spirit world does exist, it's not been possible to obtain any reliable evidence of it, so it's not amenable to scientific investigation.) But we have long known that another evolutionary process profoundly impacts human lives—the evolution of our cultures. We know that culture changes over time. It changes fast enough that we can experience cultural evolution in our own lifetimes. Our technology develops, bits of our human-built world are destroyed while other things are added, and we're often aware that our feelings and beliefs have changed.

In its broadest sense, *culture* is the monstrously complex, swirling mass of ideas, beliefs, habits, customs, fashions, and things that surround us. Our beliefs and emotions seem so personal to us, but they're connected to what people around us do. Our minds change, sometimes without our under-standing why. We can suddenly start to question something we once took for granted. It might be something trivial—like whether or not women over 30 should wear leggings—or something more profound—like whether anyone has the right to voice an opinion about what women over 30 should wear.

Some scholars have argued that "culture" is outside science's scope—more of the spirit than the physical brain. Others point out that because science is part of culture, it's impossible to have a "science of culture." They argue that a person simply can't take an objective ("scientific") view of something that he or she is inside of and part of.[21]

But a broader definition sees culture as just information—the vast amount of data that we access via our social networks. Cultural information is like a set of "tools" that we use to operate in our world. Sometimes our cul-tural information allows us to create or use physical tools—such as a knife or the warm clothing necessary to survive in a cold environment. But our knowledge about our environment—such as what plants are good to eat—is also an important survival tool. Many of the tools our culture gives us are "social tools" that allow us to interact with one another in complex ways. Language is probably the most important social tool—it's been used by our ancestors for tens of thousands of generations. But our culture provides us with many other social tools. We have agreed-upon rules for how to behave in "polite society," as well as physical objects, like money, to make it easier to trade. There are drugs, like alcohol, that can make people feel more relaxed interacting with strangers, and there are weapons to use when interactions get hostile.

In the past, separate populations each had their own cultural network. Some groups lived their lives in complete isolation. But over the last few thou-sand years, links between populations strengthened, allowing information, goods, and people to travel more easily between them. Nowadays, there's a global network and we're super-connected. Each of us has access to some parts of the network and, because people are always editing and adding data, culture is constantly changing. The change isn't random, and neither is the way we access cultural information. Finding and evaluating patterns in the movement of information is beyond the processing power of our brains. But the computers we're building are getting better and better at handling data and they're capable of being far more objective about how to handle it.

It's impossible to make precise and reliable predictions about how people's minds will change. Who would have thought that television programs about cooking would be so popular when people are spending less and less time cooking their own food? But there are patterns in culture. People who spend time together tend to think in similar ways and are more likely to agree on how "things" need to change. People who spend time together are culturally similar partly because we prefer to be with people who think like us, but also because spending time with people influences the way our mind changes—as the life of Jemmy Button clearly shows.

But even this isn't predictable. The ether that connects human minds and allows ideas to flow between them often seems to defy analysis. It seems as though it might as well be happening in a spirit world. Many of us know only too well that close family members often perceive aspects of the world very differently. The thing we call "culture" is woven into our minds. It doesn't just supply the clothes we use to decorate our bodies; it also provides the mental organs we use to digest and metabolize our experiences.

This book tells a backstory that explains how our ancestors managed to harness this culture thing—well, to kind of "harness" it. (Our ancestors used culture, and their culture used them.) The story goes back a long way. Culture in its broadest sense isn't unique to humans. Apes and many other kinds of animals learn from one another, so a group of these animals can also be said to have "a culture," albeit a simple one. Among our ancestors, culture became huge. Their efforts to survive in new habitats millions of years ago led them to use culture in new ways. As the Earth's environment became unstable, it became more and more important for human populations to have a complex culture that kept them thinking together so they could cope with new challenges. This allowed some groups to survive, thrive, and eventually prosper.

The story of the evolution of our species and our cultures begins in the next chapter, which is about our ape ancestors who lived in an African forest seven million years ago. Even during this prehuman stage, learning and being connected were important. In apes, however, only one connection is strong—the bond between a mother and her infant. We believe this had changed by about three million years ago, which is the setting for chapter 3. By this time our ancestors had evolved into the kind of ape scientists have named "australopithecines." Chapter 4 describes the lives of our ancestors who lived about a million and a half years ago. They were human, physically similar to us in many ways, and their brain was bigger than the brain of any ape, ancient or modern. But their brain was still considerably smaller than the brain of humans today. Chapter 5 is about the ancestors who lived 100,000 years ago. Physically, they were virtually identical to us, with a brain of the same size. But their culture was far less complex than any seen in humans today. By the time we get to the humans we feature in chapter 6, who lived 30,000 years ago, culture had become far more sophisticated in at least some human groups.

We cram almost all human history into chapter 7. Many stories have been told about the people who lived during the time between the end of the last ice age and today. New evidence and our new understanding of human evolution suggests new ways of looking at these times. For thousands of years, people with brains like ours walked an Earth full of natural resources, but their population and the impact they had on the planet remained small. Then, a few hundred years ago, things started to change very rapidly. Chapter 8 looks at these changes and how our species reacted to them. Finally, in a short epilogue, we speculate about the near future.

2

Ape Ancestor (About Seven Million Years Ago)

Thanks to technology that allows us to analyze the DNA of different living things, we do know the basic shape of our family tree. There are three basic kinds of great ape alive on the planet today (four if you want to include humans). The ones called chimps and bonobos, who are closely related to each other, are the ones most closely related to us. We know that at some time in the past—about six or seven million years ago—there were apes living in the forests of Africa that link us to the rest of the animal kingdom (see Figure 2.1).

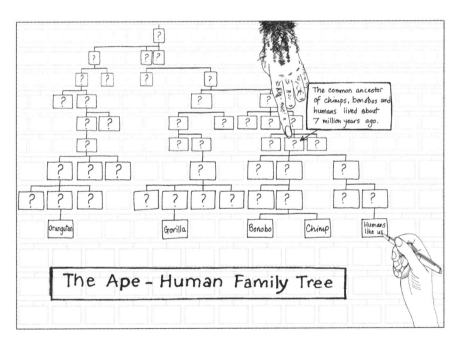

Figure 2.1 Most people know that scientists believe humans are descended from apes—or maybe that we *are* apes. But what does this mean exactly? If we tried to draw the ape family tree, it would be almost all question marks. Scientists may be sure that humans have ancestors that connect us with the ape family, but none of these ancestors have been identified for certain, and they all became extinct long ago.

These apes are our ancestors and they're also the ancestors of the chimpanzees and bonobos that live in African forests today.

To work out what our relationship to great apes might mean, scientists are putting together a picture of what this "last common ancestor" (LCA) was like.

Millions of years ago, one of your ancestors was born in a tree in Africa. Its first breath drew in warm damp air with a mildewy tropical smell. When it opened its eyes for the first time, it saw light filtered through many layers of leaves and branches. And it saw its mother's fur. If you had been born as this ancestor, you would have been an ape with a brain about a third the size of your human brain. We can't know what it would have been like to think and feel with this sort of brain. But we do have a fair idea of how you would have behaved and what your ape life would have been like.[1]

After eight or nine months of being in your mother's womb, you slid down her birth canal with rather more ease than a human newborn. When her contractions began, your mother stopped foraging and tried to make herself as comfortable as possible on a tree branch as she pushed you out of her womb. As your head started to emerge, she reached down to hold it and started to gently pull you out of her body.[2] Then she licked and licked you, removing every trace of amniotic fluid from your skin and hair. Its flavor, mingled with the smell of healthy baby, was delicious to her. When the placenta came out, she ate it hungrily. If you had been dead or unhealthy, she might have eaten you too. But you were healthy, and she felt almost overwhelmed with love for you.

Compared to human babies, baby apes are scrawny little things—about half the size and much thinner. In spite of your smaller size, however, you're more developed, both physically and mentally, than a newborn human (see Figure 2.2). In many ways, your level of development at birth is closer to that of a one-year-old human. Within minutes of your birth, you're capable of giving a mind-piercing screech, which lets your mother know how healthy you are. She nudges your face toward her nipples, trying to get you to suckle. Throughout your infancy you will make many loud shrieks and screams, alerting your mother to any tiny disruption of your comfort. Your tiny hands and feet tightly grip the fur of her abdomen. You hold yourself there for a few seconds but feel yourself slipping and start to scream. She puts a gentle hand on your back.

Eventually you'll be able to hold onto her front and help yourself to milk from her teats as she forages in the trees. During the few weeks it takes for you to fully develop this skill, you and your mother spend quite a bit of time on the ground. She carries you around, putting you down briefly to rummage for food in the leaf litter and plants.

Figure 2.2 When this baby chimpanzee was born, it was more developed, both physically and mentally, than human newborns. A baby ape's survival depends on its being able to cling to the fur of its mother's abdomen and help itself to milk from her teats. For most of their waking hours, ape mothers need their hands and arms free to move around and find food.

She's never far away and snatches you up again at the slightest sign of danger, but you screech and scream, hating the feeling of not being in her arms. She has to harden her heart, because she needs to find food to be able to make the milk you need to survive. She looks for things that have dropped from the trees—fruits, dead flowers, and baby birds and eggs that have fallen out of nests. There is also the occasional termite making its way along a tunnel, as well as snails, spiders, lizards, and caterpillars.

As soon as your mother gave birth to you, you became the center of her universe. You and she will be inseparable for the next three years at least, and you will stay very close to her for at least five years after that. You and your mother have a companion—a young ape that was your mother's baby until you were born. Now you have taken his place. He wants to be close to your mother and he is fascinated by you, but your mother makes him keep his distance. Oxytocin, the hormone[3] that triggered the contractions that pushed you from her womb, also works on her brain, triggering powerful emotions. This hormone is present at the birth of mammals, and in many of them it causes mothers to bond very strongly with their newborn. The effect of oxytocin on the emotions of human mothers seems to be weaker. Human mothers can become very absorbed with their new baby, but even the most ardent human mother proudly hands her infant over to friends, family members, midwives, and pediatricians. An ape mother is obsessed with her baby. Your ape mother would have wanted you all to herself and fiercely protected you from everyone and everything.[4]

You and your mother spend part of the time with a group of apes. It might be a small group, comprising only a few adults, but it's more likely to be larger. During the day, the group spreads out as its members move around searching for food. Sometimes the whole group comes together, and sometimes different subgroups form. As night approaches, each ape finds a suitable place in a tree and quickly weaves some branches together to make a platform—a sort of nest where it can sleep. You share the nest that your mother makes.

Finding food in an African forest

Things become easier once you're able to cling to your mother and hold on tight as she climbs and swings around the forest canopy looking for food. You're surrounded by things to eat. Researchers at the Gombe Stream National Park in Tanzania have identified 235 different things that are eaten by the chimps that live there,[5] and because the ecology of African rainforest varies from region to region, chimps in other parts of Africa find different things to eat.

The fact that your mother has survived up to now shows how well she has learned what there is to eat—and how to get at it. Her problem is that the food is mostly in very small morsels. Think of being at a buffet restaurant where, instead of the food being on display, it's hidden all around the room. When you do find a piece of food, it's usually something like a lettuce leaf, celery stick, lemon slice, or one of those little plastic packets of ketchup—things that might keep you from starving but don't make you feel well nourished or satisfied. Some of the hidden morsels are more nourishing. A cocktail sausage can usually be found underneath the cucumber slices. A pineapple chunk sometimes has a cheese cube attached. Your mom is always on the lookout for the good stuff. She knows that the best morsels are the ones that are hardest to spot and get at. Sometimes they're surrounded by a protective coating that is nasty tasting, covered in prickles, or impossible to bite through. Your mom is an expert forager. On a good day, she finds plenty to eat in this cryptic and meager buffet.

Great apes evolved to survive in this habitat. Actually, it's more correct to say that apes and their forest homes evolved together. Some of the plants in the forest are adapted to use apes and other plant-eating animals to spread their seeds. They grow their seeds inside a fruit that apes find tasty. The apes chew up the fruity package, but some of the seeds pass through their digestive system unscathed and emerge several days later, dropping to the forest floor, surrounded by moist feces that will serve as fertilizer for the seedling that germinates. The symbiotic relationship between apes and trees works well. Apes eat a lot of fruit. But it isn't enough to satisfy all their nutritional needs. Plants have evolved to be frugal, and they put just enough nutrition into their fruit to get apes to eat it. Fruits, leaves, and the other plant bits that apes eat contain a lot of fiber. Apes can't digest plant fiber any more than we can. But another symbiotic relationship helps them get some nourishment from it. Quite a lot of the fiber can be broken down by the microorganisms that live in the ape's large intestine. Apes get extra nourishment by absorbing chemicals that are made and released by their tiny abdominal tenants. This extra nutrient income is a big help, but it's still a struggle for an ape mother to produce enough milk for her growing baby.

The most nutritious foods in the forest aren't free for the picking. Small animals or eggs are a lucky find. There are beehives with honey and larvae, but these are surrounded by angry bees. Your mother knows where certain plants store energy. Some plants put a lot of nourishment in their seeds, some store it in their roots, others have a carbohydrate-rich pith inside their stems. Over the millennia, the plants and animals in tropical forests evolved to be better and better at protecting their energy stores from hungry apes. At the same time, apes became better and better at getting around these protections. Apes evolved the ability to thoroughly learn the territory they live in, to remember how it changes throughout the year, and to acquire the skills necessary to get at the nutritious morsels it contains. But being this clever requires a large brain, and the larger a brain is, the more energy it needs to function. Even when animals sleep, their brain is working. So it's a fine balance—an ape needs a big brain to find enough food in the meager tropical forest buffet, but the bigger its brain, the more food it has to find. The balance is most critical for a mother, because she has the task of finding food for herself and her infant with its brain that needs to grow and develop.

If your ancestors were like today's chimpanzees, some of them might have sometimes hunted small monkeys or other similar-sized animals. A chimp monkey-hunt usually begins when one of the males spies a monkey in a tree and starts to chase it. Other males join the pursuit, coming at the monkey from all directions, swinging and jumping through the trees trying to block its escape. If your ape ancestors hunted in this way, your mother probably didn't join in. The chase would have required a level of speed and agility that's impossible when carrying a young one. If chimpanzee hunters manage to catch a monkey, they tear it apart, each hunter trying get as much as he can. When a hunter has a piece of carcass, he's surrounded by his friends begging for a scrap of meat. He keeps the most tender pieces for himself and allows his special friends to grab the grisly bits. If one of your mother's friends had some meat, she would likely be in there begging, but she has her own ways of getting animal protein. What she lacks in speed and agility she makes up for in ingenuity and know-how.

Even so, ape mothers can only find so much nutrition in a day. The milk they produce is very watery and low in nutrition compared to the milk of most other mammals.[6] Kittens, for example, get thick creamy milk high in fat and protein, which supports rapid growth. By about six weeks of age, kittens are weaned and ready to start learning to hunt. Four months later, the females in the litter will likely be ready to mate and produce kittens of their own. Female cats can produce several litters of kittens a year. Apes hardly ever produce more than one baby at a time and even the best ape mothers seldom manage to raise more than five offspring their whole life.

The slow growth of baby apes makes it possible for them to survive on low-grade milk. Once you're about six months old, you're of a size that makes it uncomfortable for your mother to carry you on her front. So she moves you to onto her back. At first you feel insecure there and you screech and scream. Finally, you get used to balancing on her back or a nearby branch as she forages. You will gradually grow in size and strength for another decade or more before reaching your full size and becoming sexually mature. All primates—the family of mammals that includes apes, monkeys, lemurs, prosimians, and humans—have a slower metabolism than most mammals, and they grow and develop more slowly too. Great apes are extreme in this respect.[7] And humans are real outliers, strange in a number of ways. More about our strangeness in later chapters.

Keeping you alive in a hostile social world

As you grow stronger, your mother must give you a little independence, but this doesn't mean she can start to relax. When you're 18 months old, your brain has reached its full adult size and you need to play and explore if you're to learn and develop. Rough and tumble play is good, especially with your older brother. Your mother keeps watch and is still fiercely protective. The hormone oxytocin, which was released during your birth is also released each time you suckle. The physical action of your sucking on her teats sends a signal to her brain to release oxytocin and so she continues to feel the powerful emotions that bond her to you.

Observations of chimps in the wild have shown why natural selection has favored females who are such obsessively protective mothers. Youngsters are always in danger. In some groups, researchers have seen an ape tear the infant from the arms of a groupmate, kill it and eat it with their friends. In times of stress, female chimps sometimes kill the infant of an unpopular mother—usually a young newcomer to the group.[8] But it's more often males that kill infants. If you were your ape ancestor, most of the males in your group would (evolutionarily speaking) have a good reason for wanting you dead—you would be the offspring of some other male.[9] Your mother probably mated with several of the males when she became pregnant with you; any of them could be your father and no one knows for sure.

As long as you keep sucking vigorously at her teats no male will have a chance to mate with your mother. This isn't just because of the bonding emotions that keep her focused on you. Your sucking also sends signals to a gland in her brain (the "pituitary") causing it to release a hormone called "prolactin." High prolactin levels in your mother's blood have two effects. Prolactin stimulates the mammary glands in her breasts to keep making milk, and it also prevents the eggs in her ovaries from ripening. If you die, or even if you just get sick and lose your appetite, the sucking will stop, and her prolactin levels will fall. Soon after that, one of her eggs will start to ripen. Once that happens, your mother's body will no longer be focused on keeping you alive and will start working to produce a new baby. The ripening egg will trigger the release of hormones that cause her body to change in ways that make males want to mate with her. She may also want to mate—but not necessarily. The males will mate with her without regard to her wants. It has been observed that chimpanzee males who are the most sexually aggressive toward females, father the most offspring.[10] Primatologists have seen female chimps mating with a male that had killed her baby a couple of weeks earlier.[11]

Happily, none of your ape ancestors were victims of infanticide. We know that they all lived to adulthood and had offspring of their own. Their success was in a large part due to the mothering they received. This suggests that if you were born as one of your ape ancestors, your mother would have been older and quite experienced. The mortality of firstborn apes is much higher than that of their later-born siblings. First-time mother apes feel a strong bond with their baby, but that isn't enough to make them good mothers. It takes great skill to forage efficiently while caring for an infant. But as with so many of the things that apes do, they get better with experience. An older mother is also more likely to have formed the friendships and social ties that help to keep her baby safe in a group where both males and females might be hostile. An ape group isn't a herd where animals just live side-by-side; it's more like a community, with members who have feelings for one another and long memories.[12]

Their greater competence and better social connections may explain the extraordinary sexual allure of the older female chimpanzee. Primatologists observing chimp groups report that males virtually ignore the sexual swellings of a young female if she is ovulating at the same time as an older female who has already raised several offspring.[13] The males are all busy competing for access to the sagging body and wrinkled skin of the senior matron. Their preference may seem odd if you've been exposed to glossy magazines and other media promoting the attractiveness of youth. Or if you've read stories of human evolution based on earlier work that placed little emphasis on the work of females and the importance of parenting. There has been much written about what "evolutionary theory says" about choosing sexual partners. But what evolutionary theory actually says is that mating is about producing surviving offspring. If that's the case, it makes sense that natural selection would favor male apes who are keen to mate with an experienced mother. In her care, offspring are most likely to thrive.

In the three or four years that you're dependent on your mother's milk, you have plenty of opportunity to see how she lives her life and makes her living. You're curious and alert and you learn how she treats the other apes. Having friends is important and enemies are dangerous. Your mother is careful with the large males, and when she sees that they have spied a piece of food that she is going after, she drops back and lets them have it. If she gets it, they will just snatch it from her anyway. She usually forages on her own with you and your brother and, perhaps, with one of her friends. Friends make friendly sounds to each other, not talking exactly, but signaling, checking in, and letting each other know when they're enjoying food. Friends also spend a lot of time picking dirt and insects out of each other's hair. Grooming one another has the practical value of helping to keep their skin healthy, but apes, like their descendants, simply enjoy the feeling of touching and being touched. They might sometimes touch one another in ways that we would consider sexual. While you're a baby, your mother is the only one who grooms you. She delicately runs her fingers through your fine fur, and you learn to love the feel of it.[14]

Your mother spends most of her waking time foraging and eating. There's something about her chewing mouth that fascinates you. You watch her putting things in her mouth. She's a very messy eater, so her face and chest are often covered in bits of food. You pick them off and put them in your mouth, savoring the taste and smell. A colony of bacteria starts to grow in your large intestine, digesting the plant fiber that you eat. You suckle whenever you want, but as you grow bigger, the watery milk becomes less satisfying. You watch your mother forage and try to copy her so you can get food for yourself. Sometimes you're lucky, but you chew on a lot of nasty-tasting and inedible stuff as well. Anything that doesn't kill you makes you wiser. The safest food to eat is stuff your mother has just found. Sometimes she lets you take a piece of food off her, particularly if it's something that needs peeling or pulling apart to get at the edible bit. If she finds something really nutritious, she'll ignore your screams and eat it herself. This looks like greed and selfishness on her part, but this behavior is essential for your survival. If she gave you energy-rich food, you would feel less hungry and suck less vigorously on her teats. If this results in her prolactin levels dropping too low, an egg may start ripening in one of her ovaries. It's important for your survival that she doesn't give birth to her next baby until you're able to support yourself with your own foraging. Of course, you mother knows nothing of pituitaries and prolactin, but for many generations mother apes who kept all the best food for themselves had more surviving children.

If the growth pattern of your ape ancestor is similar to that of a chimpanzee or bonobo, you'll be drinking your mother's watery milk until you're at least three years old. Gorilla babies tend to be weaned a little earlier, but orangutan babies are sometimes not weaned until they're over five years old. As long as you are hungrily suckling, you will be her baby. How fast you learn to forage determines the exact timing of when her body starts working on her next baby. As you grow bigger, you need to find more and more food to supplement what you get from her milk. As your foraging skills improve, your mother starts to push you away when you try to suckle. The tipping point may come during a season when food is so abundant that even a novice forager is able to find plenty of food. Your suckling diminishes to a point that one of her eggs begins to ripen. Males in the group begin to get so close to you and your mother that they make you feel very uncomfortable, but you're too afraid to make a fuss. She allows them to get close and she mates with several of them.

Your mother starts to give you more freedom and you soon learn that you must be careful of certain adults, especially the largest and strongest ones. Everyone takes care not to make them angry. For eight or nine more months you continue to be your mother's closest companion, but then your sibling or half-sibling is born. This scrawny scrap of screeching life becomes the new center of her universe and you must now weave your own bed nest at night. You still stay close to your mother because you feel nervous and because you have quite a bit more to learn about foraging and recognizing the opportunities and risks that surround you. Your mother doesn't let you get too close to your baby brother or sister, but unlike the other adults in the group, she doesn't mind you following her around as she forages. After several years of this, you begin to become sexually mature. You never forget your mother, but you're now ready to get on with your adult life. You took a long time to grow up, but now you may have 25 or 30 years of life ahead of you.

Why do we think our ape ancestors lived like this?

We know that our ancestors who lived seven million years ago were apes, so we can assume that they had ape characteristics. These characteristics would have placed certain constraints on how they could live. Two lines of evidence tell us that, seven million years ago, not only was our ancestor an ape, but it was an ape that lived in Africa.

First, there is genetic evidence for this. You may have heard it said that human DNA is 98 percent identical to the DNA of chimpanzees—or 99 percent or 96 percent. The precise number given depends on the technique used to do the mathematical analysis of the results. All the techniques give a number close to 100, and this supports Darwin's conclusion that humans are descended from African apes. The precise degree of closeness makes it possible to estimate how long it's been since an interbreeding population of apes existed that contained ancestors of both modern humans and modern chimpanzees. The estimate often given is that this population existed between six and seven million years ago, but the possibility that a few individuals moved between populations more recently can't be ruled out.[15]

(The same method of genetic reckoning reveals that the common ancestor of humans and dogs lived about 100 million years ago, and of humans and insects about 500 million years ago. Two humans living today, even humans born in very different parts of the world, are likely to have a common ancestor that lived in the last few thousand years.)

Second, there is fossil evidence (and a lack of it). Very few fossils of humanlike animals have been found that lived earlier than about four million years ago. This suggests that our early ape ancestors lived in tropical forests. Tropical forests covered large parts of Africa until about four million years ago, and the soils of tropical forests tend to be moist, slightly acidic, and teeming with life. The bones, teeth, and tissues of animals that die in a tropical forest are rapidly broken down, and the nutrients they contain are recycled into the ecosystem. A few ancient ape remains have been found, however, perhaps because they were carried away from the forest by a flooding river or predator and ended up in a place where they were preserved.

In 1994 over a hundred fragments of bones and teeth from a single individual were found in northeastern Ethiopia. Analysis has revealed them to be from a female ape that lived about 4.4 million years ago (see Figure 2.3). Bits of bone from similar apes have also been found, and their kind have been given the name *Ardipithecus*. The female who possessed this precise skeleton was given the nickname "Ardi." *Ardipithecus* is thought to be at least distantly related to our ancestors because of the shape of several of the bones in their skeletons. They're similar to the shape of the equivalent bones in our skeleton, suggesting that these ancient apes were quite comfortable standing and walking upright on two legs—certainly more comfortable than chimpanzees and other modern apes. But Ardi probably spent most of her time walking

Figure 2.3 These are the bones of a female ape that lived about four and a half million years ago. The shape of some of the bones in her skeleton suggests that, unlike chimps and other great apes alive today, this ape walked upright, more like we do. But the shape of the bones in her feet suggest that they are adapted to climbing in trees rather than walking on the ground. This kind of ape, which has been given the name *Ardipithecus*, may have walked upright along tree branches, leaping from branch to branch.
Credit: http://www.sciencemag.org/cgi/content/full/326/5949/64/F3, Fair use, https://en.wikipedia.org/w/index.php?curid=28198561.

along tree branches rather than the ground. Her foot bones show that her feet had grasping thumb-like big toes like the big toes of today's apes. Such toes are useful for holding onto tree branches but not so good for walking on the ground. Fragments of foot bones from upright walking apes with forward-pointing big toes like ours have been found, but they lived more recently. So the evidence from fossils discovered so far suggests that our ancestors spent a great deal of their time in trees until about four or five million years ago.[16]

Primatologists have spent many years making meticulous observations of the great apes living today—the chimpanzees, bonobos, and gorillas living in Africa and the orangutans living in the forests of Indonesia. They believe that their study of these apes provides clues to the lives of our forest-dwelling ape ancestors. They've been able to work out how ape behavior is shaped by having an ape body and living with others of their kind in a diverse and complex tropical habitat. The chimpanzee is the most numerous modern ape, and it's the one that has been studied most, both in the wild and in captivity. In terms of their DNA sequences, today's chimpanzees (and their rarer relatives the bonobos) are as different from our ape ancestors as we are. And they're even more different from gorillas and orangutans. But in terms of their bodies and their lives, today's chimpanzees are probably not that different from our ancestors. They were also apes trying to make a living in and around tropical trees.[17]

We humans are very different and can make a living and create a home in almost every habitat on Earth. But information gained by studying today's apes allows us to tell the first part of our evolutionary story. It reveals the changes that have occurred over the last seven million years and shows us what characteristics didn't change because they were already present in our ape ancestors.

Living with a great ape body and social life

Some aspects of the life of your ape ancestors can be described with confidence.[18] They had a great ape body, and we know that this would have provided them with certain abilities. For example, great apes, like humans, have a throat and mouth that can be used as an instrument for making a wide range of noises. Apes aren't able to make the same speech sounds as humans do, and the noises they make aren't "words." Young apes don't learn that certain noises have a symbolic meaning that can be used to communicate complicated information. Even so, the sounds they make are an important part of ape social interactions. Their vocalizations are a "social tool." The screeches, pants, and grunts communicate feelings rather than ideas, but they're still useful. When an angry adult male chimp screams in frustration, smaller group members know to keep out of his way, and this reduces the chances of someone getting hurt.

Having a great ape body also places important constraints on how they can live their lives. Perhaps most important is their pattern of growth and development. There is a little flexibility; in both apes and in humans, for example, the speed of a youngster's growth and the age it becomes sexually mature can vary a bit, depending on how well it's nourished and perhaps also how well its mother was nourished.[19] But the basic pattern of growth and development is programmed by its genes. Compared to most mammals, apes grow very slowly. So do humans, and there can be little doubt that our ape ancestor that lived millions of years ago was also a slow grower. Because their young grow so slowly and have so much to learn, the lives of female apes are much more constrained than the lives of male apes. For most of their adult lives, female apes are caring for an infant and may have one or two older offspring hanging around as well.

The four different great ape species living today organize their social lives in different ways. Orangutans live on their own, although a mother and her youngsters live together. Gorillas live in small groups, usually made up of a male and a couple of females and their young. The apes most closely related to us, chimps and bonobos, live in much larger groups, often consisting of a few dozen adults of both sexes and their young. Mother chimps and bonobos spend a large part of each day foraging on their own with their youngsters, but the group generally meets up as evening approaches and the members tend to sleep in a group of adjacent trees.

Since we humans also organize ourselves into large social groups, it seems a fair guess that the ancestors that we share with chimps and bonobos lived in social groups similar to theirs. If so, these groups were probably organized into a hierarchy, with higher-ranked group members dominating the lower orders and getting their own way. The hierarchical behaviors seen in chimpanzee and bonobo groups look very antisocial to our eyes, but conflict is bound to occur in a community of individuals who are competing to find food in the same limited territory. Hierarchical behavior isn't cooperative, but it does tend to reduce the fierceness of the competition. Everyone is safer when it's routine for those lower down the ranks to simply give in. That way, fighting only happens when lower-ranked members think they have a realistic chance of moving up and replacing someone above them. Hierarchy, therefore, is a second "social tool" that our ape ancestors probably used to make their lives easier. Once hierarchies are established, members of the lower orders might decide to cooperate. For example, among both chimps and bonobos, males are larger and will be aggressive to get what they want. In bonobo groups the females often form alliances and support each other. Bonobo males are still aggressive toward females, but they don't dominate them to the extent seen in chimp groups.

Grooming is a third "social tool" that apes employ to improve their lives, and it's also used by many other group-living mammals. Mammals stay healthier if their skin and fur are regularly groomed to remove dirt and insect parasites like ticks and fleas. But grooming often has an important social and psychological function as well.[20] Many group-living mammals form friendships, and special friends tend to groom one another (see Figure 2.4). They spend quite a bit of their day licking, biting, scratching, or picking away at each other's skin and hair. It's stressful to be always looking for food and

Figure 2.4 As apes carefully remove bits of dirt, dead skin, and insects from the skin of their friends, they demonstrate their trust and their trustworthiness.
Picture credit: Thomas Deco.

keeping out of the way of the dominant male, especially when you also have a baby to protect. It must help to have some groupmates that you really trust. By allowing themselves to be touched, often in intimate sensitive places, grooming partners show one another that they feel trust. If a grooming session goes well, it helps cement their friendship. The partners are reminded that their trust is well placed. But there's more going on that just learning. The bodies of social mammals release chemicals during grooming that influence their mood. Being groomed feels good. Oxytocin, the hormone that causes mothers to bond with their babies, is released during a nice grooming session, so the bonds of friendship that animals create and maintain when they carefully touch one another is supported by the same hormone system that supports the fierce mother-infant bond.

Different kinds of animals have different ways of grooming, which isn't surprising, because they have quite different bodies. But there are differences, even among apes. For example, bonobo grooming involves much more touching of sexual organs than among either chimps or gorillas. What's more, some of the most active touching of sexual organs occurs between bonobos who don't know one another very well. A friend of ours who observes bonobos in the wild told us that this grooming seems more like a social obligation with grumpy neighbors than a pleasant, relaxed time with friends.[21] Some grooming may be more about keeping the peace than skin care and friendship.

It could be said that the ape body is programmed by their genes to enjoy being groomed, but this doesn't mean that grooming behavior is programmed into apes. Exactly how ape friends touch one another varies between friendships and between groups as well as between species. If grooming is a social tool that some mammals use to help them form and keep friendships, then the cleverness of apes allows them to develop this tool and add individual touches. Genes indirectly influence ape behavior by providing them with an ape body. The structure and functioning of its body define what an animal can do and feel, and what its limits are. But, like us, apes gather information throughout their lives, and this learning influences how they behave. Short-lived animals with small brains are known to respond to certain situations in set ways,[22] but apes are long-lived and have large brains.[23]

Apes are also able to get around some of the constraints placed on them by their body. For example, a female chimp can't help advertising to the males around her that she is becoming fertile, and the males can't help wanting to mate with her. When an egg is ripening in one of her ovaries, a female chimp can't prevent the tissues around her vaginal opening absorbing water so that they swell up and start to protrude several centimeters behind her crotch. As the egg becomes ripe, the skin of the swollen area absorbs extra blood, becoming warm and red. Male chimps find this extremely attractive and, because males are much stronger, females don't really have a choice about mating. During her fertile period a female chimp may mate 30 or more times and with several different males. But this doesn't always happen. Several primatologists have reported observing a female disappearing with a single male for several days. She seems to spend her whole fertile period with him. (We say "seems" because it's impossible to know for sure. Primatologists following chimps around their habitat can only see what they manage to see, and it's possible that other males did see and mount the fertile female when the human voyeurs had lost track of her.) But what these observations show is that female chimps don't seem to be "programmed" to have sex with many males. We can't conclude that our female ape ancestors were either.[24]

We also can't know if our female ape ancestors who lived seven million years ago had sexual swellings or gave any other sign that they were ovulating. Chimps and bonobos signal their fertility with sexual swellings, but these aren't a general ape characteristic. Neither gorillas nor orangutan females advertise their ovulation in such an obvious way, and, of course, neither do human females. Nor can we know about the structure of the groups that our ape ancestors lived in. Primatologists describe chimps and bonobos as "male philopatric," which means that males generally spend their whole lives in their mother's group (which is also usually their father's group). Their sisters tend to move to a different group. We can't conclude from this that male chimps are genetically programmed to stay near their mother and females are programmed to want to leave her. It may just be the most reasonable thing to do. A female is usually accepted if she moves to a new group, while a male venturing into the territory of another group is likely to be killed by males from that group. Some female chimps do choose to stay near their mother, but they pay a price. Many of the males in the group are their close relatives. A baby conceived through mating with her father or brother is less likely to be healthy than a less inbred baby. A female has the best chance of reproducing successfully if she goes to another group's territory and tries to fit in with them. That's what most female chimps and bonobos do. It's interesting to try to imagine the feelings and perceptions of a young female chimp as she decides to leave her mother and venture into an unknown community. But we can't know what it's like to experience life with a brain so different from our own.[25]

Given what we said earlier in this chapter about mothers being obsessively bonded to their offspring and entirely responsible for their care, it's important to point out that there is also some flexibility here. Some mother chimps have been observed to allow certain trusted individuals to hold their baby, such as their own mother, an older offspring, or even a human primatologist. And, although males don't usually care for infants and youngsters, males have been observed walking around with an orphaned infant clutching onto their abdomen. Males have even been spotted giving food to a hungry orphan. If a mother chimp dies before her infant is weaned, its chances of survival are slim. But orphaned juveniles aged five or six who still need to follow their mother around are sometimes "adopted" by another more experienced female, and sometimes by a male. The adopter is most likely to be an older sibling or a friend of their mother.[26]

Culture introduces more flexibility

The behavior of the apes and humans living today is strongly influenced by the cultural legacy they receive, and this was almost certainly also true for our common ancestor with chimpanzees and bonobos that lived seven million years ago.[27] When people think about what nonhuman animals inherit from their parents, they tend to think only about genes. We humans might inherit houses, old-fashioned furniture and some photographs from our parents, but what do other animals have to pass on? In fact, many young animals need to be endowed with a lot more than genes if they're to get a foothold on life. They need food, care, and protection while they grow, and they need to have the experiences that will allow them to develop their abilities.

Many abilities do seem to develop just as a matter of course. A baby ape, like all baby mammals, is born with a body that is able to move in certain ways and experience lots of sensations and feelings about sensations. These develop as a result of the genetic inheritance it received. The nose of a newborn ape can detect chemicals in the air, its brain processes the information, and some of the chemicals trigger feelings. Some chemicals trigger feelings of hunger. Others trigger feelings of fear. The eyes of a newborn ape can detect light, and its brain can perceive patterns in the information coming from the eyes. Some patterns trigger comfortable feelings. Others make the newborn feel uneasy. The newborn's skin, ears, and internal organs are also sending information to its brain, and this is being processed too. Its genetic inheritance creates all this. It also produces a baby ape that is ready to learn and is poised to develop further. But when it's very young, it mostly just tries to be close to its mother's abdomen and drink her milk.[28]

As with most mammals, it's the mother ape who provides virtually all the extra nongenetic endowment that her youngster needs. Male apes rarely pass on anything but their genes, while females spend all, or nearly all, their adult lives caring for at least one offspring. When the eight or nine months of pregnancy are included, a mother ape spends between three and seven years in intimate contact with every baby she successfully raises. Her efforts give her baby the time it needs to grow and develop. And if it is to develop skills, she needs to give it more than food and protection.

Once her youngster is past the clinging stage, a mother ape begins to pass on her expertise. In the course of her endless foraging, she demonstrates again and again what her baby will need to do to get food. There's very little evidence that mothers *consciously* teach their youngsters. It's more a matter of helping them to learn for themselves. Being near her increases the chances that a youngster will learn complicated foraging techniques that an ape (or even a human) would have trouble thinking out for itself. Each mother gained the knowledge and skills she possesses while she was being cared for by her own mother, so mothers aren't just passing on their own personal expertise, they're passing on a family legacy. To succeed in life a young ape must receive a "cultural inheritance." Apes may not have a culture of human-grade complexity, but their cultural legacy does include traditions, survival secrets, and the "tricks of the trade" of being an ape that makes its living in a particular set of habitats.[29]

One of the best-studied chimp foraging skills is "fishing for termites."[30] Termites are extremely nutritious (about 20 percent protein, rich in vitamins, and full of the kind of fatty acids that a growing brain needs). It's almost certain that some of your ape and human ancestors ate them. There are about a thousand different kinds of termites in Africa. The ground underneath a tropical forest has large communities of them, with a queen and different "castes" doing jobs like "worker" and "soldier." They build large nests or "mounds" with tunnels and chambers. Despite their ant-like lifestyle, they're actually related to wood-eating cockroaches. They consume the wood and leaves that fall to the ground, getting nourishment from them with the help of microorganisms that live in their tiny digestive systems and inhabit chambers in their mounds.

Catching termites is difficult. Their underground life makes them hard to get at, and, not surprisingly, an animal that is such a nutritious food item has evolved elaborate protection. Your termite-eating ape ancestors may have used the same technique to get around their defenses as chimpanzees use today. Termite mounds are tough and difficult to dig through, but they need a ventilation system, and this is what makes them vulnerable. Chimpanzees make a flexible probe, sometimes more than half a meter long, by tearing up a leaf, blade of grass, or twig. They gently push this tool down a ventilation hole and deep into the mound. If the probe gets down to where termites are working or guarding the nest, the insects grab onto it. Then, by gently pulling the probe out, a chimp pulls out the attached termites, nibbling them off the probe as they arrive at the surface. Termite "fishing" is interesting to scientists studying human evolution because they believe it showcases some of the most human-like features of chimp behavior. It's an intricate tool-based skill that is difficult to learn and that most chimps can't figure out for themselves.

Since cultural inheritance is so important for chimps, it's hardly surprising that, once primatologists began to study chimps in a systematic way, they found cultural differences between the groups.[31] Each chimp learns what it needs to know to make a living in the habitats where it forages. Groups that live in rainforest areas have a body of cultural knowledge that's different from groups with a territory that includes drier forests and savanna. But what chimps know is also determined by their group's history—what former members of their group learned to do and what they passed on to their offspring. Termites and their mounds are found all over Africa, but the chimps in some groups have never been seen to try to catch them, presumably because they've never learned how to do it. They may not even know that that it's possible to get nutritious morsels from the hard, earthy structures that they see every day on the forest floor. It's not because chimps in these groups are stupid. Non-termite-eating groups have their own foraging tools and intricate skills—they may use rocks to crack nuts or sharp sticks to stab sleeping bush babies—but termite fishing isn't part of their cultural legacy.[32]

The wide diversity in chimp behavior is important to note because, as Darwin explained, variation is the engine of evolution. The more diverse and flexible a population is, the greater the possible directions for change. Having a great ape body places limits on how they can live, but the flexibility of their behavior partly compensates for this. They can adapt, and their ways of adapting can sometimes be passed on to their offspring. The other important component of Darwinian evolution is inheritance. If a mother ape discovers a new way of getting food in the meager forest buffet, her knowledge can be passed on to her own offspring, and, as long as this knowledge continues to be of value, it is passed on down the generations.

There's no reason to believe that the cultural inheritance received by our ape ancestors that lived seven million years ago was any more complicated than the culture of today's apes (or monkeys or dolphins or any of a number of animals that have a simple culture).[33] But our side of the ape family tree was destined to evolve cultures of dazzling complexity. The beginnings of this were present in our ancestors who lived seven million years ago. Their apish ability to learn, and especially to learn by watching others, eventually began to increase.

This increase didn't occur in the branches of the ape family tree that stayed in the tropical forests because, however good they are at learning, their cultural inheritance was limited by the way they raised their young. Each young ape is cared for entirely by its own mother, so it benefits only from the knowledge and skills that she has learned. Apes may learn a few things for themselves as they go through life, but once they've left their mother's side, they will no longer have the chance to closely observe another ape foraging. If a female develops some new bit of know-how, there's a good chance that her offspring will learn it from her. But if a male chimp learns something for himself—if he discovers a new food item or invents a new way of getting at a food item—his innovation and insight will likely die with him. Every aspect of ape culture is maternally inherited, so the cultural matrix of an ape group is supplied entirely by the females. And it can never be more than a flimsy matrix because there are few cross-links between matrilines.

At some point, perhaps about four million years ago, our ancestors began to spend more and more time outside of the tropical forest. This, we believe, made it necessary for them to change the way they raised their young. The ancestor featured in the next chapter was a great ape, but it was quite different from any great ape alive today.

3
Apes That Walked Upright
(About Three Million Years Ago)

Millions of apes lived and died between the time when the human lineage split from the branch that led to chimps and bonobos and the time when the first humans appeared. A tiny fraction of them left behind physical remains—bones and teeth—that were preserved. By studying these remains and comparing them to bones from humans and apes, scientists have been able to draw some conclusions about the animals they came from. We know roughly when they lived, how they moved, what sort of habitat they lived in, and how closely they were related to our ancestors (see Figure 3.1).

Among them were some apes that were "bipedal"—they walked upright on two legs. Unlike earlier apes, they spent quite a bit of their time outside tropical forests in drier and more open habitats. The first evidence of them was a fossil skull found in 1924, in what is now South Africa. The fossil was given the name "Taung child" because it was small and still had some baby teeth in its jaws. Its discoverer realized the skull belonged to an extinct ape species never seen before. He made up a name for the species by putting the Latin word for "southern" (*austral*) with the Greek word for "ape" (*pithekos*). Since then, hundreds of "australopithecine" bone fragments and teeth have been found in many parts of southern and eastern Africa. Some of their footprints have even been found.[1] Paleontologists agree that some australopithecines must have been our ancestors.

Your australopithecine ancestors born three million years ago opened their eyes on a sunnier landscape than the ape in the last chapter. This landscape had some trees, but it wasn't heavily forested. These ancestors might have been born on the ground rather than in a tree. Like all non-human apes, australopithecines were born at a more advanced stage of development than humans, and their newborns had a smaller and thinner body. But they had at least one very human feature. If you had been born as this ancestor, you would be revealing it not long after your birth: you would be pulling yourself up onto two little legs, and before long you would be walking upright alongside your big brother, holding his hand (see Figure 3.2).

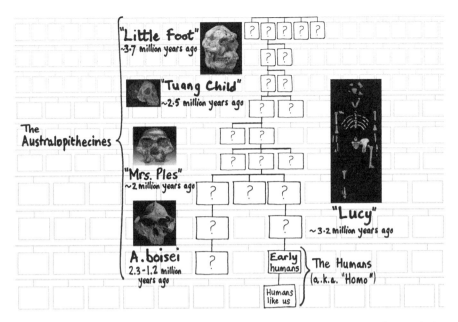

Figure 3.1 The ape-human family tree is still full of question marks, but we know a little more about our ancestors who lived between two and four million years ago. And we even have some pictures for the family album.

If you were born as one of your ancestors that lived three million years ago, you wouldn't just be physically different from the ape in the last chapter. You would also be eating different foods and learning to forage in a new habitat. Your upbringing would also be different, because you would be cared for by a number of different apes, not just your mother. This way of caring for youngsters was part of a new kind of social life for apes. It required new kinds of social interaction and different responses to emotions such as anger and internal feelings such as hunger. These changes were important because they eventually changed the way our ancestors evolved.

Imagine growing up as this australopithecine ancestor.

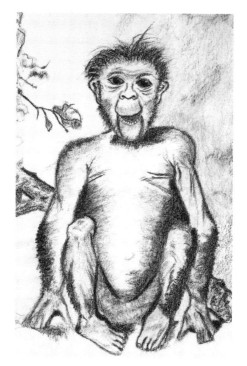

Figure 3.2 The shape of their bones tells us that australopithecines walked upright on the ground more or less as we do. But their bones also reveal that the body proportions of australopithecines were much more apelike than human. They had longer arms, shorter legs, and much bigger bellies.

For the first six months of your life, you sleep a lot, but when you're awake, you do your best to explore the interesting things around you. Most interesting are the others that you live with. At first you just see them as large, warm, and cozy moving things with hair that you can clutch when they hold you or when you crawl and clamber over them. But you also notice that they have faces that move in interesting ways and make noises. You soon learn to recognize each one of the others you live with. One of the large ones is your mother and you get most of your milk from her, but other large ones also have milk and will let you have some if you're hungry and can't find your mother. When you feel hungry, uncomfortable, or worried, you make a quiet whimpering noise, and someone comes and pays attention to you.

As you grow older, you learn more about what's going on around you. During the day, mothers and some of the larger youngsters go away for a time. They come back later, bringing things. Sometimes your mother goes away too, but you're never alone. There are always bigger ones nearby. You don't like to be too far away from the others, but as you become more mobile you sometimes forget and just keep wandering. Someone notices and brings you back before you get into danger. The group stays in one place for a while—perhaps a few weeks—and then everyone goes on a journey, sometimes an uncomfortable journey, to a new place. When you're very little, bigger apes take turns carrying you on journeys, but the bigger you get, the less you get carried and the more you have to walk for yourself. Everyone is happy when you get to the new place. There is water to drink and you can rest and explore new things.

Lots of different kinds of animals come around, and some of them make everyone very nervous. Sometimes, just their smell makes you want to hide. They try to sneak up on youngsters, to snatch them into their jaws and haul them away. When someone sees an animal creeping up, they scream, and then everyone screams and runs to climb into trees or onto a boulder. As they go, mothers and older youngsters often pick up sticks and stones to throw at the animal. Danger can also come from above. Some birds of prey are big enough to pick up youngsters in their talons and carry them off.[2] When someone spies a large bird of prey, they make a squealing sound similar to the sound made by these birds as they fly around the sky in the springtime.[3] Shorter, gruffer sounds are made to warn someone if they are too close to some danger they don't see, like a snake or stinging plant. Everyone has their own special sound, and if a little one hears its own sound, it stops what it's doing. You first learn to recognize the warning sounds, and then you learn to make them yourself.

You're not the victim of a predator or a snake. All your ancestors survived to adulthood and had offspring of their own. This was thanks to the care they received not just from their mother, but also from the others who lived with them.

Every so often, strange apes come by that are larger than the mothers and smell different. If they hang around the group, the mothers don't get nervous, but they sometimes get flustered. Sometimes one of the mothers goes with a stranger and mates with him.

It's always very exciting to see a new baby come out of its mother. The whole group wants to touch it and smell it. At first it stays with its mother or with one of the other mothers, but after a while it starts to move around on its own. All the youngsters want to play with a new baby, but sometimes their play is so rough that it whimpers. This causes the mothers to make warning noises in an angry way. You play more gently. But a baby can be rough too, and sometimes it pulls your ear or your lip so hard that it hurts. This makes you want to hurt the baby back, but you don't. You bit a baby once and it really screamed. The mothers were very angry and one of them hurt you. For a while, none of the mothers would be friendly or let you have milk. But then, just in time, because you were starting to get very hungry, things went back to normal. The whole experience made you frightened, and you don't forget it. You never hurt a baby again and always think twice about hurting smaller group members, even if they make you angry. As you grow bigger, you take a turn carrying a baby during the journeys, and you also help to look out for the babies. You notice if one of them crawls too far away, and you carefully bring it back. When you do this, you may get a cuddle from a mother and some milk. This makes you feel very good.

As you grow older, you also become more interested in what the mothers and larger youngsters bring back after being away. At first, it just seems like a jumble of woody stuff, but the older ones are very interested in it. In time you realize that they're interested because there's food in the jumble and they're working to get it out. They seem to just fiddle with things using their fingers, lips, and teeth, or they pound and rub at things with a stone. You pick up stones and try to fiddle with them like they do. It's fun to play at being like them, and the mothers seem pleased to see you doing it. You keep watching and, in time, you start to be able to pick out the details of what they're doing and understand the purpose of each action. At the same time, you're getting better and better at controlling your limbs and fingers and lips. Eventually, you start extracting bits of food from the stuff. After that, it all comes together quite quickly. You have become a worker like the mothers and the bigger youngsters. And you're very glad to get the extra food because the mothers aren't letting you have as much milk as you would like and you're hungry.

The older youngsters like to go out foraging with the mothers, and when you're old enough to be able to run, your mother lets you go with her. You need to be able to run fast in order to get away from danger, but mostly you walk on these trips. As you walk along, your mother is always looking out for danger and for things to eat. On these trips, you learn about all kinds of new and marvelous foods—eggs, berries, insect larvae, honey. Then you help collect tubers, nuts, and stems. These need to be worked on to get the food out. Together you take them back to the group and put them in a jumble on the ground. You want to go out foraging every day, but sometimes you must stay back and help with the little ones. You often go out with your own mother, but sometimes you go with other mothers. This is interesting because other mothers sometimes go after foods that your mother doesn't bother with. You notice that one of the other mothers keeps a thorny twig stuck in the hair on her shoulders. This means she always has a tool handy if she sees a fat juicy grub that needs to be fished out of a narrow crack or crevice. You find your own thorny twig and carry it in the same way. In fact, you decide to carry two of them in case one breaks. You get better and better at foraging for food, but there's a lot to learn. Every time the group journeys to a new place, there are new areas to explore and new kinds of food to learn about.

You keep growing bigger and, when you're about 10 years old, you start to become an adult. If you're a female, the transition doesn't bring such big changes. One difference is that the strange males who visit start to be interested in you and you feel shy. The mothers protect you from the males for a while, but when you become less shy, they let some of them be with you and you mate with them. Eight or nine months after the mating, a baby comes out of you and you are now one of the mothers in the group and you make milk. You want to take special care of your own baby, but you continue to care for other babies and youngsters of the other mothers, and they help to care for your baby.

If you're a male, reaching adulthood brings big changes to your life. You keep growing and growing until you're bigger than the mothers. You're strong too, and could really hurt someone if you felt like it. You've learned not to hurt little ones, but they still feel nervous around you. You're so big and you smell different, more like the strange males that sometimes visit. It's also because you seem angry. You're not so much angry as always hungry. Now that you're bigger you need to eat more, and you no longer feel frightened to be away from the mothers. You start to feel less comfortable staying with the group than when you're away on your own finding food. You start to spend more and more time away and visit your mother less often. You find some other males and hang around near them sometimes. You sometimes follow an older male when he visits a group of mothers and you watch him mate with one of them. No one minds as long as you watch from a distance. Eventually you decide to visit groups of mothers on your own, and after a while some of them choose to mate with you. Even though you're bigger and stronger than females, you know from experience that mothers don't like bullies, and besides, you know you're no match for a group of several angry females. So it is the females who decide if they want to mate with you. Pleasing them is more successful than using force. Using force has negative consequences.

Why do we think our australopithecine ancestors lived like this?

We know that at some point in our evolutionary story, perhaps about four million years ago, our ancestors started to live in different habitats. Earlier apes and the apes that live today—chimpanzees, bonobos, gorillas, and orangutans—are adapted to a tropical forest habitat. There are groups of chimpanzees living on the edge of forests, and they often forage in less densely wooded areas. Sometimes they venture out into drier grasslands for a while. But our australopithecine ancestors traveled farther from the forest, stayed longer in the grassland, and eventually began to live in the habitat called "savanna." This gave them new opportunities for making a living. The tropical forests were slowly shrinking as the Earth's climate cooled and became less rainy.[4] As the climate changed, organisms on the edge of the forests attempted to survive in more exposed lands that didn't have such abundant and regular rainfall. Some were able to survive and reproduce, and their descendants created new habitats with a wide diversity of new species. The new habitats had some trees but were more shrubby and grassy than the woodlands and forests that had been the home of apes for millions of years.

The evidence that australopithecines were out in these new habitats comes from their remains—the many bones and teeth that have been found. The places where these remains were found is one clue. Because many of them fell on drier ground, they weren't rapidly dissolved in the moist acidic soils of the tropical forest. A few ended up in places where they were preserved for millions of years. In many cases, these remains have been found among animals and plants adapted to drier habitats. The shape, size, and strength of australopithecine teeth provide further evidence that these apes were living in drier habitats. They are the teeth of an animal that grinds up hard and tough food, like the seeds, tubers, and plant stems common in grasslands[5] rather than the fibrous but relatively soft fruit eaten by apes living in the trees of tropical forests. The chemical composition of the teeth also provides evidence that the animal that grew them was living in grassland rather than tropical forest.

Australopithecines, or at least some of them, might have still spent a good deal of their time up in trees, but they had adaptations that would only be present if they also spent a lot of time on the ground and outside the forests. Unlike other apes, they didn't have a thumb-like big toe that could hold onto branches and make it easier to climb around trees. Their big toe pointed forward like ours. It was a big toe that made it easier to balance on two legs and push each foot off the ground as they walked and ran.

Solving the problem of life in the grasslands

Adapting to the gradually expanding African plains wasn't just a matter of having pluck and an adventurous spirit. We believe that a specific problem kept earlier apes from fully exploiting this habitat, and the same problem prevents today's apes from spending a lot of time away from wooded, forested, or swampy habitats.[6] For apes, raising young is difficult in drier and more open habitats. The most obvious problem is predators. It's easier for predators to spot and run down their prey in more open environments. Youngsters and mothers carrying babies are slower and most likely to be the ones caught. The second problem was surviving in a habitat where clumps of food and sources of water are far apart. Again, this was more of a problem for mothers and young ones.

We think that the solution australopithecines found was for females and their youngsters to group together and help one another.[7] Upright posture would have given them a chance to detect stalking predators at a reasonable distance, and being in a group would improve their chances. The bipedal australopithecines had arms and hands free to use stout sticks as clubs and to throw stones. If caught in the open and unable to make a dash for the trees, a mob of club-wielding and rock-throwing mothers and juveniles would not be easy prey.

In all the ape species living today, mothers raise their babies almost entirely on their own,[8] and their lives are extremely busy. It's only possible for them to satisfy their baby's needs because, in a healthy tropical forest, they're surrounded by small bundles of food and water. As long as they know how to find and collect the morsels and can put in the hours necessary to collect enough of them, they will survive and there's a reasonable chance that their infant will too. This isn't the case in drier habitats. Items of food in grassland and savanna habitats are often larger and more nutritious than the forest morsels, but they're farther apart. In one location, for example, there may be an underground tuber weighing a couple of pounds and full of starch. Half a mile away there might be a tree full of nuts, and half a mile in another direction there might be a hive full of honey and bee larvae. Meanwhile, the lake, which is the only source of water in the area, might be half a mile in a third direction. Drier habitats are also more changeable than forests. They change more with the seasons and they change more from year to year. A mother might be knowledgeable and hard-working, but if she's unlucky she might still not find enough food. After a couple of unlucky days in a row, she and her baby would be starving and dehydrated.

What's more, simply being out in the hot dry sunshine away from the cover of trees would have been dangerous and uncomfortable for an australopithecine mother and her infant. The heat and the dryness would be more of a problem for mothers than for males. Every time her infant suckles, an ape mother's body releases a considerable amount of water along with the nourishment. This isn't a problem when the mother can easily replace the water her body loses from a stream or a tree full of juicy fruit. But it is a problem if she is carrying her suckling infant half a mile away from a source of water.

These problems could be overcome if mothers banded together. They could find a central place, a sort of "home base" that could be defended and where there was cover and a source of water. The infants could stay there with some mothers and older youngsters while others took turns to go out to forage. There would be risks. To leave her infant, a mother has to trust that another female will protect and maybe feed it while she is away. But the risk of leaving her infant has to be traded off against the risk of carrying it while she forages in a hot dry habitat. Mothers also have to trust that some of the foraged food will be shared. If one of them isn't able to find much food during a foraging trip, chances are that other mothers will be luckier on that day. Many of the food items they find would need to be peeled, cracked open, ground up, or processed in some other way to get at the edible bits, but australopithecine foragers didn't need to do this out in the hot sunshine. They walked on two legs, and so their arms and hands were free to carry food back to the home base where it could be processed in more comfort. When the area surrounding the home base was depleted of food, they could make a journey that would take them to a home base in a new area.

It's possible to think of other ways that australopithecines could have organized infant care. What to do with infants when you need to go out for food is a problem that many mammal mothers face. Hiding them in a burrow or nest is one option. That's what a cat does with her newborn kittens. But this option wasn't available to australopithecine mothers. Newborn apes (and also human newborns) are too developed and active to be just hidden away somewhere, and they are dependent for years, not just a few weeks.

In a few mammal species, males and females pair-bond and the father helps to care for the baby.[9] It has been suggested that our ancestors may have pair-bonded and that infants might have stayed with their mother in a safe place while their father went out to get food to bring back for the family.[10] In the case of our ancestors, this scenario is less likely than mothers banding together.[11] One reason is that pairing up with a male would be riskier for the mother and her infant. It would make the infant's survival dependent on a single male. If he died, the infant would die, and its mother might as well. Leaving his mate and their infant to bring back food would also be risky for a male australopithecine. Examination of their remains suggests that males were larger than females, so the female would be unable to protect herself and her infant from other males. If a pair-bonded male left his mate and offspring to get food for them, another unbonded male could kill the infant and rape the female. In all the known mammals that form pair-bonds, the male and female are inseparable, and the male is the same size or slightly smaller than the female. It is generally observed in nonhuman mammals that when the males grow to be larger and stronger than the females, the males compete to mate with as many females as possible.

It seems reasonable to us that australopithecine mothers trying to raise their young in drier habitats would have developed an infant care solution similar to that used by sperm whale mothers. Sperm whale females have a similar problem raising their young. Like young apes, sperm whale calves are very slow-growing and born about five years apart. But sperm whales are predators and younger calves can't accompany their mothers on hunting trips. The main food of sperm whales is squid that live at depths greater than 300 meters. Hunting these squid involves diving, sometimes to depths of more than a mile, swimming rapidly, and spending more than half an hour at depth. To solve the problem of what to do with their calves while they hunt, female sperm whales live in stable groups of between five and ten mothers who are often but not always genetically related. While a mother is away hunting, her infant stays on the surface with other group members. The mothers schedule their hunting so that there is always an adult present on the surface. It's difficult to get close to groups of sperm whales in the open ocean to observe the details of their behavior, but observations of mothers who are keeping an eye on another mother's calf suggest that the babysitters sometimes allow the calf to have some of their milk.

Male sperm whales are born and grow up in their mother's group, but as they approach maturity, they set off to live in the colder water near the poles. Male and female sperm whales have different deep-water hunting grounds. Females tend not to go into colder water, probably because their smaller calves can't thrive in the low temperatures. By leaving the warmer water, males avoid competing with the females, so there is more food for them and their offspring. Young males may briefly join a bachelor group but mostly they live alone, heading to warmer regions occasionally to visit female groups to see if any of them want to mate. Because males don't have to put effort into producing and caring for offspring, they can put more effort into growing and maintaining their own body. Large size is less of a handicap for animals that live in water, so male sperm whales grow to become huge—the largest are over two or three times the size of the females, at over 20 meters long. Females prefer to mate with large males and it's easy to see why natural selection would encourage this preference. All animals living in the wild go through hard times when food is scarce or they're fighting an infection. These difficulties may not kill them, but they do tend to make them smaller. By choosing to mate with larger males, females are choosing healthier and more competent males. This increases the odds of their offspring inheriting genes that will help them avoid bouts of malnutrition and disease.[12]

The larger size of australopithecine males is probably explained in the same way. Larger males would have had more chances to mate, and there were also practical reasons for being as large as possible. Predators were less likely to tackle a bigger male, for example, and longer-legged males could run faster. As is the case with male sperm whales, male australopithecines would have been able to forage in areas where females couldn't go. Moving away from the females and foraging in those areas would have increased the amount of food available for the females and youngsters. The males likely also experimented with foraging techniques that the smaller females would be less inclined to try, such as scavenging meat by frightening predators away from an animal they had killed. Perhaps groups of male hunted together for large prey.

An important difference between sperm whales and australopithecines is that young whales only need to learn how to find and catch their prey. They don't process their food. Much of the food australopithecines found would need processing in some way, such as peeling, pounding, or cracking open. We think it likely that mothers brought these food items back to the safety of their home base, allowing the young to start to learn the processing skills they would need.

Other adaptions to life outside the forest

To survive in more open landscapes, where items of food were further apart, our ancestors needed to be able to move around the ground in a way that allowed them to cover long distances efficiently. Walking on two legs is quite an efficient way of traveling, and it takes less energy to move along the ground than to climb up and down trees.[13] Even so, our ancestors had to adapt to eating different foods and cope with new dangers. Examining their remains reveals how their teeth and bones had adapted. Because they could walk on their two hind limbs, their front limbs could evolve to be more specialized. Apes need nimble fingers to forage effectively, to grip, manipulate, and carry things. But the hands of chimps, bonobos, and gorillas need to serve a dual purpose. The first two knuckles of each hand have to do double duty as a sort of hoof. Their index and middle finger have to take their weight and absorb shocks as they gallop along the ground on all fours. The need to walk on their hands constrains the evolution of their fingers. They can't evolve to be that nimble.

Evolving the ability to walk and run balanced on their hind legs involved many other changes to their body. Changes to their pelvis meant that our australopithecine ancestors gave birth like humans rather than like chimps. Being born and giving birth is an arduous procedure in all primates, but for humans (and probably australopithecines) it's literally less straightforward. Primates that walk on four legs have a pelvis that's wider in the front-to-back dimension, and this allows the baby to pass straight down from the womb and emerge from the vagina facing front. The pelvis of humans and australopithecines is shaped for balancing on two legs and is widest on the side-to-side dimension. To pass out of the womb and into the world, their baby must rotate so its head can fit into the birth canal and then rotate again to allow its shoulders through. What's more, a large bone at the base of the mother's spine (the sacrum) sits at the back of the birth canal. This means that babies are usually born facing their mother's back, which makes it difficult for the mother to help get her baby out. When other primate mothers give birth, they can reach between their legs and pull their baby out. A chimp mother who is about to give birth goes away to be on her own throughout the ordeal. In human cultures, mothers usually have help delivering their babies, especially the first few times. This may well have also been a tradition in australopithecine groups.[14]

The remains of australopithecines can also provide clues about the muscles that were attached to their bones and the organs that the bones supported and protected. The size of their skulls tells us that their brains were no larger than chimpanzee brains. Their ribcage was the same shape as the ribcage of chimpanzees, and this tells us that it sat on top of an abdomen that was, in proportion to the rest of their body, considerably larger than ours. It suggests that the end part of the australopithecine digestive system was like that of chimpanzees. They had a very large colon where the fiber of plants was digested with the help of bacteria. Even if their diet was different from that of a forest-dwelling ape, we can be reasonably sure that foods eaten by our ancestors three million years ago still included plants with a high proportion of the plant fiber that mammals can't digest, and which were fermented by bacteria living in their digestive system. They weren't able to cook their food but might have pounded tubers and other fibrous plant matter with stones to help break down plant tissues.

There are, of course, plenty of things about australopithecines that can't be deduced from looking at their bones and teeth. But knowing what their bodies had to cope with in their new environment makes it possible to make some informed guesses. Random changes in the genes that organize embryo development are continually introducing subtle variations in all parts of our bodies. In the new environment, different variants would be favored.

For example, as they spent more time away from the cover of trees, they had to deal with strong sunshine and greater extremes of temperature—hotter days and cooler nights. Individuals born with skin that developed more "eccrine" sweat glands could keep cooler in the sunshine during the heat of the day.[15] Eccrine sweat glands release watery fluid onto the skin, and as it evaporates, the skin cools. Being able to perspire more would have increased our ancestors' chances of surviving by making it possible for them to put more effort into looking for food when other animals were resting in the shade. Sweating would have been more effective in individuals born with fewer hair follicles and/or follicles that produced finer hair, because their sweat would more often evaporate directly from the skin rather than being soaked up by hair. So natural selection favored individuals with thinner hair. The downside of having thin hair was that their skin got exposed to harmful ultraviolet radiation. Natural selection therefore favored individuals who developed more "melanocytes"—the skin cells that produce the skin's natural dark-colored sunscreen. In a species that walks upright, the top of the head is most exposed to bright harmful sunlight. Natural selection favored individuals who continued to grow a natural "hat" of thick hair on the top of their head that gave extra shielding from the sun.

When our ancestors began spending more time on the ground away from the cover of trees, they faced an increased risk of being caught by lions, packs of hyenas, and other predators. The youngsters were also at risk of being picked up by large birds of prey. To reduce this risk, they would have to be able to control the noises they made. Those who survived best were the ones who could learn to save their screeches and screams for real emergencies to signal the presence of danger. Even as babies, they would have to be quieter and more furtive in their behavior than apes living in the safety of trees.

If mothers lived in groups that shared the care of their young, they would have had to adapt to a very different *social* environment. The strong mother-infant bond seen in forest dwelling apes had to be weaker in order to allow a group of females to help each other care for their young. A mother had to be willing to leave her infant with a babysitter and also willing to babysit the infants of other mothers. Unlike their forest-dwelling ancestors, young australopithecines were in close contact with several adults. Growing up together and grooming each other regularly, the whole group may have felt closely bonded.[16] Nights might have often been cool in the new less-forested habitats, so instead of sleeping in separate nests, australopithecines may have huddled together, with the adults taking turns to be on guard. They would have needed to develop a different set of social tools than those used by the apes living in a forest. Some groups may have developed vocalizations that carried more information, the beginning of a simple language that allowed group members to warn the others of danger. Vervet monkeys, which live today in the same kinds of African habitats as our australopithecine ancestors, make warning calls, and the call they give differs depending on the kind of danger they detect.[17]

The experiences that animals have when they're young affects the way their brain develops, and this has a lifelong impact on how they perceive and respond to the world. If australopithecine mothers worked together to raise their young, youngsters would have been competing for the attention of their carers. To be successful, they would have to be sensitive to what others were feeling, trying to please them, and choosing the right moments to seek attention. The experience of growing up in this new social environment would therefore have had a direct effect on their behavior. The responses they learned would become "second nature" to them. But this social environment would have also had a less direct long-term effect on brain evolution. Individuals with genes that gave them a brain better able to learn how to behave appropriately in this kind of group would raise more offspring. Thus, natural selection would have favored genes associated with brains that became well-adapted to living in a community that shared food and the care of young.

Genes play an important role in organizing brain development (see Figures 3.3 and 3.4).[18] Our brains have the same basic components as the brains of apes, but our human genes influence the way our brain develops, and so does being connected to a human body and experiencing the world with that body. Some scientists believe that the inside walls of australopithecines' skulls can tell us something about the structure of their brains. Some of the ancient skulls that have been found have a sort of imprint that was left by the brain that grew there. These imprints suggest to these scientists that the brain in some australopithecine skulls may have been the same size as a chimp's brain, but it was closer to a human shape.[19] There's certainly no reason to believe that having a chimp-sized brain meant that our australopithecine ancestors behaved, thought, or felt like chimpanzees or like their close relatives, bonobos. As we pointed out in the last chapter, chimps use hierarchy as a social tool for reducing aggression. In chimp groups, there is a male dominance hierarchy and bonobo groups have a female dominance hierarchy. There are friendships in these groups, but the only reliably caring relationship is between an infant and its mother. Otherwise, individuals have to look out for themselves and be careful to control their behavior based on whether they're interacting with someone of higher or lower rank in the hierarchy.

In groups that share the care of the young, there is still competition to survive and reproduce—in nature there is always competition to survive and reproduce. But being competitive in a group that raises its young together requires more complicated social tools and a subtler control of emotion. Fitting into such a group isn't just a matter of knowing your rank in a dominance hierarchy. If our australopithecine ancestors raised their children together, they would have had to be less aggressive in their day-to-day interactions. But, at the same time, they couldn't be pushovers. They had to be able to work out

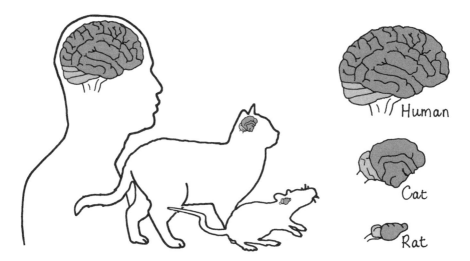

Figure 3.3 Brain Evolution.
The human brain is larger than the brain of most other animals, and it is shaped rather differently. This isn't because it contains a special organ for thinking in a human way. All mammal brains are made up of the same basic structures, but they're in different proportions.

Many scientists believe that the expansion of certain brain areas was crucial in human evolution, but they're a long way from agreeing to a story of how differences in our brain relate to differences in our behavior. Simply having a larger brain with more brain cells increases the number of neural connections that can be made. This allows more information to be stored and worked with, and a finer control of the body and emotions can be achieved.

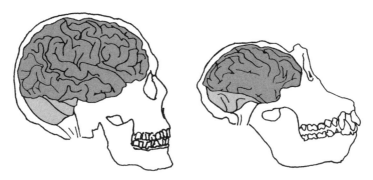

Figure 3.4 The evolution of our brain from that of our ape and australopithecine ancestors didn't require an entirely new set of genes. Changes in the size and shape of organs and body parts can result from alterations ("mutations") in sections of DNA that control how genetic information is followed during embryonic development.

To understand the evolution of the human brain, it's important to think about how the new conditions our ancestors faced might have favored the development of a slightly different kind of brain.

when it was right to stand their ground, especially when they became mothers themselves and had to deal with rowdy youngsters. And, of course, they needed to be very aggressive when their group was being attacked by predators or, perhaps, by another group.

The hierarchical social structure of their group obliges chimps and bonobos to also learn to control their behavior. But, if australopithecine groups had more complex caring relationships, they would have to be more flexible in how they act on their emotions.[20]

Learning to control your behavior can be tough, but humans seem to be born with an inbuilt advantage compared to chimps. During the 20th century, a number of scientists investigated the difference between humans and our ape relatives by trying to teach apes (mostly chimpanzees) language and other human skills. A few even tried raising a chimp in their home as if it were a human child. In the 1940s, a two-day-old chimp called Viki was taken home by a married couple who were psychologists and wanted to compare a chimp's progress to that of their own baby son.[21] The behavior of the two babies was similar at first, but huge differences emerged before they reached their second year. We complain about our toddlers being uncooperative and prone to tantrums because long experience with raising children has taught us that their behavior will eventually improve. We expect them to want to please us, to be distressed when they've made us annoyed and be delighted when we praise them. Viki the chimpanzee didn't care what her foster parents thought of her. Rewards had to be in the form of an edible treat; punishments had to painful. She wanted what she wanted and could not be deterred from using all her strength to get it. She was much stronger and more athletic than a human of the same age. As she became more mobile, rooms in her adoptive parents' home had to be turned into cages to keep Viki from helping herself to food and treating all their possessions as toys. Chimpanzees raised by humans are unquestionably very different from chimpanzees raised by their mother in the wild, but they can't be turned into humans. It's not simply that human children are more intelligent. Their emotional responses and motivation to learn are completely different.

Undoubtedly humans can be violently aggressive and capable of great brutality, but during our day-to-day interactions, we tend to be more docile and compliant than almost any other large animal. The difference between human and ape behavior can be seen at a very early age, long before children learn to talk. The genetic changes that caused our ancestors to evolve a brain inclined to such tolerance began when our ancestors began to raise their youngsters together, and it was linked to another important change—a change in the way that culture worked for them.

A selective cultural inheritance

As we described in the last chapter, apes living in the wild are experts at finding food in the habitats where they forage because they have a "culture" that is passed from a mother to her offspring. This cultural "inheritance" changes over time because the foraging knowledge that a young ape learns by watching its mother changes as the environment changes. For example, if environmental change causes a plant to be scarce, a youngster might not get the chance to learn how to recognize and process it. Even if its mother knows the trick of removing the spines from its seed coat, if her youngster doesn't see her removing them, it won't "inherit" the knowledge and its cultural legacy won't include this knowledge. This is cultural change (or evolution), but it's not "Darwinian" evolution, because there's no selection involved. There just happens to be fewer plants so the youngster didn't happen to see its mother remove the spines.

But selection could begin to influence cultural change in groups where mother apes raised their young together. Each mother would be bound to do things in slightly different ways, and youngsters would be able to watch most or all of them. This allowed them to be exposed to more overall information and to sometimes choose what to learn. For example, youngsters might see that while most mothers use a stone to crack the hard shell of a nut, one of them is quicker because she has discovered that if she places her teeth at a certain weak spot in the shell, her jaw muscles are strong enough to crack it. A youngster's mental ability would have influenced how it used the extra knowledge.[22] Those who were clever enough to make comparisons and select techniques that work best would be more successful than those who simply got confused by seeing multiple techniques. The more mentally able ones would be better nourished and have more offspring, so, on average, the group would become a bit cleverer. Each new generation would inherit better foraging skills, and because the better-nourished mothers would have more offspring, the genes that helped them be better foragers would be inherited by more youngsters. Over time, the group would become better at improving its shared body of knowledge and skills. By selecting which bits of culture to inherit—"downloading" and "uploading" what they learned to a shared body of knowledge—the group's culture would become more and more useful.

Also, in a group that raises its young together, knowledge about how to behave toward one another and how to raise offspring would be part of the group's evolving culture. New mothers wouldn't be complete novices because they would have had experience helping to care for babies in the group. And there would be more experienced mothers around to help. The chances of a baby surviving would be greater than in groups where mothers raised their young on their own. Successful youngsters will learn the parenting practices that contributed to their own survival and apply them to their own young.

The most successful groups would be those with an effective parenting culture that included customs of rewarding helping behavior and punishing harmful behavior. If mothers intervene in fights and punish youngsters who hurt those younger and smaller than themselves, then youngsters will learn to control the way they respond to feelings of anger. If mothers punish those who take food from their younger and smaller groupmates, youngsters will learn to control how they respond to feelings of hunger. The skill of controlling the way they respond to feelings will likely stay with them all their lives. This gives the groups social tools to help them live together. In a way, it's similar to the hierarchy social tool that, as described in chapter 2, helps to minimize conflict in chimp and bonobo groups. But it has the opposite effect. For a hierarchy to work, weaker individuals must learn to control their behavior to avoid conflict with stronger individuals. In groups that work together to care for their young, the stronger individuals have to learn to control their behavior so as not to harm younger ones.[23]

Groups with the custom of punishing bullying and ending cycles of retaliation will have fewer disputes and be more cohesive. These groups will be more likely to survive and grow so that, in time, they become so large that they need to split. The two subgroups will need to make journeys to different home bases, but both will be supplied with the cultural knowledge of the "parent" group. And so groups become like organisms. The information that makes them successful is stored in the culture of the group, just as information is stored in the DNA of an organism.

Back in 1865, Alfred Russell Wallace observed that humans are kinder, gentler, and more generous than would be expected from a simple interpretation of the theory of evolution by natural selection.[24] And Darwin himself observed in his 1871 book *The Descent of Man* that these mysterious characteristics could have evolved in our species because groups full of noble and high-minded people were more successful than groups of greedy, selfish people.[25] The question is, how were such groups originally formed? Today's very large societies are composed of individuals with a wide range of (sometimes rather vague) ideas of what constitutes misbehavior. Our ideas about justice and punishment are perhaps even more diverse. As a consequence, we have developed complicated (and not always very effective) institutions to identify, apprehend, judge, and punish individuals who behave in ways believed to cause damage.

Today's justice systems evolved from much simpler systems of justice. But long before a human existed who could conceive of something called "justice," mothers and their little ones needed to work together to survive in difficult conditions. It was necessary to control responses to emotions, including feelings of love, in a way that was good for the group. If her own offspring was being disruptive, a mother had to harden her heart. A parenting culture that favors behaviors that make it easier to raise offspring together and to thrive as a group doesn't just change the behavior of group members during their lifetime. It can also influence the frequency of genetic traits in future generations. To be successful, a group that raises its offspring together needs youngsters who *are capable* of learning to control the way they respond to feelings of anger, hunger, and lust.[26] Youngsters who can't, or won't, cooperate handicap the whole group and the offender might have to be eliminated. Groups that evolved the custom of casting out those who repeatedly misbehave are able to maintain an environment for raising offspring who are a credit to the group. And this custom also creates an environment in which future generations will be better behaved. Difficult youngsters who have been cast out of their parental group are very unlikely to reach adulthood and produce offspring, so the genes associated with their damaging behavior will become less common in the population. As a result, generation by generation, group members will become less aggressive and competitive and more group-oriented.

It was in this way that, even millions of years ago, the groups that contained our ancestors didn't just have members that looked a bit more like us. The behavior of their members was also a bit more "human." They had a lot more changes ahead of them, but they had taken some important steps. If we could spend time with our australopithecine ancestors, we would see that becoming more human wasn't just a matter of becoming cleverer or becoming gentler and more compliant. It involved becoming able to learn to behave in more complex, variable, and controlled ways.

Selecting a gentler temperament

An experiment that began in Russia more than 60 years ago has taught us what can happen to a population of mammals when gentleness and tolerance are selected. It's not yet possible to know how relevant the specific findings of this experiment are to the changes that occurred in our ancestors, but the findings are important to all evolutionary stories because they demonstrate how powerful, wide-ranging, and sometimes bizarre the effects of selection can be.

Part of the aim of the Russian experiment was to make life easier for foxes raised in captivity. In the 1950s, fur coats made from the skins of foxes with black and grey hair were fashionable, and farms had been set up in Siberia where foxes domesticated and bred for their fur could be raised. Even though they had been raised in captivity for many years, being kept in side-by-side cages and so close to humans frightened many of the foxes, and the fear made them behave aggressively. Their human keepers' response to their aggression reinforced their fear. It made practical sense to try to breed foxes that could cope better with captivity. But the scientists also wanted to address a broader question: To what extent would the parents' temperament be passed from parents to offspring? Foxes at the experimental farm in the Siberian city of Novosibirsk were given a simple test of their reaction to the approach of a human. Generation after generation, the ones that showed the least fear and aggression were chosen as breeding stock. The others were sent to the fur-coat factory. But later, so they could make comparisons, the scientists took a second group and bred a lineage of foxes that showed more fear and aggression.

It took only a few generations to see differences in the descendants of the more laid-back foxes, but the scientists also saw differences they weren't expecting. The selected descendants weren't just more tolerant of humans and other foxes; they were physically different as well, and they behaved in ways that had never been seen in foxes before. They wagged their tails as humans approached, licked them, and whimpered and barked to attract their attention. In short, they behaved more like pet dogs than wild foxes that needed to be kept in a cage. They also looked more doggy, with floppy ears, curly tails, and cuter, less foxy-looking faces. It was as if a pet dog was somehow hidden in the genes of foxes and it just took a little selective breeding to get it to emerge.

International teams of scientists are now working together to learn how the genetic information in the cells of the friendly dog-like foxes differs from that of their fearful and aggressive cousins. It has long been known that similar differences can be seen between tame and wild varieties of all kinds of animals, from rats to cattle.[27] It's called "domestication syndrome" (see Figure 3.5).

The surprise was how quickly the changes could happen. Is it possible that our own ancestors experienced similar changes? If they were forced to live together, might they have "domesticated" themselves? The selection for "tamer" people might have resulted in people who didn't just tolerate one another but also felt a bond with a whole group as strong as the mother-infant bond that's seen in apes.

It will be possible to say more about this as the research progresses and more evidence is available. But we can be sure of one thing. There aren't simple genes for "tameness" or "tolerance." Evolution doesn't "design" living things

Figure 3.5 It's possible to see the similarities between a young wild boar and its distant relative, the domesticated piglet, and also the differences.

Figure 3.6 As it became a farm animal, the pig underwent some of the same changes as other (unrelated) domesticated mammals—floppy ears, curly tail, and loss of coat color.

the way a human would. There isn't a set of genes devoted to temperament, another devoted to hair color, and a third to ear shape. It's hard not to think of "evolution" as a designer or architect logically going about building a living thing the way a human would. But that's not the way it happens. Biologists have found that bits of genetic information are sometimes used in the making of several different cellular systems, to produce bits of molecular machinery that work in different ways in different parts of the body and different stages of development.

That's why selection for one characteristic often has many unexpected effects like floppy ears, a curly tail and changes in coat color.

4

Early Humans (About 1.5 Million Years Ago)

Scientists who study ancient bones and teeth don't entirely agree on how to classify the long-dead animals they were once part of. But they mostly agree on which of them are "human" enough to join us in the *Homo* genus and which should be considered australopithecines. For the purpose of this book, we're going to use brain size to determine which apes to call "people." In our definition, "humans" are ape-like animals with a brain that is larger than the brain of a chimpanzee.[1] On this basis, none of the australopithecines found so far are human. Some of the animals whose ancient bones look quite human in other respects only just make it into the category. A skull has been found of a male who lived about 1.8 million years ago (see Figure 4.1). It held a brain that was 25 to 30 percent larger than a chimp's brain. By our definition, he was definitely a human, and by 1.5 million years ago there were probably more humans with a brain as big as this. But humans with smaller brains also continued to exist, and so did australopithecines.[2]

Your ancestors born a million and a half years ago were larger and had a slightly different shape than the australopithecines that came before them, but they began their lives in pretty much the same way. When they were born, they became part of a group and they were looked after by their mother with the help of other mothers and older youngsters. Some, perhaps all, of your ancestors who lived at this time had a brain that was quite a bit larger than an australopithecine brain, and, because of this, they probably had to be born at an earlier stage of development. The less-developed babies were born weaker, less coordinated, and took longer to learn to walk.

If you had been born as one of these less-developed babies, caring for you would be more work, but your mother and her helpers could see that you were healthy. Your eyes were bright, and you were soon smiling back at them. Your mother had also been one of the weaker babies when she was born, but she survived and soon caught up with the other youngsters.

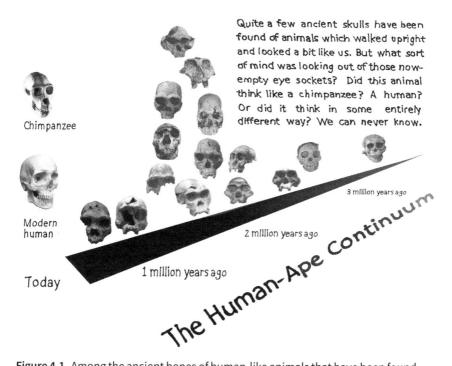

Quite a few ancient skulls have been found of animals which walked upright and looked a bit like us. But what sort of mind was looking out of those now-empty eye sockets? Did this animal think like a chimpanzee? A human? Or did it think in some entirely different way? We can never know.

Chimpanzee

Modern human

Today

1 million years ago

2 million years ago

3 million years ago

The Human-Ape Continuum

Figure 4.1 Among the ancient bones of human-like animals that have been found, more human-like ones start to appear around two million years ago. The bones, teeth, and stone tools they left behind give us no more than glimpses of the transitions taking place.

Your bright eyes can't focus very well at first, but your ears are sharp. You're surrounded by noises. The sounds you notice most are the voices of your carers. You pay special attention to voices because they sound familiar to you. You had been listening to the sound of your mother's voice since before you were born. As you grow older, one of the first parts of your body that you get control of is your mouth and throat, and you try to copy fragments of their voice sounds with your own voice. Soon you're trying to put together a long stream of voice sounds like the others do. The youngsters who play with you love it when you do this, and they make sounds back at you. The mothers and the fathers do too. An important difference between the family you are in and the earlier australopithecine families is that males don't just visit. Fathers are part of the family. They don't have milk, but they do help to carry babies and little ones on journeys.

When you start to get your first teeth, your mother gives you your first "kiss food" (see Figure 4.2). This is food that she has chewed up until it's ready to swallow. She puts her mouth up to yours as if she is going to kiss you, but instead of just kissing you, she squirts a bit of food from her mouth into yours. It's such a shock the first time

Figure 4.2 To grow a larger brain, early human babies needed more nutrition. Giving them small amounts of food before they were weaned would have made them healthier and more content. Since these babies were only just beginning to get teeth, their carers may have chewed their food up for them and transferred tiny portions into their mouths. "Kiss-feeding" has been observed in groups of foragers living today.

because it tastes and feels very different from milk. You suckle on her breast afterward and the food goes down with the milk. You learn to love kiss food, and after you get more teeth you sometimes give kiss food to babies.

About the time that you're able to walk around you start to understand what people want you to do when they make certain voice noises. Sometimes they want you to stop doing things or to stay where you are. Or they might be encouraging you to come to them, copy them, or bring them something. They often point and make a voice noise. You find that if you copy them, and make the same noise, they get very pleased and make the voice noise again. You love to copy, and you love to make them pleased. In time, you can reach out to a certain thing, make its noise, and they give it to you. Different foods and lots of other things have a special noise—a name. There is also a name for every member of the group, including you.[3]

Your mother gives birth to a new baby, a sister for you, and so you spend more and more of your time playing with the other youngsters. You mustn't go far away from the mothers, but you and your friends make pretend collection journeys, taking turns at being fierce animals that attack and need to be chased away. You make toy tools out of sticks and stones and play at doing food preparation work. You also go on real journeys with the whole group. Sometimes the journey is a move to a new camp, and sometimes you travel to a place where there are lots of nuts or berries. In these wonderful places, even the little ones can collect lots of food and everyone has more than enough to eat.

Once you grow bigger, you can be a real help with the food preparation. You work on the easy things at first, like stripping leaves from branches. When you're old enough to be relied upon, you're shown how to do jobs that require more skill. You have to be able to understand that sometimes things have to be done in just the right way even though it may seem easier to take short cuts. Because you learn fast and prove to be a good worker, you're taught some skills that not all youngsters learn. You also start to go along on real collection journeys. The older children in the party show you what to look for—not just food, but useful things too. You must collect certain bones, stones, and teeth because they can be used to make tools. The large bone from the leg of a buffalo can be used to break open a termite nest. Some birds lay eggs that are so big and strong they can be used to carry water, if the egg is opened very carefully.

And then something happens that means that you do less food preparation work and go on more collection journeys. One of the fathers falls and hurts his leg while out collecting. The others in the party have to carry him back to the camp. After that, it hurts him to walk so he has to stay in camp and just do food preparation. No one knows if he will be able to walk when it's time to journey to a new camp. Everyone is worried, because there will be one less strong father to collect food. You feel happy that you'll now be able to go on more collection journeys, but your mother says that even though she understands that you're pleased about this, you should not be happy that the father is injured.

Every time you go out on a collection journey, you spy some plant or animal that you've never seen before. On one journey, you get your first glimpse of the animals that walk on two legs. They look a bit like people but they're quite a bit smaller, they can't talk, and they have a funny snouty face. It would be fun to capture one of their young ones to play with it and learn more about them, but they're clever and would be hard to catch. You love learning about new things, and if you come across something that no one in your party knows about, you pick it up and take it back to camp to show the injured father. He tells you what he knows about it and if it yields food at certain times of the year. The injured father is soon able to walk while leaning on a stick or someone's shoulder. The group can make the journey to the new camp, but the father is still not able to make collecting journeys.

Collecting is tiring but it's so much more interesting than food preparation. You often see carrion birds circling, and if they behave as if they have spotted a carcass, members of your party run around looking for it. Predators fiercely protect the carcass of the animal they've killed while they're still hungry, but after they've eaten for a while, they get lazy, especially when the sun is high and it's hot. If they're feeling lazy enough, it's possible to frighten them away and get some meat.

One day, your collecting party finds the carcass of a large buffalo that still has plenty of meat on it. It also has three hyenas on it, but the adults in your party decide it's worth trying to frighten them off. This is dangerous because a hyena can easily kill even a large father,[4] but they can also be easily fooled. The party of small people must act like a single very large animal. Everyone must forget the fear they feel and be part of a huge angry monster running at the hyenas, throwing stones, roaring, and shaking sticks. If just one of the hyenas gets scared and runs away, the others always follow. Your party tries it and the trick works, the hyenas abandon the carcass and the party swarms over it. The fathers use their hand axes (see Figure 4.3) to hack at the joints of the buffalo and cut off pieces that can be carried back to camp. Everyone else

Figure 4.3 Our ancestors who lived 1.5 million years ago made stone tools like these, which could be held in the hand and used to chop and cut. Making them was skilled work. The earliest stone tools found so far, which are estimated to be about 2.6 million years old, are much cruder.
Photo credit: José-Manuel Benito Álvarez.

surrounds the fathers, waving sticks and roaring. It's important to keep up the large angry animal act until they've got as many pieces of carcass as can be carried. Then you all escape as one group carrying off enough meat for a wonderful feast. When you get back to camp, everyone is very pleased.

There are good days like this, but in general things are not good. With one of the most experienced fathers injured and not able to collect, there is often a shortage of food. Then there is a long period without rain and the sources of water start to dry up. Collecting is difficult, and everyone becomes thinner and thinner. Your mother has had a new baby, a brother this time, but he dies, and so does one of the other babies. One good thing is that the injured father gets better. He can walk quite well again, and he says that he no longer feels pain. But he walks slowly and can't run. He will never again be a good collector.

The fathers and mothers start talking about trying to find their brothers and sisters. After a lot of discussion and looking at the sky, they decide that they should do it. When it's time to make the next journey, the group sets off in a new direction. Life continues pretty much as before. You still journey for a day and then make camp, stay for a few days, go collecting and rest, and then journey for another day. But the feeling in the group is different. The fathers and mothers are looking forward to seeing people they haven't seen for a long time, but they're also worried. If they can't find them, your group will feel very alone. And if they do find them, the fathers and mothers know they will still feel some sadness because they're sure to find that some of the people they love have died.

The group keeps moving, the weather improves, the countryside starts to look very different, and it becomes easier to find food. You still see plants and animals that you know, but you also see many new things that only the fathers and mothers recognize. One evening you make camp in a new place and, on the ground, you see waste from the food preparation of other people. The next day, when you're out collecting, your party finds more signs of people. A pile of dry bones from an old carcass has cut marks that could have only been made by a hand axe. The mothers and fathers start to feel hope that they will meet another group. Everyone is happier.

When you do finally meet the other group, they are the ones who find you. A party of their fathers walks toward your camp calling out and holding their hand axes. Your mother calls out when she sees that one of the fathers is her brother. This is the beginning of one of the best and most important times in your life. The two groups camp near each other for a few days. There's so much for the adults to talk about. At first, you feel shy with the children from the other group, but the mothers are friendly, and their babies are very sweet. During the next few days, you get to know the older children and you go out food collecting with them and some of their fathers. It's different from what you're used to, but they know the area very well and you learn a lot. You also do food preparation work together and find that this group does things slightly differently.

With so many people collecting in the area, the food is soon gone. The fathers and mothers talk about where to journey next and when it might be possible for the groups to meet again. You feel sad about leaving the people you've met, and your mother feels especially sad about leaving her brother. Then the mothers and fathers of both groups decide a wonderful thing together. There is going to be some swapping around. You and your mother and sister will move to the other group along with the old father who injured his leg. Meanwhile, some young fathers from that group will move into your old group.

You soon start to feel like you belong to your new one, even though you miss the people of your old group. You know that when you're older you will mate and have your children with your new group. A year after you join the new group, the old father with the injured leg dies. When your mother has a new baby boy the group decides to give him the same name as the old father.

Why do we think our early human ancestors lived like this?

Several lines of evidence provide clues to what early humans were like and how they lived their lives. A good place to start is with a skeleton[5] found in 1984 near Lake Turkana in Northern Kenya (see Figure 4.4). Dating techniques have revealed it to be between 1.5 and 1.6 million years old, and examination of the bones has determined it to be the skeleton of a boy. He was about 160 cm (over 5 feet) tall when he died, considerably larger than the largest australopithecine males, but the bones of his limbs had not yet finished growing and he didn't have all his adult teeth. He may have grown a few more inches if he had lived. He was too young to have fathered any children when he died, and so we know that he couldn't have been one of our ancestors. But his skeleton is so similar to ours that it's reasonable to conclude that he was related to our ancestors.

It's possible to tell from his skull that the face of this boy was more human-looking than the face of an australopithecine. The australopithecines' nose was part of an ape-like snout, but this boy had a proper human nose that stuck out of his face like ours does. His teeth were also more like ours, smaller than australopithecine teeth. But he didn't have much of a chin, bony ridges stuck out over his eyes, and, above these ridges, his forehead sloped back more steeply than ours. Beneath the forehead was a brain with a volume of about 880 cc (cubic centimeters). Most modern humans have a brain of more than

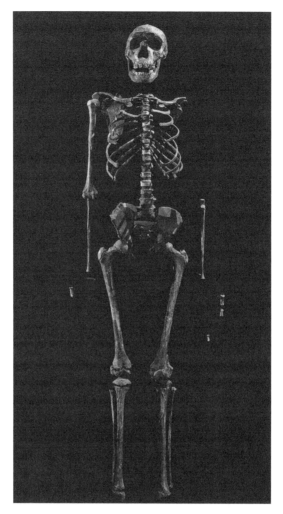

Figure 4.4 This is the skeleton of a boy, often referred to as "Turkana Boy" who died between 1.5 and 1.6 million years ago. It is similar in many ways to the skeleton of humans living today.

1300 cc, so his brain was considerably smaller than ours, but it was about half again the size of a chimp or australopithecine brain if they had been his size. Even though he wasn't fully grown, his whole body was larger—an australopithecine male would have only come up to his shoulder. But his larger brain wasn't simply a consequence of his being a generally bigger animal.

This is the only nearly complete early human skeleton found so far, but many other human and human-like remains have been found from this period of our history—bones, teeth, and stone tools. They show that between one and two million years ago, there were still australopithecines living in

Africa alongside early humans. Not all the humans living at that time would have looked as much like us as the boy who died by Lake Turkana. The remains that have been found suggest there was quite a bit of variation, including the shape of the skull and the size of the space that held the brain. Some of the human-looking skulls held a brain that would barely qualify its owner as being "human," at least according to our definition of humans needing to have a brain that is larger than a chimp's brain. Their brain was larger, but if their larger body size is taken into account, the difference is barely noticeable.

It's impossible to know how having a larger brain affected the way early humans could think and behave, but we do know that by this time the stone tools they were making were more finely crafted than those that had been made earlier. The makers of hand axes needed a brain capable of learning a complex skill and planning how they were going to produce the finished article. It's possible that early human groups were made up of people with different-sized brains. Mixed groups might have fared better than groups whose members all had the same-size brain. If a group had some members who were able to learn faster, plan further ahead, remember more, and have bright ideas, the whole group would have benefited. But smaller-brained members might have also been valuable. They would have been hardier and less hungry.

The newborn babies of larger-brained humans might have appeared slow or poorly developed compared to the babies of those with smaller brains. A baby has to be born before its head has grown too big to squeeze through the opening in its mother's pelvis. In order to end up with a bigger brain, a baby must be born when its brain is at an earlier stage of growth and development. Then it can grow larger after the baby's head has made it through the narrow pelvic opening. In baby apes, brain growth continues after birth, but it's much slower. Because of this slowdown, baby apes are able to survive on milk alone until they're old enough forage for themselves. The brain of a modern human baby continues to grow rapidly for a full year after it's born, and by about six months of age it needs to start receiving extra food on top of the milk it's getting. Our babies don't thrive unless they're given some extra calories. At six months old, they aren't able to chew food for themselves, so the food they're given has to be specially chosen or prepared.

The brains of early humans were considerably smaller than ours, but mothers probably still had to deal with less-developed newborns and hungrier babies. It's likely they would have tried giving them mushed-up food and found that it made them quieter and less demanding. What better way for a mother to mush up food that by chewing it herself? The "kiss-feeding" of babies is part of the culture of many hunter-gatherer groups today.[6] Food that has been chewed by the mother is already partly digested by enzymes from the mother's saliva, and it will also contain healthy bacteria from the mother's mouth to inoculate the baby's developing digestive system. Early human mothers who invented the practice of feeding their babies would have had healthier, quieter babies. And they would have also had more babies. If a baby is receiving kiss food, it will be less hungry and suck less vigorously on its mother's nipples. As a result, the pituitary gland in its mother's brain will release less prolactin, the hormone that prevents the ripening of eggs in her ovaries. Giving food to babies deactivates the natural means of birth control that served our ape ancestors for millions of years. Mothers who give their babies food are likely to become pregnant with another baby long before their older baby is ready to find food for itself. This would be a huge problem for an ape mother, who has to do all the work of finding food for herself and her dependent offspring. We believe it wasn't such a problem for early human mothers because they were in a group made up of males and females of different ages who helped each other.

New social tools for helping families work together

Few accounts of early human evolution depict males being fathers or sons, and yet every single one of our male ancestors was both of those things. That's not to say that early human fathers put as much effort into hands-on child care as early human mothers did. But once the size of our ancestors' brains began to expand, it's likely that adult males began to live with mothers and children, and groups worked together to collect and prepare food. It seems to us that mothers would have needed help from adult males to get enough food for themselves and their larger-brained children. Chimp mothers are stretched to capacity finding enough food to keep their larger-brained babies alive. We reckon that australopithecine mothers could have managed if they worked together and got help from their older children. But it's hard to see how mothers could find enough food to supply the calories needed to support the growth of even larger-brained youngsters while at the same time supporting their own larger brain. Males were the only possible source of the extra nourishment.

Very few adult male mammals help provide food for their young, and male apes certainly don't. Most put all of their "fathering" effort into competing to be the male who gets to mate with the most females. This is perhaps why the early human "apemen" were so often depicted as violent warring figures. But if the transition to being human is growing a larger brain, it seems likely that the "apemen" had to start becoming family men. In all human populations, as far back as records go, males have lived in families and provided parenting help, even if they didn't do much hands-on caring for babies. In most cultures, men are expected to help raise the children they have fathered, but this isn't universal. In a few cultures, mothers get help from their brothers instead. In extreme cases, men hardly help at all.[7] It's impossible to know at what point in our evolutionary history males began to provide parenting help, but if it began about two million years ago, it would have been part of the transition from australopithecines, with a chimp-sized brain, to humans, with a larger brain. Both genetic and cultural changes must have accompanied the adoption of this style of parenting. In the next chapter, we'll talk more about how this could have evolved into the institution we call "marriage."

The early human children in this chapter's story developed mental skills necessary to take part in their family's work as they grew up. They learned to make "voice sounds" that had a meaning and weren't just a signal of their emotional state. They played together, imagining themselves in different roles. These types of mental skills make it possible for a family to act together. They can put aside their fears and imagine themselves to be part of a ferocious animal that frightens hyenas off a buffalo carcass. We know that humans today possess these mental skills, and we know quite a bit about how children develop them as they interact with adults and other children.[8] We don't know when our ancestors first had a brain that was capable of developing them, but we think it likely that they were at least beginning to emerge by a million and a half years ago.

The complexity of the stone hand axes early humans made suggests that families shared a culture that was more complex than the culture of australopithecine families. Hand axes and other stone tools are very durable, so they provide abundant evidence, but early humans must have made tools from more perishable materials too, such as sharpened wooden digging sticks, animal hide containers, and hollowed gourds for carrying water.[9] Australopithecines knew their environment and were able to make a living there, but the knowledge and expertise of the early humans must have been greater. And they would have passed on to their children the social tools that they would need to fit into their family. Family life runs more smoothly when youngsters are brought up to respect their elders' traditions and agree on the sorts of behavior and feelings that are appropriate.

The early humans in our story had the very important social tool of language. It is, of course, impossible to know when humans began to communicate with language, and there are some scholars of human evolution who argue that the capacity to use language only evolved within the last 100,000 years or so.[10] Scholarly discussions of language evolution can get rather heated because it's such a puzzle. If you think of evolution as being driven by individuals competing with one another, it's hard to imagine how language could have evolved. Language allows humans to communicate useful information to one another, but it also allows us to lie and deceive one another. So why would our ancestors have started to use language? How could they have benefited by giving useful information to their competitors? And why would they have listened to anything a competitor said to them?

Language could have only evolved if our ancestors were part of a group that wasn't competing with other members, or at least not competing very vigorously. Members of the same group needed to trust one another in a way not seen in chimpanzees and bonobos. It's true that chimps and bonobos form friendships and demonstrate trust by allowing their friend to groom sensitive parts of their body. But trusting that someone won't hurt you isn't the same as trusting that he or she is a friend who has your interests at heart. Primatologists report that chimps often make a series of noises that seem to say "Hello, friend. Join me on this branch and we can enjoy the fruit together." But this isn't real generosity on the part of the grunting chimp. It's just recognition that a few pieces of fruit aren't worth fighting over. A chimpanzee would be unlikely to bring some fruit to an injured friend who was having trouble moving around the tree. Chimpanzees are selfish compared to humans. Apart from mothers and young, they don't share food or give each other information about how to get food. It's not surprising that they didn't evolve the ability to use language.[11]

A 1.8-million-year-old skull and lower jaw found in the Caucasian mountains north of Turkey (see Figure 4.5) provides strong evidence that humans were different. They are the remains of a man who had lost all but one of his teeth. There's evidence of healing, which means that he continued to live on for several more years after losing the teeth. Whether his tooth loss was due to old age or disease, being toothless must have been quite a severe disability. Survival would have been impossible without help from family and friends.[12]

Whenever our ancestors evolved language, we think that the fact that they did is strong evidence that the last few million years of human evolution were different from the evolution of other animals. The evolution of other animals can be explained in terms of competition between individuals and their genes.[13] But in human evolution, another layer of complexity must have emerged that created conditions in which the selfish behavior seen in chimpanzees was no longer favored by natural selection.[14] As we argued in the last chapter, this layer of complexity may have emerged when our ancestors began living as part of a family that shared food, information, and care of young ones.[15]

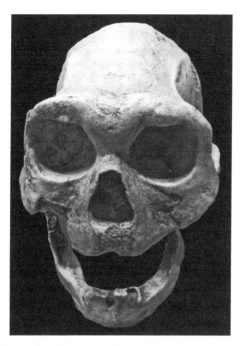

Figure 4.5 This skull, dated to about 1.8 million years ago, is from an individual who survived for some time with only one tooth. This is strong evidence that by this time humans were living in groups whose members cared for one another.

We believe the evidence reveals enough about the practices and accomplishments of early humans to suggest that their lives were quite complex, and that by 1.5 million years ago individuals were connected and helped one another. Their stone hand axes were far harder to make than earlier stone tools, and they passed on the skill of making them down the generations as they traveled out of Africa, right across Asia and up into Europe. It's easier to imagine them doing this if they had a social tool like language that would allow them to teach, give advice, discuss problems, consider options, and agree on courses of action.

This new way of living didn't stop competition, but it changed it enough so that individuals with different kinds of characteristics started to be favored by natural selection. Family members didn't behave like robots, thinking nothing of themselves and just working for the good of the collective. There was deceit and there were squabbles. But in the more successful families, members cooperated well enough so that most of them were able to raise more offspring than they could have on their own. Sharing food with other members didn't reduce how much an individual got for herself and her baby, because, by working together, and especially putting men to work, the family was able to get more food and raise more babies than the individual members could by living and working alone. Families that worked together best got the most food and more of their offspring survived. Individual family members grew old and died, but new members grew up to replace them. The family and the cultural information used by the family lived on. If inventing "voice noises" that had meanings allowed members to communicate better and work together more effectively, then families who had language would survive while families who lacked this social tool died out.

The brains of early humans were considerably smaller than our brains, so if learning and using language requires a brain as large ours, then our ancestors couldn't have evolved language until a few hundred thousand years ago. But why should it take a massive brain to begin to use language? When a pair of cognitive scientists looked at language learning as simply data storage, they calculated that the words and grammar remembered by the average person could be stored as about 1.5 megabytes of information.[16] This may sound like a lot, but this amount of data can be stored on a microchip far too small to see with the naked eye. Processing the language data to produce and understand speech takes extra brain power, of course, especially when there is complicated information to communicate. But the language ability of our early human ancestors was probably quite a bit simpler than ours. Their lives were difficult, and they faced big challenges, but it might have been possible to discuss their problems and plan courses of action using language much simpler than ours.

Language and other cultural practices that made families successful were inherited by each new generation. The less successful families died out. Children in the more successful families inherited their parents' genes as well as their culture, so they had a good chance of developing the sort of brain that would allow them to learn language and fit into the family. In this way, human psychology evolved, creating the foundations of our uniquely human "mental life." The children in our story enjoyed imitating and pleasing others. They played imaginatively, taking turns pretending to be fierce animals and practicing responses that they would use when they were older. A child's misbehavior would sometimes attract punishment, but carers may sometimes have tried to explain why some behavior and some emotions aren't appropriate.

A new diet, too

The skeleton found near Lake Turkana suggests that, by a million and a half years ago, our ancestors were eating a diet that was more like our diet today than the diet of our australopithecine ancestors that lived a million and a half years earlier. Take a look at the Turkana boy's skeleton and note the way that the bottom ribs curve in just above his waist. Our ribs do the same thing. The ribs of the two nearly complete australopithecine skeletons found so far are quite different: they form a cone shape similar to the ribs of today's apes. This suggests that their ribs sat on top of a large apish digestive system where bacteria fermented prodigious quantities of fibrous plant material that is indigestible without bacterial help. The diet of modern humans, even vegans, contains far less indigestible fibrous material than the diet of apes. The plant food we eat is made up largely of parts of the plant where nutrients are stored, like seeds or special underground storage organs.[17] We do still eat fruits and the structural parts of the plant, such as the leaves and stems, but they aren't such an important part of our diet as they are for apes. And we often make our food more digestible by cooking it or processing it in some other way. Although the ways we process food can destroy or remove some of the nutrients it contains, it can also destroy and remove the toxins and indigestible stuff that food often contains. As a result, our digestive system handles food that contains, weight for weight, considerably more calories and nutrients than the food that runs through an ape digestive system. Less of the stuff we swallow goes through to our colon, the final part of our digestive system, where it's fermented by the bacteria that live there. Our abdomen is smaller because our colon and its population of fermenting bacteria are smaller than the equivalent organ in apes.

The shape of their skeleton tells us that our australopithecine ancestors three million years ago had an abdomen similar to an ape's, while our early human ancestors had an abdomen similar to ours (see Figure 4.6). The best explanation for this change is that their diet gradually changed during the intervening time. The shape of the teeth supports this. They are less suited to grinding up tough plant food than australopithecine teeth. These changes seem to us to be an example of cultural evolution driving the evolution of the genes that organize body shape. Our ancestors learned to recognize and obtain food items with a higher nutritional content, and they developed ways of processing food to make it safer and easier to digest. They passed this knowledge and know-how down the generations, introducing changes as appropriate. They added to or refined techniques as their environment changed or as they discovered better ways of doing things, and they forgot techniques that were no longer useful. As their diet became richer, they no longer needed to carry around such an elaborate digestive system with room for a very large population of bacteria. Natural selection favored those whose genes directed the development of a more compact and more economical gut. Learning,

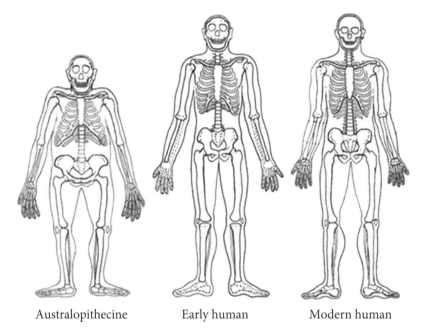

Australopithecine   Early human   Modern human

Figure 4.6 Comparing the shape of the skeletons of modern humans to the skeletal remains of human-like animals of the past provides clues to the ways that our ancestors' diet changed.

remembering, and improving techniques for getting and processing food required quite a bit of brain power. Natural selection would, therefore, favor those whose genes directed the development of a larger and more elaborate brain that could maintain a larger body of cultural knowledge, even though it would require more energy to run such a brain.

We can be reasonably confident that by a million and a half years ago our ancestors were finding and preparing a relatively nutrient-rich diet for themselves, but the evidence found so far doesn't tell us what they were eating or how they were preparing it. Could they have been cooking? Expertise on making fire and using it to cook food has been part of all human cultures for tens of thousands of years. Cooking makes many of the things we eat more nutritious and safer. It's by far our most important food processing technique.[18] But we don't know when our ancestors first learned to make and control fire. A burning fire leaves traces that an archaeologist can find; high heat alters sand and clay in characteristic ways. Traces of ancient fires have been found near human remains dating back beyond 1.5 million years, but it's impossible to conclude from this that humans made the fires. They might have just been wildfires caused by lightning strikes. Pieces of burning wood from a wildfire may have been collected by some early human groups and used to set fire to fuel they collected. In this way, groups may have kept a fire going for a while and found ways of using it. But to use fire regularly to cook, keep warm, and protect themselves from predators, they would have had to know how to use friction or strike a spark to set fire to kindling.

Better evidence that our ancestors were using fire regularly would be charred bones or fire-blackened stones in the shape of a hearth. This sort of evidence has been found, but so far only in archaeological sites of a much more recent date. The oldest sites are about 300,000 to 400,000 years old.[19] Our ancestors who lived 1.5 million years ago may have used fire to cook their food, or they may have cooked when they happened to have fire. At this point, we don't know. But we can be quite sure that if they didn't have fire, they must have used other techniques for preparing food to make it more digestible. They may have cut it up, pounded and pulverized it, allowed it to be fermented by safe nutritious microorganisms, or marinated it in plant extracts. No evidence has been found so far of exactly what they did. Once our ancestors could control fire and cook the food they collected, there was less need for other food processing techniques, but we still use quite a few different techniques to prepare our food.

There must have been times when finding food was easy, such as during the seasons when plants produced a large number of edible fruits and nuts. But it's likely that our ancestors living a million and a half years ago spent a lot of their waking hours collecting and preparing food. The most successful groups were those who cooperated best, sharing food and the labor of collecting and preparing it. Group members who were less mobile, like children, mothers with young children, and those who were injured, would do more of the food preparation.

Adapted to rely on tools and to be on the move

Our ancestors who lived a million and a half years ago were slow and feeble compared to the animals they were competing with. Their bodies lacked inbuilt "weapons" like fangs, claws, or hooves on the end of legs that could deliver walloping kicks. Their relative weakness was probably another side effect of having a larger brain. With so much energy needed to support a large brain, there is less available to send to other parts of the body. Having a smaller digestive system saves energy, and so does having smaller, weaker muscles.[20] The advantage of a larger brain was that it made it more likely our ancestors could invent ways of compensating for their lack of speed and strength. Their hands were capable of a wide variety of minutely controlled fine movements and could turn sticks and stones into tools and weapons. They discovered that, by practicing, they could become better at using tools and weapons. Also, having a larger brain would have allowed early humans to compete by being better at finding ways to take advantage of unexpected events and unpredictable situations. Several separate lines of evidence tell us that this was a time when such a strategy would have paid off.

At about the time early humans appeared, the Earth's climate was starting to go a bit crazy. A new geological epoch that scientists call the "Pleistocene" began about two and a half million years ago. Evidence for the way this changed living conditions on the planet can be found everywhere, from sediment at the bottom of the oceans to the ice on the top of mountains and at the poles. The planet-wide changes, which had been driving the gradual shrinking of forests and the formation of new drier habitats, reached a tipping point during the Pleistocene. After that, there were periods, thousands of years long, when the Earth was cooler, had less rainfall, and temperatures were more variable. During these "ice ages," large volumes of water froze and became part of huge sheets of ice that covered the poles and mountain tops. Sea levels fell, and weather and rainfall patterns became much less stable.[21] Habitats changed and shifted position. The organisms living during times of change did their best to cope.

Volcanic activity added to the instability in eastern Africa, where so many of the australopithecine and early human remains have been found. The occasional eruptions of ash, cinder, and noxious gases were hazardous in themselves, and they also caused rivers to change course and lakes to dry up. Unstable conditions would have constantly presented both new challenges and new opportunities. They may have often given our larger-brained ancestors an advantage over animals that weren't such quick learners and relied more on genetically evolved instincts.

The body of modern humans has adaptations that allow us to be more mobile than other large animals. Even though we aren't fast runners compared to horses, lions, and chimpanzees, we can walk for miles without tiring, and we can, with training, become better at running long distances than any other mammal. Some of our muscle fibers are actually built differently than those of the animals we are most closely related to; our muscles are built for efficiency and endurance rather than speed and strength.[22] We can run marathons in the blazing heat, as long as we keep drinking water. The skeleton found at Lake Turkana suggests that our ancestors that lived 1.5 million years ago already had uniquely human adaptations for endurance. Their longer legs would have allowed them to stride along faster at a walk than australopithecines, and the dimensions of the muscles and tendons attached to the leg bones would have given their legs the same spring and strength that makes modern humans such efficient runners.

It's likely that early humans also had our ability to stay active, even in the heat of the day. Other large mammals that live in hot climates tend to find a shady spot and lie quietly during the hottest part of the day. Keeping their activity to a minimum reduces the amount of body heat they generate and means they don't need to breathe in large lungfuls of hot air. We know from his skeleton that, like us, the Turkana boy had an air conditioner sticking out from the middle of his face—a nose. When we breathe through our nose, the air passes through a chamber where it receives treatment. The damp walls of the chamber moisten the air and adjust its temperature while hair growing from the walls trap sand and dust. Our nose makes it less likely that the air we breathe will damage the delicate membranes inside our throat and lungs.[23]

The Turkana boy's skin was probably also adapted to helping him stay active in the heat. As we mentioned in the last chapter, the skin of australopithecines was probably sweatier and less hairy than chimpanzee skin to allow them to stay cooler while foraging in the hot sunshine. If the legs of early humans were adapted for foraging at a faster pace, then their skin would need to be even better at cooling them off. They may have already had the elaborate cooling system that we have, which sends blood running just below the surface of our skin where the evaporation of our sweat cools it. The cooled blood then travels from the skin and through our muscles and organs to cool them. Humans can run in the desert sunshine because we can release about a liter of water an hour onto our skin. People who are very experienced at working in hot weather might produce as much as three and a half liters of sweat an hour.

Early humans would have probably preferred to rest during the heat of the day, like other large animals do, but their adaptations meant that they could be active during this time if they had to be. They had to burn more calories to keep their larger brain going but, thanks to the "mental life" provided by this brain, they could make good use of adaptations that allowed them to walk long distances and be active in the heat. An ability to plan ahead and imagine outcomes would have shown them the need to make containers for carrying water on journeys and to choose routes that took them past sources of water. Imagining outcomes would have also allowed them to strategize. Pushing themselves—and each other—to be active when other animals were taking a rest would have made it possible to survive even when they were up against animals that were more physically able. The large hyenas that hunted on the African plains a million and a half years ago would be more than a match for a group of humans most times of the day. But when the sun is beating down, a well-fed and sleepy hyena may decide to give way to a noisy bunch of stone-throwing humans.

Our early human ancestors needed to keep moving and adapting to new environments, and they were well adapted to being on the move. It isn't surprising that some of them moved right out of Africa. Human remains older than a million years have been found (so far) near the Black Sea on the eastern edge of Europe, in China, and on the Indonesian island of Java. The presence of early human remains on Java suggests that they were already making their way across Asia nearly two million years ago. To get to the island from mainland Asia today, it's necessary to cross the Strait of Malacca, which is over 500 miles wide, to get to Sumatra, and then cross another 20-mile stretch of water to get to Java. But the sea level is much lower during ice ages, when much more of the Earth's water is frozen into sheets of ice at the poles and on mountain tops. During the ice age that occurred around 1.9 million years ago, the sea level was 200 feet lower than it is today. The islands of Java and Sumatra were attached to the mainland at that time, and early humans could have simply walked to them.

More evidence of early human wandering will no doubt be found in the future, but such finds will never give us a complete picture of where they went and what they did. They must have walked across and died in many parts of Africa and Eurasia leaving no trace that can be found today.

Are we descended from these ancient travelers? Based on the evidence we have so far, it's impossible to tell how much of our ancestry is made up of people who traveled out of Africa more than a million years ago. The close genetic relatedness of all humans that are alive today suggests that all of us are descended from a few small populations—perhaps less than 20,000 people in total—that were living in Africa less than 100,000 years ago.[24] But we don't know the history of that population. It might have contained people whose ancestors had traveled great distances out of Africa and then came back. On the face of it, it seems extraordinary that our ancestors could have managed to travel on foot with their children all the way from Africa to China—a distance of some 12,000 miles. Could they really have gone all the way there and found their way back to Africa? It doesn't seem so extraordinary when you take into account the amount of time that they had to do it. If they averaged just 60 miles of progress per generation, it would have taken them less than 50,000 years to travel from Africa to China and back again.

5

Humans Like Us
(About 100,000 Years Ago)

The ancient human remains that have been found so far tell us that there were at least five different kinds of human walking the Earth 100,000 years ago. The ones that look most like us were living in Africa and in the area east of the Mediterranean where Africa connects with Asia. Most, or perhaps all, of your ancestors were living in Africa 100,000 years ago (see Figure 5.1).

If time-traveling aliens abducted a baby from a family of these ancestors and brought it forward in time to be adopted by 21st-century parents, the adoptive parents probably wouldn't suspect there was anything unusual about their new family member. The baby's skin would be dark, adapted to spending a lot of time in the African sunshine,[1] and it might be one of those children who seems to catch every infection going around. Most babies born today inherit an immune system adapted to respond quickly to viral threats. The ancestors of babies born 100,000 years ago didn't have frequent contact with strangers, so their immune systems probably wouldn't be adapted to tackle such a wide range of infections. But, with the help of vaccines, antibiotics, medical care, and good nutrition, a baby from the past would likely grow up healthy in the modern world. He or she would probably have the same sort of problems growing up in the modern world as the children of 21st-century parents, as well as the same chances of living a happy productive life.

If those time-traveling aliens took a 21st-century baby back to a mother in ancient Africa, the modern baby would also have a reasonable chance of fitting in and surviving, although its life would be physically tougher and its chances of surviving to adulthood would be much lower than if it had stayed in the 21st century.

Part one

Imagine being born as one of your ancestors who lived 100,000 years ago in Africa. But first imagine the day before your birth when three women who

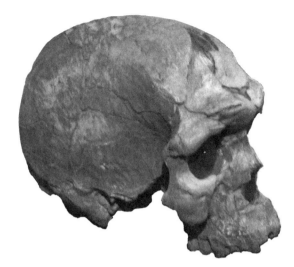

Figure 5.1 On the basis of its size and shape, this could be the skull of a person living today. But it's the skull of a man who lived in east Africa about 160,000 years ago.
Credit: By Alessandrosmerilli—Addis Ababa national museum, Public Domain, https://commons. wikimedia.org/w/index.php?curid=25048966.

would be very important in your life are waiting and worrying. They are your grandmother and her two daughters. The older daughter will be your very special aunt and the younger one is your mother.

Like every woman in the final stage of her first pregnancy, your mother is feeling exhausted but also restless. She's lying down with her head and shoulders propped up against a pile of branches covered with a sling made of antelope skin. It's the sling that your great-aunt used to carry her baby, and it's the sling your mother will use to carry you. But, at this point, she doesn't want to tempt fate by imagining herself using the sling. Of the four births she's seen in her life, only one of them was good. She hopes that her birth will be like her mother's sister and not like the terrible torture her own sister had to endure last year. She rubs her face against the sling for luck and breathes in the baby smell still clinging to it. Her belly and breasts seem huge to her. She hasn't felt like herself for weeks. Your father has been working hard doing most of her share of the work, collecting and preparing food and gathering firewood. Your grandmother sits down beside her, putting a hand on her belly. She thinks the pains in your mother's back are a sign that her labor is beginning. Your mother hopes so, even though she's very scared.

　Your grandmother also has hopes and fears. If this birth goes well, she will have her first surviving grandchild. She has experienced many births, including giving birth to five babies herself. Three of her own babies survived—her two girls and a son, who has now joined his uncle's family. She thinks about the different births, looking for reasons why some went so badly while others went well. The freshest memory still has the power to make her stomach turn over—the torture that your aunt suffered. She labored for two nights and everyone feared she would die. Finally, the baby was born. It started to breathe but was too weak to cry, and its body seemed floppy. The labor had been too long, and all the women had warned that the baby probably wouldn't survive. Your grandmother held the baby tight with its face pressed against the skin of her stomach until it stopped breathing. It's better for a baby to die quickly than to have to watch it slowly weaken and die.

While your grandmother carried her dead grandchild to a place where the carrion birds would take it, she tried to cheer herself up by imagining smothering the baby's father. He was a young man who had come up to them one evening and asked to travel with them. He pronounced some of his words in a strange way, but they could understand him. He told them the name of his family. They had heard of the family, so, as was customary, they showed him hospitality. He explained that he needed to get away from his family for a while because he'd quarreled with his older brother. Your grandmother remembers not liking the look of him, but your grandfather had said that he "looked like a fine young fella." The stranger certainly smiled a lot and was very agreeable. He was also tall and strong and a good worker. The family started to like him and were happy for him to stay. But, unfortunately, your aunt had liked him too much. It had been planned that she would soon marry your father. But she secretly mated with the stranger and became pregnant. If the stranger had kept his promise to her and married her, the family would have accepted him, but instead he sneaked away in the night. Your aunt was devastated and, as her belly grew larger with his child, the family decided she had been punished enough and started to be kind to her. Your grandmother shudders at the memory of smothering the baby that her poor daughter had labored so long to bring into the world, but she's glad it's dead.

Your aunt hadn't minded when it was decided that your father should marry her younger sister. She was sore for a long time after the difficult birth and wasn't interested in marriage. But her breasts were making lots of milk and she had willingly helped suckle her aunt's baby. Everyone knows that babies do better if they have milk from more than one woman. She now plans to help feed you if your birth goes well.

Your mother's labor continues throughout the night, but shortly after the sun rises next morning, you give your first loud and lusty cry. Your aunt holds you as your grandmother cuts the cord with a flint blade and sears the ends with a glowing piece of wood brought from the fire. She then holds onto the end that is still attached to the inside of your mother's womb. The placenta comes out and your grandmother inspects it to make sure none of it is left inside. She has heard of mothers dying because part of the placenta had remained inside. Then she slices up the placenta with her blade and skewers the pieces on sticks to be roasted over the fire. It's the custom for the baby's mother to eat the placenta to help replace the nourishment she has given to make the baby, but your mother says she wants to share it with her sister because her milk will help feed you.

Your mother gazes at your face and smells your body. She feels so happy, perhaps happier than she has ever felt in her life, but she tries not to show her joy out of respect for the feelings of her older sister, who'd had such an awful time giving birth and is now unmarried and childless. Her sister has been so good, showing no sign of resentment or jealousy, only love and caring. She carefully puts you in your aunt's arms tells her that she feels so tired that she's afraid she might drop you. While your mother sleeps, your aunt patiently teaches you how to latch onto her nipple and you get your first taste of milk. *(To be continued.)*

Why do we think our ancestors were like this 100,000 years ago?

We think these ancestors were very similar to us for two reasons:

1. Bones among the remains of humans living in Africa even 160,000 years ago look very similar to the bones of people today. Scientists have categorized these remains as being from members of our own species— *Homo sapiens.* (This name for our species was coined around the beginning of the 19th century using the Latin words for "wise" [sapiens] and "man" or "human" [homo]).

2. It hasn't been possible to recover and analyze DNA from these bones or any of the really ancient bones found in Africa. We therefore can't know precisely how similar the genes of people who lived in Africa 100,000 years ago were to the genes of people today. But 100,000 years is only about 4,000 generations in our long-lived species. This isn't enough time for radical genetic change to have occurred. It has been possible to recover and analyze DNA sequences from bones of humans who lived tens of thousands of years ago in Europe. These are probably less closely related to humans today because they were physically different. Even *their* DNA is very similar to ours.

If humans who lived in Africa 100,000 years ago were very similar to people living today, we think it's reasonable to use evidence gained by studying people living today to inform our ideas about what our ancestors were like 100,000 years ago. But it's important to keep in mind that even if these people were similar to us physically, their cultures were very different from ours. In many ways, their lives were probably not that different from those of the humans who lived over a million years ago. They likely lived in small groups that moved around from area to area foraging for their food.[2]

The best sources of information about the lives of nomadic foragers are the reports of anthropologists and explorers who've studied people who live like this today. Nowadays, only a tiny proportion of people live in what anthropologists call "small-scale foraging societies," and today's foragers have more advanced tools than our ancestors who lived 100,000 years ago. In fact, most of today's foragers have regular contact with modern people, wear manufactured clothes, and use metal knives and cooking pots. Some keep in touch using mobile phones. But information about the way they live their lives can still provide clues about the lives of our ancestors. Also, explorers and anthropologists have been living with and writing about foragers for well over a century, so there are plenty of reports of people living with simpler technology. The "ethnographies" they have written give us a glimpse of the wide variety of ways that humans can make a living. They also reveal patterns in the way foragers solve the day-to-day problems that all humans share, such as the need to give birth to and raise large-brained children who require a great deal of care.

Our ancestors living 100,000 years ago likely found solutions similar to those used by today's foragers. For example, when people live in small mobile groups, it isn't easy for young adults to meet up and find mates, so it makes sense for parents to arrange marriages between their children. This practice is still quite common in many cultures today. It's also common for women in foraging groups to help each other by suckling another woman's baby, especially the baby of a sister or daughter. And, because the life of a nomadic forager makes it impossible to care for a very sick or severely disabled child, newborn babies that are unlikely to survive are sometimes not cared for and allowed to die or killed as painlessly as possible.

Brains and climate

By 100,000 years ago, our ancestors had brains that were, like ours, more than twice the size of the brains of the humans described in the last chapter. There isn't much physical evidence of what these ancestors were accomplishing with their large brains. We know that they were using fire and that their stone tool kit had become larger. Evidence of these have been found. Their hunting weapons included small sharpened bits of stone that might have been attached to the tips of spears and darts. And archaeological digs into ancient mud where the growth of microorganisms was inhibited have yielded long wooden spears with fire-hardened tips that humans had made several hundred thousand years ago.[3] No doubt our ancestors made many other tools of biodegradable materials. They sometimes attached wooden handles to their stone tools, but they were still making tools by hitting pieces of rock together just like their smaller-brained ancestors that lived over a million and a half years earlier.

It was once thought that evolving a large brain would have given our ancestors a huge advantage over all the less intelligent animals they were competing with. But the evidence says otherwise. Humans have been on the planet for well over a million years, but only a small number of human remains have been found. This suggests there weren't many of them compared to the prehistoric horses, cats, baboons, and many other kinds of mammals that were living at the same time. A vast array of bones and teeth from these species have been found. Clearly, these species were thriving while our ancestors were just managing to hang on. It isn't surprising that these smaller-brained animals were doing so well. Their bodies had adaptations that made them faster, stronger, and, in some cases, far more dangerous than humans.

Scientists now realize that there are considerable *dis*advantages to having a large brain,[4] but there must have also been advantages, because large-brained humans were definitely doing better than humans with smaller brains. Based on the ancient human remains that have been found so far, there were at least three kinds of large-brained humans living 100,000 years ago, and they were living in many parts of Africa and Eurasia. Humans may not have been common, but they were surviving. The ones that looked like us were mostly living in Africa, and some of them had moved into the eastern Mediterranean, where Africa connects to Eurasia. It seems that smaller-brained humans were almost extinct. Some had continued to live in Africa until about 300,000 years ago, but, so far, no more recent remains of small-brained humans have been found except on islands south of Asia.[5]

The larger-brained humans may have been doing better because they had a better chance of surviving periods of extremely variable climate conditions. Scientists studying evidence of past climates believe there may have been several times during the last million years when the Earth's climate was extremely unstable (see Figures 5.2 and 5.3).

It's been known since the 1800s that northern parts of North America and Eurasia were once covered in vast sheets of ice, and scientists now believe that over the last two and a half million years, the Earth has gone through many alternations between warm and icy conditions. Until about a million years ago the fluctuations were dominated by a 41,000-year time scale. For

Figure 5.2 Mud that settles and lies undisturbed at the bottom of the ocean or a lake builds up a record of past times. By carefully drilling into the sediment, scientists can retrieve a sample of mud layers going back tens of thousands of years. (This is known as a "core" sample.) Analyzing chemicals in layers that were laid down year after year yields information about changes in temperature, atmospheric conditions, plant life, and much more. Information about conditions that existed even further back in time can be obtained by analyzing samples from sedimentary rock that formed millions of years ago.

the last million years the dominant cycle has been 100,000 years. Between the "ice ages," there were warm periods that also lasted up to tens of thousands of years. During the warm periods, conditions were pretty much as they are now, with average global temperatures of about 18 degrees Celsius (64 degrees Fahrenheit).

During the ice ages, sheets of ice at the Earth's poles and on mountaintops expanded, the sea level fell, and there was less rain. But these changes, on their own, wouldn't create conditions that ecologists see as harsh. Scientists once assumed that conditions weren't particularly harsh—that the decrease in rainfall and temperature would have just caused a slow expansion of deserts and grasslands, a changed coastline, and a shift in the position of habitats. Results of more recent research suggest otherwise. Superimposed upon the main cycles are periods of more rapid variation. During some of these periods, average temperatures in at least some parts of the planet were bouncing up and down, changing by more than five degrees Celsius over a few decades. There were times when our ancestors experienced temperature fluctuations in a single lifetime that were several times greater than the temperature change that has occurred over the last 10,000 years.[6] Although fossil evidence is

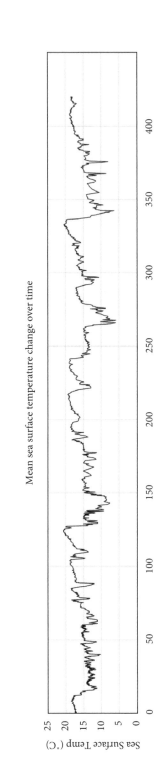

Mean sea surface temperature change over time

Figure 5.3 This is a plot of temperature estimates going back over 400,000 years. It's based on an analysis of layers of mud from the ocean floor off the Atlantic coast of southern Spain. Samples collected in different parts of the world differ slightly because they reflect local conditions. But they all suggest that the Earth experienced long periods of climate instability over the last half million years or more. There were several long ice ages, and during these periods, temperatures weren't just lower, they were much more variable as well. The climate that our planet experienced during the last 10,000 years has been very stable in comparison.

Data from Belen Martrat, Joan O. Grimalt, Nicholas J Shackleton, Lucia de Abreu, Manuel A Hutterli, and Thomas F Stocker, "Four Climate Cycles of Recurring Deep and Surface Water Destabilizations on the Iberian Margin," *Science* 317 (2007): 502–507.

inevitably patchy, these increases in climate variation seem to coincide with increases in human brain size and cultural sophistication, suggesting that complex culture and costly large brains are adaptations to an increasingly variable environment.[7]

Temperature fluctuations of this magnitude wouldn't simply have caused habitats to shift position. They would have caused chaos. When environmental conditions are stable, natural selection drives living things to become better adapted to those conditions. But when conditions become unstable, the habitats are destroyed, and many of the creatures adapted to the habitat become extinct. There are, however, animals that specialize in taking advantage of habitat destruction. We see many of them thriving today in areas where humans have moved in, destroyed the ecosystem, and dumped their waste. It doesn't take long for the animals we call pests to arrive—animals like sea gulls, pigeons, rats, and mice. These animals do well in disrupted environments because they're adaptable and fast learners. They're able to find food and make homes in a range of environments. When there's food, they devour it and produce many offspring. When the food starts to run out, their descendants move off in all directions to look for new environments to exploit. Animals like these would have done well during the ice ages, and so would the animals able to catch and eat them.

Having a larger brain might have helped our ancestors find ways of making a living during the times of climate chaos. Their ability to cooperate, share information, and rapidly adapt their culture to new conditions would have also been valuable. During these desperate times, even the strongest and smartest individuals would have been better off sharing with a family group than trying to live independently. Through working together, discussing ideas and outcomes, and learning from mistakes, they could be adaptable, like rats or sea gulls. They could never reproduce as quickly as these animals, but if the family groups included males who were ready to devote their ingenuity and greater physical strength to helping females to protect and feed the family, they could manage to raise young.

A hundred thousand years ago was not a time of climate chaos. Some of the evidence of past climates suggests that it was much like today. But that doesn't mean life was easy. In fact, during periods like this, our ancestors would have been struggling to compete with animals that were better adapted to stable habitats. Men carrying spears might have been quite successful hunters in the chaotic times, but in times of climate stability, lions, tigers, saber-tooth cats, hyenas, and wild dogs would be back in force. Their successful hunting would have left fewer animals for humans to hunt, and there was also the danger that they would be prey for these top carnivores. Times were stressful and yet, at the same time, their lives would have held few big challenges. In a more stable climate, families weren't faced with very many formidable new problems or surprising new opportunities. It was just constant worry and there was much less advantage to having quick wits and working together.

If our ancestors that lived 100,000 years ago were like people today, the lack of exciting new challenges and the endless stress of keeping children safe and fed would have made them less inclined to cooperate. Males in the prime of life may have been tempted to go it alone and not saddle themselves with a bunch of woman and children. If they did, their lives would have been easier, and their survival chances would have been greater. But going it alone would have decreased the chances of *their genes* surviving. Young humans could only be brought up within cooperative groups that included adult males who were willing to contribute to the effort of feeding and protecting them and being a role model for their sons.

Some men, like the tall smiling stranger in our story, might have managed to get women pregnant without being a loyal member of a family. A few of the babies of these men may have survived. But men who stayed with a family and helped to raise children had a better chance of passing on their genes, and they would also pass on their cultural knowledge and beliefs. Young boys growing up in a family of men and women would be able to learn the skills they needed to survive, and they would also see how people worked together and raised their young. These groups of humans might not have been "families" in the sense that they were all genetically related, but they were families in the sense that they shared the purpose of raising new young members. Families that failed to raise enough useful new members inevitably died out.

Growing up to fit into a family

To raise their offspring, our ancestors had to work together and coordinate their efforts. Their culture included important practical skills like starting fires and making tools. But to work together successfully, they also needed social skills and social tools that would allow them to agree on goals and settle disagreements peacefully. The genes that governed their brain development coevolved with the social tools they invented that made it easier to work together. It made sense to believe that males and females should have different roles and do different work. Since women are biologically equipped to provide food for infants, it was efficient to have women take on the main responsibility for baby care. Ideally, the other tasks they took on would be ones that could be done while pregnant, carrying a baby, or keeping an eye on young children.

Males could do other work, including the riskier work and work that took advantage of their superior physical strength.

Child labor also made sense. It was essential that children were as small a burden as possible and that they start contributing to the work of the family as soon as they were physically and mentally able to help. This gave children the chance to learn on the job and contribute more as they grew older. Sickly and lethargic children couldn't be indulged. Caring for children was an investment in the future, and it didn't make sense to invest in a child if there was little chance of it growing up to be an asset to the family. Everyone had to do their bit, and older children had to help to look after younger ones, as they still do today in many cultures.[8] We're descended from the children who survived in this challenging environment, who were stimulated but not overwhelmed by it. They met and sometimes surpassed the expectations of their elders, but they didn't overdo it. Children who were too willing, passive, or obedient risked being worked to death. Successful children got the balance right and grew up to be team players who expected those around them to also play their part.

Most people living today are experiencing a cultural environment very different from this, and beliefs about how children should behave and be cared for are changing rapidly. Because we have different experiences today, our brains will develop to be rather different from the brains of our ancestors, but the baby brain we're born with is pretty much the same as that of humans going back well beyond 100,000 years ago. This doesn't mean that all human babies are born with identical brains, then or now. Variation exists in all populations, and families struggling to survive 100,000 years ago benefited from having members with a mix of mental perspectives, just as they benefited from having members with a mix of physical abilities. But to survive, a human baby born 100,000 years ago needed to be born with the capacity to develop into a child, and then into an adult who was willing and able to fit in and be part of a group that worked together. We're descended from successful babies, so we have inherited a brain that can adapt to life in the Stone Age. It also happens to be a brain that can adapt reasonably well to life in the 21st century.

Our brain is sensitive to the actions and expectations of people around us. Psychologists studying baby behavior have found them to be primed to soak up information from what surrounds them. A few weeks after they have come into the world, they are already meticulous observers of behavior. After watching a short play involving puppets or a cartoon, babies a few months old can recognize which characters were being helpful and which were being "bad." When they're a few months older, babies can show, by reaching for the puppet, which character from a puppet drama they prefer. They like the helpful puppets and puppets who agree with them about what food to eat.[9]

Experiments have confirmed what most parents have long believed: babies understand a lot of what they see around them long before they can talk about it. They can work out the purpose of an adult's behavior and notice when things don't go to plan. A good way of getting a laugh out of a baby is to pretend to be very frustrated about something that keeps going wrong. Babies and young children seem to be obsessed about the purpose of our behavior. They want to know "why" and are keen observers of mistakes and bad outcomes. Among the first words children use are those their carers say when something goes wrong. English-speaking children tend to learn words like "uh-oh," "oh dear," and "whoops," and, if you're unlucky, they pick up words like "damn" and "oh shit."

But, even if they appear to take delight in catastrophe, toddlers can actually be very helpful. Psychologists demonstrate youngsters' ability to understand what others are planning to do by putting them in a situation where an adult is trying to do something but can't. On seeing someone struggling to reach something or trying to open a door with his hands full, children as young as 14 months toddle over and provide help.[10] So far, most of this research has been done in Western populations. Babies and young children living in non-Western cultures have similar capacities for observing and understanding behavior, and a similar desire to be helpful and to play and work with others. But the much smaller amount of research performed on babies and children living in non-Western cultures has revealed that different ways of parenting have important effects on the behavior of babies and children.[11]

Thousands of YouTube videos of five-year-olds singing and dancing like the latest pop icon provide ample evidence of how good children are imitating.[12] But learning how to behave isn't just a matter of imitating. It may be fun to play being different people, but small girls aren't really supposed to behave like pop icons. Children have to develop some sense of who they are in order to work out who to copy. This may be more difficult for children in modern societies than for children born 100,000 years ago. Children today are exposed to so many potential role models, and many of the characters they see aren't even real. But most find a way. Psychologists studying children in the United States have found that by the age of five, most have developed ideas about whom they should pay most attention to. They're more likely to remember what strangers say if they're like people the child is familiar with. Interestingly, even though American culture often seems obsessed with appearances, what strangers look like was found to be less important to the children than what they sound like. A person speaking with an accent like theirs made the biggest impression on the youngsters. How they dress and the color of their skin were less important.[13]

If you think about the lives of our ancestors that lived many years ago, it's possible to see how being born with a social and team-building brain like ours would have given them the best chance of surviving and raising children in an uncertain world. A group of people of different ages had to live and work together. A brain that just made them behave like passive groupies would have been no good. Everyone needed to try to strike the best balance between helping their family to survive and surviving themselves. And families needed members capable of being both leaders and followers. It would be in the best interest of younger, less experienced members to follow their elders, but, ideally, they would start developing knowledge, understanding, and skills they would need to be elders one day. If older children and adults fell into line, it was rarely because they feared punishment. It was because they understood the goal of the work and accepted it as their own goal. They worked toward an immediate purpose—climbing the hill to reach a cave—because they understood how it fits into a bigger purpose that they agreed with—keeping the babies and children safe.

Humans today are pretty much the same. We can be good team players, but we perform much better if we understand "the game" and believe the goal is worthwhile. We can't help having opinions about how the game is being played. Being part of a discussion about such matters is part of what it is to be human. To be left out of it feels bad and life seems without purpose.

Part two: Imagine growing up as one of your ancestors living in Africa 100,000 years ago

As you grow up, family members often remark that you are more like your aunt than your mother. Most say that it must be because her milk was the first milk you tasted. But some wonder if the soul of her dead baby moved into your body. You hate that idea because that might mean you have some of the evil trickster in your soul. You're determined that you will prove to everyone that you are not evil or a trickster, but it's not easy to be good all the time. By the time you're old enough to help with the family's chores, everyone knows about the evil trickster who betrayed the kindness of your family and made your aunt pregnant with a baby that died. Your grandparents' friends and relatives had spread the terrible story among all the families and warned them about the evil one. They had heard stories about a tall young man who had been seen dying in agony from a snake bite. According to other stories, he had been struck by lightning. No one really knows what happened to him.

About the time your last baby tooth falls out, you experience your first hunger. You have, of course, often gone to bed hungry in the dry seasons, but this is your first real hunger. It begins when your family travels to a place that elders remember as a riverside, with many trees that produce nuts during the dry season. On the way, you meet another family going to the same place, so you travel together. As you get closer, it becomes clear that something isn't right, and when you arrive, you find that the river must have changed course. It's now only a trickle, and the nut trees are nearly dead. The parents of the two families decide to travel together to another place, but after several days of walking, you find that the river here is also nearly dry. The elders decide to follow the old riverbed upstream. It still has a small stream of water, which means you won't have to fill and carry water containers.

You trudge along, taking your turn carrying a baby who is so still that you keep checking to see if she's still breathing. You remember the stories of families nearly on the verge of collapsing when they are saved by a cloud that suddenly appears on the horizon. The storytellers say that that some clouds turn into swarms of delicious locusts that will provide more than enough food for starving travelers. The lesson of this story is that you must never give up hope. You keep scanning the horizon for unusual clouds and try not to notice the carrion birds that circle overhead from time to time. You feel on the verge of collapsing but still you walk on, late into the night, until you realize that the mood of the people around you has changed. Then you sense it too—the smell of water and the sound of a flowing river. You have reached the place where the river changed course. There are nuts and fruit here and plenty of firewood. As dawn breaks, you start to feast on what you can find. The next day men from each family go out to hunt the baboons that are starting to come to help themselves to the nuts and fruit. Everyone is worried that there will be very little game. If a group of lions is working in the area there won't be much left for the hunters. But it seems as though their run of bad luck has ended. One of the parties arrives back with two young baboons and the other has killed a medium-sized antelope.

That night, everyone has a full stomach for the first time in ages and the members of both families laze around one fire listening to the older ones talking about the hungers they had experienced. They soon start telling the stories of the time long before any of them were born, when there were terrible hungers but also times of great plenty. The elders from the other family tell slightly different versions of the stories about the long droughts, massive floods, and terrible windstorms. They describe the endless restless water at the end of the land and mention that it isn't ordinary water. They say that it's undrinkable because it's so salty. This seems strange to you because you've heard that there is food growing on the rocks at the beginning of the endless water. How can food grow in undrinkable water?

As the fire starts to die down, the elders talk about lessons learned from this hunger. A grandfather from the other family says that everyone deserves credit for the way they made decisions together and didn't blame each other for misfortunes that were no one's fault. Your grandmother then makes the children very pleased by praising their behavior during the long hard journey. She says that it's much easier for the adults to keep their temper if the children are quiet and don't complain. Then a father from the other family says that he thinks that there is some fault to be reckoned because some other families must have known about the river changing course and they didn't spread the word. It is evil, he says, to withhold information that can mean life or death to families.

Everyone goes silent. "Evil" is a strong word and it frightens people. You and several other members of your family look at your aunt because she has strong opinions about using the word "evil" carelessly. It takes a moment for her to realize that people are expecting her to speak. Then she sits up a bit straighter and says, "I was once called evil and I wondered myself if I was an evil person. This was because many years ago, a stranger convinced me that my wishes were more important than my duty to my parents. The consequences were bad for me and my family but, even if there hadn't been bad consequences, my behavior would have still been foolish, selfish, and wrong. The shame I feel for what I did is the worst punishment I can imagine. This shame has convinced me that I myself am not evil. If I were evil, why would I punish myself in this way? I think that we should tell other families about our terrible journey, so they won't make the same mistake. Any families that withheld information will hear about our suffering. I think they will be punished by their shame. But, if that isn't the case . . . if they're happy or amused by the hunger we've endured, then we'll certainly hear about that and then we'll have good reason to think them evil."

Members of the other family look at your aunt with new respect and you feel proud. You also feel relieved because what she has said makes you think about evil in a new way. You know the feeling of shame very well. There have been times when you were naughty or silly. You might have thought at the time that you were being clever, but now remembering them makes you feel horrible, especially if there were bad consequences or someone saw you. If these terrible feelings of being ashamed are a sign that you are not evil, it means that, even if there is some of the evil trickster in your soul, you are not evil.

Many years later, when you are one of the old ones sitting around the fire, surrounded by your children and grandchildren, you sometimes tell the story of your first real hunger. If they ask for it, you also tell them a story about the life of your wise aunt and how she taught you the importance of feeling shame.

Social tools and social emotions

It's useful to think of our brain as having two different kinds of capacity—the capacity to "think" (interpret information, figure things out, make predictions, have ideas, formulate plans) and the capacity to "feel" (hungry, achy, cold, angry, frightened, sexy). How we feel minute-to-minute motivates and colors our thinking and influences what we decide to do. Emotions like "pride" and "shame" influence how we behave toward others.

As humans evolved, they became smarter in their dealings with their physical and social environment, and this is because they were getting better control of what was happening inside their heads. Over the generations, our ancestors' feelings—both physical and emotional—became more intricately entwined with their thoughts and memories. This made them less likely to act on impulse and increased their ability to behave in calculated ways. They could, for example, suppress feelings of anger or sexual desire if acting on them would get in the way of carrying out their longer-term plan. And they could even *conjure up* feelings of anger or sexual desire if these feelings might help them achieve their goals.

Scientists comparing the anatomy and workings of human and animal brains are beginning to work out the details of how our brain changed so that we could develop this new control. The brain of our ancestor who lived 100,000 years ago was more than twice the size of the brain of the ancestor who lived three million years ago. A great part of the expansion was in the part of the brain called the "prefrontal cortex"—the part that sits above our eyes. The nerve cells in this part of the brain extend their long thin arms (or "axons") to other parts of brain. Axons are a bit like information-carrying wires. They connect the prefrontal cortex to the different mental organs within the brain—the organs that receive sensory information, generate emotions, store memories, and organize actions. These mental organs are similar in all mammals. What's different in humans is the way they're connected. Inside our brain's large prefrontal cortex, a vast amount of information is integrated and processed in ways that can't happen in other animals.[14] This makes our experience of the world and our behavioral options very different. Inside our frontal lobe, our feelings are sliced and diced, mixed together and flavored with our memories.

We may experience our mental life privately, but it evolved to be shaped and shared by the social network we're part of. We rarely make decisions by thinking things through for ourselves. Even if we don't discuss a decision or ask for advice, we can't help thinking about what other people would do and how they might react to our decision. This is because, to survive and reproduce, our ancestors had to rely on and be relied upon by others. We've inherited a brain that evolved to be connected to other brains, and we've inherited a culture with a vast number of social tools, such as language, that allow us to function within a network and bring up children that will also be able to function.[15]

Seven million years ago, our ape ancestors were social animals with a few social tools. As they evolved, they took "being social" to completely new levels. The most important social relationship in a chimp group is the bond between a mother and her baby. Through the action of neurochemicals in her brain, a female ape becomes attached to her newborn. In humans, thanks to the labyrinth of connections that can be made in our brain, this basic bonding mechanism can be triggered by many things. We can feel briefly attached to a person we sit next to on an airplane. We can feel bonded with our friends, our pets, and the singer of a song that moves us. We can really care about someone like Elizabeth Bennett (who never actually existed) and feel great relief when Mr. Darcy proposes to her again and she agrees to marry him. We can love our new phone, our country, and the ideals that our country stands for. And we can suddenly feel at one with half the people in a stadium because someone on "our" team scored a goal.[16]

We come into the world ready to start relationships and, as we gain control of our body, we're keen to take part in games and tasks that involve working with others. In this way, we're so different from young chimps. Experiments have shown chimps can understand collaborative tasks perfectly well, but they only bother to take part if they can see how it will result in their getting a piece of fruit or some other reward. Humans, by contrast, often work together just for the joy of it.[17] Experiments have shown that working with others affects children's behavior. Afterward, they're more generous in sharing any goodies the experimenters give them—as if working with others has put them in a better mood. It seems unlikely that children's greater willingness to share is simply the result of learning that they should pay people for working with them, but the way we feel about everything is strongly influenced by the experiences that shaped the development of our brain. Our childhood observations of others don't just help us learn how to behave; they help us understand how we're supposed to *feel*. And then, as we learn language, we're able to understand the goals and values that our family expects us to feel strongly about.

Once children are judged to be old enough to know how to behave, they're expected to feel bad when they do something wrong. Most of them do. We might call this feeling "guilt" or "shame." We may be "remorseful," "embarrassed," or "humiliated." English isn't the only language that has a lot of words for this bad feeling. People of all cultures know the feeling and see it as a complicated thing. Such feelings are vitally important social tools. The bad feelings we have when we remember the foolish, hurtful, or odious things we've done in the past make us less inclined to do them again. This is good for both ourselves and the people around us.[18]

Our huge kit of social tools also includes good feelings, such as the "pride" we feel when we do the right thing, even if no one even notices. But it's usually best when our goodness is noticed. If our athletic prowess, piano playing, or business acumen receives recognition, we feel "thrilled." If a soldier dies hero-ically in battle, his mother feels honored when she is given his posthumous medal. Even as we compete, our social feelings keep us bound in the web of rules and judgments of the groups we belong to. And if we come to believe that we don't belong—if we see ourselves to be of no use to anyone—we may no longer want to live. We (or most of us[19]) were born with the capacity to have social feelings, but the details of what we feel, when we feel it, and how we respond to the feelings are shaped by our experiences.

Because they had feelings of guilt and pride, our ancestors could put more effort into finding food and keeping safe. Such feelings made our ancestors more trustworthy and they didn't need to keep an eye on each other and punish wrongdoers. To a large extent, everyone policed themselves. Behaving in the expected ways became part of the way they thought and felt. Behaving "abnormally" felt "unnatural" to them. They felt fine while behaving in ways everyone around them thought was "normal," and they felt bad when they did something that would be considered weird or selfish. Social scientists call this the "internalization of social norms."[20] Modern towns and cities tax their residents to invest in policing and punishment systems that work to "keep the peace." But a population can only feel safe and comfortable when its mem-bers are part of a community and have similar ideas about right and wrong behavior. If people who live in the same neighborhoods have internalized the same "neighborliness" norms, police don't need to patrol very often. If everyone on the street hates loud parties and feels disgusted by public drunk-enness, then weekends will be quiet in their community. In neighborhoods full of people who love a good party, weekends will be noisier, but not neces-sarily less peaceful.

Our modern way of life relies on internalized social norms as much as that of our ancient ancestors, but it has also revealed problems.[21] We now often live among people from different cultures, and we sometimes have to work hard to avoid feeling disturbed when we see them behaving in a way that seems shameful or even disgusting to us. It's hard not to see such behavior as "unnat-ural" or "inhuman." Another drawback is that, even when social norms don't make sense, we often still punish ourselves in painful ways when we infringe them, even if the infringement is accidental. Why should a woman feel mortified if a little menstrual blood seeps through her white jeans? People and populations differ in the norms they internalize and how they feel about these norms, but caring about what others think of us is a vital part of being human. There are individuals, sometimes called "psychopaths" or "sociopaths," who reportedly don't feel bad about doing wrong, and we find them scary.

Why did our ancestors become parents?

From a genetic evolutionary perspective, the most important thing that a living organism does is produce offspring. But nowadays many people are choosing not to become parents, and perhaps some of the people who lived 100,000 years ago also avoided having children. Pregnancy was just as uncomfortable for them as it is for us, and without the medical options available today, giving birth would have carried a far greater risk of pain and death. What's more, after all the suffering, the chances of a newborn reaching adulthood were often not much more than 50 percent. So why did our ancestors decide to have babies?

It's relatively easy to explain why our *ape* ancestors had babies. Apes feel like mating and they mate. Mother apes feel like caring for their baby and they care. Father apes don't have this feeling and they don't care. Baby apes inherit their parents' genes and develop a brain that generates the same feelings, so it all happens again with the next generation. But if our ancestors living 100,000 years ago had a brain like ours, then life for them wasn't just a matter of doing what they felt like doing. They may have often felt sexual desire but kept their behavior under control. It seems very unlikely that they would be ignorant of how babies are conceived or how to satisfy their sexual desire without conceiving a baby. There must have been some accidental pregnancies and sexual assaults 100,000 years ago, just as there are today, but children conceived in this way probably didn't fare too well. Most of our ancestors were conceived because their mother chose to endure pregnancy, face the risks of giving birth, and spend the prime of her life raising young ones. And they survived to adulthood because their fathers, uncles, and other relatives helped to care for them. But why did the mothers become mothers and why did the others help her care for her baby?

If our ancestors were like the people who live in small-scale societies today, they put a lot of effort into parenting because they saw it as the natural and normal thing to do. Culture is, after all, a lot like genes. If a population doesn't produce children, their culture will perish as surely as their genes will perish.

As well as obstetric and parenting skills, and parenting tools like slings and cradleboards, our ancestors' culture included a set of beliefs and customs about the value of children. Many cultures still have the custom of giving higher status to a young woman once she has become pregnant, and especially high status to a woman who has given birth to several healthy children. Because our ancestors internalized these norms, they may not have perceived that they had a choice about parenthood. They did what was (to them) obviously "the right" thing to do. The idea of parenthood being a personal choice only became widespread in the last 150 years. Prior to this, even though people understood the link between sex and pregnancy, most seemed to believe the creation of a new human life wasn't something that a mere human should try to control.[22] Their feelings might have been similar to those expressed by Kahlil Gibran in his 1923 poem *The Prophet*: "Your children are not your children. They are the sons and daughters of Life's longing for itself."

This doesn't mean that our female ancestors felt compelled to be pregnant all the time. Far from it. *Surviving* children were what made a family thrive. Conceiving a child if it didn't have a reasonable chance of survival was highly inefficient. So was allowing a newborn to live if its chances of survival were low.[23] When times were hard, young adults (both men and women) may have had to postpone having a baby of their own and concentrated instead on helping to care for the children of older, more experienced parents. In really hard times, some might never be able to have a child.[24]

Over the course of human evolutionary history, there may have been some independent-minded women who thought things through and decided to avoid the pain and risks of motherhood. These women are not our ancestors. There may also have been families that decided to do away with the rules and customs that encouraged the raising of children. Our ancestors didn't belong to families like this. Our ancestors were part of families that believed in the importance of children and worked hard to produce the next generation. That's why we exist.

And why did they get married?

The custom of formally recognizing a link between a couple that is having sexual intercourse is a social tool that is part of a larger social tool of kinship. It's impossible to know when our ancestors first started getting married, but there are two reasons for believing that the custom of marriage was already well established 100,000 years ago.

The first is that the idea of marriage is so widespread among humans.[25] Even though cultures are often vague about what is meant by "family," most languages include words that meticulously describe how people are related, either through "blood" or through marriage. In many populations, marriage is part of the cultural glue that binds individuals together in families and families together in wider clans and tribes. It helps regulate births, establish who should help raise a child, and who should benefit from the help of that child if it survives. Many cultures believe that a woman should only become pregnant if she has the security of knowing that her baby will have roots in her husband's family as well as her own. That way, if her husband dies, there's a good chance that his family will help provide the care her baby needs.[26] If marriage was a recent invention, you'd expect to find at least a few cultures with no idea of it. But you don't. The basic belief that it's important to know who fathered a child likely began hundreds of thousands of years ago when our ancestors lived in Africa.

The second reason for believing that marriage has a long history in human populations is that the genes we have inherited provide us with a body that is better adapted to a life of restricted sexual relationships rather than one in which males and females have many sexual partners. The size of men's testicles is a reliable indicator of the kind of sex life our species is adapted to. Comparing the testes of different mammals yields a clear pattern. In species like chimpanzees, which have quite open sexual relations and ovulating females are likely to mate with several males, the males have larger testes (see Figure 5.4). Not only are these larger testes able to release more sperm and seminal fluid with each ejaculation, the sperm they release are supercharged compared to human sperm. Each carries a large supply of fuel that allows it to swim long and hard once it's been released inside a female. Producing vast quantities of well-equipped sperm takes a lot of energy—gram for gram, the tissue in testes requires about the same number of calories as brain tissue. And carrying around very large testes is inconvenient. But it's easy to see why big bollocks evolved in chimpanzees. If an ovulating female mates with several males, the males with the most numerous and athletic sperm have the best chance of being the one who fathers her next baby.

Human males have the much smaller testes typical of mammals in which females only mate with a single male each time they ovulate. The sperm they produce has the minimum amount of fuel. In human females, the fertilization of eggs is more of a "fun run" than a marathon.[27]

More sobering evidence that the human body is poorly adapted to unrestricted sexual relationships is the vulnerability of women's reproductive

Figure 5.4 In the right hand is the brain of a chimpanzee. In the left is one of his testicles. Female chimpanzees usually mate with several males when they are ovulating, which means that the sperm of the different males compete to be the one that fertilizes her egg. To have a chance to father offspring, chimps have to produce large numbers of extremely vigorous sperm. To do this, they need large testicles.

organs to infection by sexually transmitted bacteria and viruses. Rates of sexually transmitted disease rise rapidly in populations when a cultural change leads to more open sexual relations. Chlamydia infections are the most dangerous from an evolutionary perspective, because infected people experience few symptoms and can continue to spread the infection for years. Infections don't threaten a person's survival, and nowadays they can be cured with a dose of antibiotics, but while a woman is infected with chlamydia, the bacteria can cause a mild inflammation in the tissues that line her reproductive system. This often results in scarring that permanently blocks her fallopian tubes so that the eggs she produces can't be fertilized and travel to her womb. Today, women with blocked fallopian tubes can sometimes be helped to become pregnant with in vitro fertilization, but, until recently, getting a sexually transmitted infection could permanently destroy her ability to reproduce. If behavior that spread these infections was common in the past, we would be descended from ancestors with a body that wasn't so vulnerable to the consequences of such behavior. And we wouldn't have inherited a body that is so at risk from sexually transmitted diseases.

Some people argue that to understand why people marry it's more important to look at the evolution of genes than the evolution of culture. They suggest that human sexual behavior is influenced by a genetically evolved tendency of humans to desire to have mates and to form pair-bonds.[28] If this were the case, the institution of marriage would be simply adding legal and sometimes religious rituals to a set of behaviors that humans are genetically programmed to carry out anyway.

One problem with this idea is that human marriage is nothing like the kind of pair-bonding that occurs in some other mammals.[29] The pair-bonding mechanism in mammals is usually triggered by mating, which causes the male to become attached to the female he has mated with. If having sex triggered the same kind of pair-bonding emotions in human males, then there would be no need for marriage and no possibility for a society with a complex culture. Pair-bonding emotions make males behave in ways that would make human social life impossible. A pair-bonded male gibbon does little or no parenting, apart from helping to defending the territory that he, his mate, and their offspring live in. He wants to be near his mate all the time and is aggressive toward all gibbons other than his mate and their offspring. A human male may get annoyed at seeing another male flirting with his girlfriend, but he would risk arrest if he showed the sort of obsessive and violent behavior characteristic of pair-bonded male gibbons.

The feelings of love that modern couples experience are real and involve the same brain mechanisms as those that trigger pair-bonding in other mammals. But what we call "love" usually involves other feelings as well, such as lust. And these raw feelings are processed through our frontal lobe so that anything we feel is heavily influenced by past experiences, including what we have learned about how we "ought" to feel. Over the last couple of centuries, an astonishing amount of creative effort has been devoted to describing the feelings (good and bad) that people have when experiencing romantic love.[30] At the same time, there has been considerable cultural evolution of beliefs and customs regarding marriage, particularly in Western populations. In the West today, the institution of "marriage" can seem irrelevant when it comes to raising children. Plenty of married couples choose to remain childless, and it's common for a single parent to raise children with the help of relatives, government-funded benefits, and/or paid employees. Divorce, remarriage, and the possibility of people of the same sex getting married and raising children adds even more complexity.

But in all cultures, "marriage" is more than just a formal agreement and a ceremony. It is a set of thoughts, feelings, beliefs, and habits that we learn from those around us. This was as true for our ancestors that lived 100,000 years ago as it is for us today. Couples that live happily together without a "piece of paper" to say they're married aren't kept together by gene-driven raw emotions any more than married couples are kept together by their piece of paper. The stability of our relationships is provided by the way we're shaped by our experiences as we grow up in a family and see how parents behave. Nevertheless, a bond between two people gets further stability from being publicly recognized whether or not the recognition is accompanied by legal or religious rituals. As well as the feelings they have for each other and their shared children, a couple also has the feeling of not wanting to look bad in the eyes of the people who expect them and their children to stay together in a family.

Hungers

Over the last few million years, many of our ancestors lived through periods of desperate food shortage. The epidemic diseases like plague and cholera that devastated human populations in more recent times were less of a problem for our earlier ancestors because they lived in small mobile groups and such infections were less likely to spread. But maintaining a healthy immune system takes a lot of energy, so when our ancestors were starving, they were more vulnerable to infections from all kinds of microorganisms and parasites. Everyone was at risk, but babies and young children were the most likely to die. If food was limited, it made more sense to give it to the older children and adults, who were able to go out and find more food. Allowing a baby to die or smothering it when its mother's milk dried up must have been painful, but our ancestors survived because they had a brain that allowed them to override their emotions and choose behavior that was informed by their experiences— both their own experiences and the wisdom of those they respected. They survived because they made the difficult choices that gave their family the best chance of survival in the long term.

Two lines of evidence provide clues to how severe these hungers were. The first is revealed by research into the environments they lived in. The unstable climate during the ice ages brought times of prosperity when people had enough to eat and the population grew. But these ended abruptly, leaving our ancestors with the problem of figuring out how to make a living in a new set of environmental conditions. Even in the warmer and more predictable periods between the ice ages, dry seasons and drought years would have put their survival in jeopardy.

The second line of evidence suggesting that our ancestors experienced cycles of feast and famine is that our bodies are so well adapted to this kind of life.[31] We start out life with about four times as much fat as the average new-born mammal and then gain more as we grow. When there's enough food, we tend to eat more than we need and store the extra energy as fat. Humans have a fatty layer lying under our skin that makes our skin look smoother and feel more cushiony than the skin of most other mammals. We also have special sites around our body where fat is warehoused. Carrying extra weight means that we have to use more energy to move around, but we cope easily. It's normal for women to carry more fat than men, and women's bodies function well even if fat makes up over a third of their body mass.

The human body's fat-storing system probably began to evolve when our ancestors started spending more time outside the forest in habitats with rainy seasons when food is relatively abundant, followed by dry seasons when it's scarce. Before the move to these habitats, the ability to build up stores of fat wasn't necessary. Food is seldom abundant in a tropical forest, but a clever and knowledgeable ape can almost always find something to eat. Also, for an ape that makes its living climbing around in trees, the extra weight of a fat store is much more of a burden than it would have been for an Australopithecine that spent most of its time walking on the ground. The bodies of chimpanzees and bonobos are extremely lean—only about 1 percent of the male chimpanzee body mass is fat. Females carry a little more (see Figure 5.5).

Many mammals that live in seasonal habitats put on fat during the times when food is most plentiful, but the amount of fat in human bodies, particularly woman's bodies, is extraordinary. It's many times greater than that seen in any other land animal. The high fat levels aren't just the result of living in modern societies with sedentary habits and an abundance of unhealthy foods. Women living in foraging societies with a lifestyle similar to that of our ancestors 100,000 years ago are nearly as fat—between 19 and 24 percent of their body weight is fat. What's more, women's bodies don't function normally unless they have high levels of fat. The monthly cycle of ovulation and menstruation stops if a woman's fat store falls much lower than 20 percent of her body mass.

It was in the 1940s that some scientists began to point out that the only other animals that routinely carry so much fat are mammals adapted to life in the ocean, such as whales. This observation led to the intriguing suggestion that the human body might have adaptations left over from a life in the water. A fat body isn't a burden to a swimming animal. In fact, it helps them float. The "aquatic ape" hypothesis seems to explain a lot of the odd things about humans.[32] Much of our fat is in a layer just beneath our skin. Whales also have a layer of fat or "blubber" beneath their skin, which helps to regulate their temperature. Perhaps a fat layer served a similar purpose in ancient humans.

Figure 5.5 Jambo, a chimpanzee living at the primate center of Twycross Zoo in the English county of Leicestershire has a rare condition that caused him to lose his hair. This provides an opportunity to see how lean and muscular chimpanzees are compared to humans. Fat makes up not much more than 1 percent of the body mass of male chimpanzees, even those living in zoos. Humans have smoother skin than other land mammals, because we have a layer of fat lying beneath our skin.

Humans are also relatively hairless, like whales, and of course some whales also have large brains. Might it be that a large part of our evolutionary history was spent on the edge of lakes and seas from which our ancestors swam out to forage for aquatic and marine plants and animals?

It's a neat idea, but the answer seems to be "no." Scientists believe that the evidence for the aquatic ape hypothesis is what a lawyer might call "circumstantial." There's no sign of our body having been adapted to spending long periods of time in the water. For example, our skin is very vulnerable to attack from microbes and parasites that live in water. And there are many signs that our skin is extremely well adapted to a life surrounded by air. For example, our amazing ability to keep our body cool by sweating is only effective on dry land and wouldn't be necessary if our ancestors spent their lives in the water.

Most physical anthropologists now agree that the fatness of the human body is an adaptation that allowed our ancestors to survive times of severe food shortage despite having a large energy-hungry brain.[33] Brains need to consume calories all the time, so to survive times of food shortage, animals need some calories held in reserve. The larger the brain and the more severe the food shortages they face, the more calories they need to hold in reserve. Over tens of thousands of generations, our ancestors were the ones who survived the hungry times, and this is partly because their bodies had the largest fat stores. As a result, we have inherited a body that is not just good at storing fat, but also tends to hang on to its stored fat as long as we're getting enough to eat. If times of hunger were frequent and severe, women would need larger fat stores than men. Our female ancestors not only survived the hungry times themselves; they were the ones whose babies were most likely to survive. It would have been risky for a woman to become pregnant if she didn't have a good store of fat. When the hungry times came, her baby would likely die and the physical effort she had already invested in that baby would have been wasted. This is why we've inherited a body that stops ovulating if our fat stores are low due to stress or illness.[34]

The evolution of the unique human life course

Our extreme fatness isn't the only odd feature of human biology. Biologists studying an animal species report on its "life history." They track how long it takes for them to reach maturity, how their reproduction is organized, and how long they generally live. Other useful models from evolutionary biology address such things as how to forage optimally. Students of human behavior who specialize in the application of such models call themselves "human behavioral ecologists."[35] Rough patterns exist in these life histories. For example, mammals that are larger tend to take longer to reach full size. They usually produce fewer offspring and live longer lives. Humans more or less follow the pattern for a large mammal, but with a few odd quirks. Thinking about how these quirks might have given our ancestors a better chance of survival gives us more clues to what their lives were like.

Animals evolve to grow at certain rates. A puppy grows to maturity in less than a year. For a human baby, it takes well over a decade. This is because we're descended from apes and, as explained in chapter 2, a mother ape can't find food fast enough to support an infant that is growing rapidly. Like other animals, we grow fastest right after we're born, and then our growth slows down. By age three, we're growing slowly but steadily, and then, at around age seven, our growth slows again, and we grow more slowly for several years. Then, as puberty begins and we start to develop an adult body, our growth speeds up again. We're the only animal that has this adolescent growth spurt. Why did we inherit a body that saves some of its growth for later? How might this have helped our ancestors survive childhood?

Research into the energy requirements of young children suggests that, like our fatness, this unusual growth pattern evolved as a way of coping with the trade-offs our ancestors faced thanks to the huge brain sitting on their shoulders. The brain of a five-year-old is as big as an adult's, and it's a lot busier. It's developing, learning, and constantly tending to its connections—forming new ones and dismantling old ones. Measurements have shown that a five-year-old's brain absorbs energy at nearly double the rate of an adult's brain.[36] During the seasons when the habitats they lived in produced the least food, the children who survived best must have been the ones whose body invested more energy in brain development and less in the growth of their body. Smaller children were easier on their parents just as smaller dogs are easier on their owners. They ate less, took up less space, and were easier to control. Staying small longer also had benefits for the child. It gave them more time to observe, to play, and to learn to use their increasing size and strength effectively. Adults expected less from them, so they didn't have to work so hard. As they rapidly got smarter and slowly got bigger, children could make an increasing contribution to their family and deserved a larger share of the food. Once their brain ceased to need so much energy, their body could devote more energy to physical growth.

It could also be that this unusual pattern of growth gave our ancestors extra flexibility to deal with the times of feast and famine they lived through. If there was food available, it was better to be big. Bigger bodies are stronger and faster, so they're better at getting that food. But bigger people need to eat more, so they're less likely to survive times when food is really scarce. The human body is genetically adapted to be adaptable. Our final height is influenced by the genes we inherit from our parents but also by environmental factors. The precise size and shape of our bodies depend on the quality of the nutrition that we and our parents received and how badly we and they have suffered from infectious diseases and parasites like intestinal worms. The slow growth during childhood provides a time for the body to sample the world it has been born into. If the body experiences no famines and little sickness, it may signal the body to have a bigger spurt of growth before reaching maturity. In many parts of the world, people are now better fed and, in general, much healthier than their ancestors who lived only a few generations ago. They are also tending to grow to be taller than these ancestors. The trade-offs involved in human growth are fairly well explored.[37]

The other odd thing about human life history is our long life span.[38] If our bodies followed the mammalian pattern of life span being correlated with body size, then our maximum life span would be about 40 years—a bit longer than chimpanzees, but not as long as an elephant, which can live 70 years, or the largest whale species, which can live to be over 100. Many of the babies born 100,000 years ago may have only had a 50-50 chance of surviving to adulthood, and until about 150 years ago, infant and child mortality was high in all human populations. But we know that our ancestors not only survived to adulthood, but they also had children and grandchildren who survived. It's likely that quite a few of them lived well into their late 60s and early 70s and helped to raise their grandchildren.

It's easy to see that having grandparents around would be useful, but the fact that older humans can be useful isn't enough to explain why our bodies last about twice as long as expected. Biologists have learned quite a lot about the evolution of life spans and why the patterns they observe exist. As with so much in evolution, it comes down to offspring. Traits are favored if they increase the number of surviving offspring an animal produces. The longer an animal lives, the more time it has to produce offspring. But to have a longer life, animals must slow down their *rate* of reproduction. A long life requires a lot of investment in body maintenance; energy must be spent in fighting infections, controlling parasites, and fixing damage from everyday wear and tear. This leaves less energy for reproduction. The life span that evolves will be around the balance point between time to reproduce and the rate of reproduction. It will be different for each kind of animal, depending on how it makes its living and the stresses and strains of its environment.

When our ancestors started to live in groups that shared food and knowledge, they created a social environment within their natural environment that shifted the balance point. It made investing in bodily repair yield better returns. That's because, in this environment, adults in the prime of their lives had a good chance of surviving damage that would be fatal for other animals. Perfectly fit and well animals are sometimes unlucky. They sprain a joint or an insect bite gets infected. Such minor things could heal within a week, but a week is a long time in nature. An animal that's less mobile is at more risk from predators. It's also less able to forage for food and may not be able to get the energy it needs to heal. Once humans lived in groups that fed and protected members who had minor injuries and infections, these members could recover and go on to produce more offspring.[39] The individuals who happened to have a body that invested a little more in maintenance had a slightly better chance of recovering, and these were also the ones likely to produce more offspring. As a result, the next generation had more members who inherited the trait of investing more in body repair and had a slightly longer life span. A few hundred generations of slight increases in body maintenance and life span could result in a substantially longer life span.

One curious effect of the longer life span is that women experience "menopause." Their ovaries cease to release eggs, their hormone levels change, and they're infertile for the last third or so of their lives. This is another unique feature of the human life course, and it's not easy to explain why it evolved. If women continued to be fertile, they could use their longer life span to produce more offspring. Instead, a woman's fertility ends at about the age her life would be predicted to end based on the size of the human body. Why didn't the ovarian life span increase as much as the life span of the rest of a woman's body? The simplest explanation is that stopping producing children didn't reduce the number of surviving descendants that women had. Rather than putting her body through the rigors of pregnancy and childbirth yet again, she could concentrate on making sure the descendants she already had would be successful. Once our female ancestors reached menopause, they probably had several children of various ages, including some who were having children themselves. Working to make sure her children and grandchildren survived was a more effective use of her time. Humans need to be raised by parenting teams, and not all the female team members need to actually produce babies. Teams need coaches and managers. Who better to take on this role than experienced older women and their husbands?[40]

DNA evidence of the extended human family tree

There is strong evidence that a large part of our evolutionary story took place in Africa. But techniques that allow scientists to determine the precise structure of the DNA in the genes of living humans and recover the DNA from the remains of humans that lived in the past have shown that the ancestry of many of us includes some humans who were living outside Africa 100,000 years ago.

The climate and geology of some parts of Europe are very good at storing evidence of past time. Because of this, we have far more clues about what humans were doing in Europe during the last few hundred thousand years than we do about what was going on in Africa. The limestone caves that are a feature of many European landscapes must have provided shelter for many generations of humans, but what's more important about these caves, from an archaeological perspective, is that their roofs and floors occasionally collapsed. Such collapses often trapped whatever (or whomever) was in the cave behind or underneath a pile of rubble. This protected their bodies from the carnivores and scavengers that would have crunched up their bones and scattered their teeth. As a result, quite a few clues to what these ancient Europeans were like remained safe for tens of thousands of years.

It wasn't until the mid-19th century that Europeans started carefully examining the ancient human bones that were occasionally discovered by people exploring caves or working in quarries. The examinations revealed that some of the old bones belonged to people who looked rather different from themselves. The skulls held a brain as large as, or even larger than, their own brain, but these ancient heads were a different shape, with a flatter forehead and a face that jutted out a bit. The rest of their body was different too: they were stockier with shorter legs, and their bones were heavier. News of these strange people spread fast and their discovery was seen as evidence of human evolution. But many European scholars of this time thought of "evolution" as "progress," imagining simpler forms of life being replaced by superior forms.[41] The bones of a now-extinct form of human seemed to be a perfect illustration of this progress. They saw the more robust bones as being typical of a "primitive" form of human and imagined the body to be ape-like compared to their own more delicate bodies. The German biologist Ernst Haeckel suggested that the extinct human be given the name *Homo stupidus* to contrast it with the much wiser *Homo sapiens* that replaced it. He was overruled. The extinct Europeans ended up being called *Homo neanderthalensis*, or "Neanderthals," after the Neander Valley in Germany where some of the first discoveries had been made. (*thal* is German for "valley").

The physical differences between Neanderthals and humans like us are so clear that even an amateur can spot them (see Figure 5.6), and it was on this basis that biologists decided that Neanderthals were a different species of human.[42] But there's no evidence that Neanderthals living 100,000 years ago had less sophisticated ways of thinking and living than the humans that looked like us and lived in Africa. And the evidence from analysis of DNA extracted from ancient human bones is raising questions about whether Neanderthals should even be considered a different species of human. It's generally thought that the different species are biologically separate types and that matings between animals of different species won't produce surviving offspring, and yet the DNA evidence reveals that humans from Africa that looked like us mated with Neanderthals many tens of thousands of years ago. A few of the offspring that resulted from these matings survived, and sections of Neanderthal DNA can be found in the genome of most humans.

In the cool dryness of many European caves, DNA from the cells inside the ancient bones degrades more slowly than it does in hotter, moister

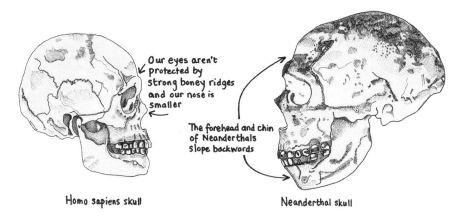

Figure 5.6 *Homo sapiens* and Neanderthal skulls compared.

environments. Human bones found in Europe that are estimated to be 400,000 years old have yielded DNA evidence. Scientists have had less success extracting intact samples of DNA from bones found further south. So far, DNA evidence has only been obtained from the remains of Africans who died less than about 15,000 years ago.

But what sorts of things can actually be learned from DNA analysis? Fictional accounts of scientists interpreting DNA evidence often give an exaggerated idea of what it can reveal. It's true that the structure of the DNA of genes carries "genetic code" for building and maintaining living bodies, but that doesn't mean that it's some kind of "program" that a computer can decode to reconstruct what a person looks like. The technology that allows geneticists to determine the precise structure of the molecules of DNA in the cells of living things has provided biologists with a whole new batch of clues about the evolution of all living things, including humans, but the analysis involved isn't like decoding a program. It's more like trying to learn about baking by analyzing a loaf of bread. Knowing about each molecule of each ingredient inside a loaf of bread can tell you quite a bit about bread-making—about the ingredients used and how they were mixed together and heated.

The problem with using DNA analysis to learn about humans is that we're all very similar. We may look and behave differently, but we're all closely related. That means that, genetically speaking, humans living today are like nearly identical loaves of bread. We all have pretty much the same genes, and we actually share most of our genes with other animals too. But because small errors sometimes occur when genes are copied and passed from parent to offspring, there are slightly different versions of these genes and the expression of these genes is slightly different. It's the precise combination of the versions of genes that we inherit that makes us human rather than chimp. Different versions and expressions of genes are also found in the same species, and it's these small differences that makes all humans genetically unique, except for identical twins. Geneticists analyzing the DNA of genes have linked some of the different versions of genes to a tendency to develop certain diseases, such as breast cancer, and certain characteristics, such as red hair. So the DNA collected from the bones of a long dead woman may reveal whether or not she had red hair, but it can't reveal what her face looked like. Faces are developed through the interaction of many genes, and these interactions are not yet well enough understood to reconstruct a face.

Going back to the loaf of bread analogy, the genetic differences between humans are at the level of how many grains of flour came from which variety of wheat and which precise strains of yeast were used to make the bread rise. This information provides useful clues about the movements and matings of our ancestors. Knowing the precise structure of the DNA inherited by people living today and comparing it to DNA extracted from the skeletons of humans who died long ago has made it possible to construct rough family trees showing the genetic links between people. Analyzing human DNA isn't like downloading a story of our species' past. It just yields a load of data. Making sense of it involves complicated mathematics and many hours of number crunching by computers to work through all possible combinations of migration and mating events that could account for the pattern of variation found. The aim is to resolve a "model" of the most probable human family history—what combination of migrations and matings best fits the patterns in the observed data. More data is being obtained as more samples of DNA are analyzed. Most of these samples come from living people, but samples of DNA are also being obtained from more and more dead people. The precise shape of the human family tree model changes as new data is added to the calculations.

In 2008 a sample of DNA was extracted and analyzed from a human finger bone found in a cave in Siberia known as the Denisova Cave. The outcome of this discovery was a substantial rearrangement of the limbs of the human family tree. This DNA was found to be different enough from both Neanderthal and *Homo sapiens* DNA to suggest that the bone came from a person who was part of a population that had separated from the Neanderthal population about 400,000 years ago. This population was given the name "Denisovans." So far, only a few fragments of bone have been found that are known to come from Denisovans, so it's impossible to work out what they looked like. But DNA extracted from those remains has established that, tens of thousands of years ago, surviving offspring were produced from matings between Denisovans and humans of African (or mostly African) ancestry. As a result, even though we think of the Denisovans and Neanderthals as having "died out," they actually have a place in the family tree of many people who are alive today.[43]

According to current models, it's likely that the humans living today are descended from at least seven separate populations of humans. These populations were made of families who exchanged members as people mated and produced children. Hundreds of thousands of years earlier, they had all been part of a single population that split as families went their separate ways. Of the seven human populations, two lived outside Africa and we know them as the Neanderthals and the Denisovans. The five populations living in Africa might have been part of a population that split up around 300,000 years ago. Although the populations were separate, the DNA of people living today carries evidence that their members did meet and sometimes mated. Children produced by these matings survived and have descendants living today.

Information about temperature obtained from mud cores suggests that changes in the climate may have been the cause of the split of the African populations that occurred around 300,000 years ago. It roughly coincides with a period when the Earth's temperature was extremely unstable. This also seems to be around the time when some human populations became extinct, such as smaller-brained humans whose remains have been found, known as *Homo naledi* and *Homo erectus*.

The relatively stable climate that existed 100,000 years ago was gradually becoming cooler. Eventually temperatures would become unstable again, and our ancestors would have to change their ways in order to survive.

6

Ice Age Humans (30,000 Years Ago)

Thirty thousand years ago, the planet was in the grip of an ice age. Large areas of land were either covered in ice or they were so dry that little plant life could grow. Where there was water, plant productivity was limited by low carbon dioxide concentrations. For 50,000 years extreme temperature fluctuations had caused periods of climate chaos. The period between about 60,000 and 12,000 years ago seems to have been the most highly variable of all the ice ages.

By 40,000 years ago, humans that looked similar to us had spread out of Africa, and by 30,000 years ago the other kinds of humans were probably extinct. No evidence has so far been found of Neanderthal or Denisovan humans surviving beyond about 37,000 years ago. Descendants of the *Homo sapiens* who had been living in Africa 100,000 years ago had taken their place. Genetic evidence shows that descendants of Neanderthal and Denisovan humans were still surviving, but they were surviving as part of *Homo sapiens* populations and most of their ancestors were descendants of the humans who had left Africa about 20,000 years earlier.

By 30,000 years ago, many *Homo sapiens* populations were not simply surviving the climate chaos. They were thriving. Their numbers had increased, and they had colonized parts of Europe, Asia, and Australia. Their culture had been transformed. Not only were they making better tools and using new materials; they were producing what we would call "art." They carved figures from stone, bone, antlers, and ivory, and shaped them from clay. They made jewelry and musical instruments and painted the walls of caves.

Our ancestors who lived 30,000 years ago were scattered over a wide area. Each isolated group evolved its own language, toolkit, skills, customs, and beliefs as they met the challenges and opportunities that faced them. Based on the evidence found so far, the people from this time that we know most about are the ones who lived in Europe and Asia. They hunted animals that lived on the frigid grassy plains that covered much of that continent. Genetic analyses suggest that many of us are descended from these people.[1]

Figure 6.1 This human figure (about the length of an adult's index finger) is believed to have been carved around 30,000 years ago. It was found in 1988 in northeastern Austria near an outcrop of the shiny gray-green serpentine rock from which it was made.

Photo credit: Creative Commons license. The figure is known as the "Venus of Galgenberg" and is on display at the Museum of Natural History in Vienna.

Part One

If you had been born into a family of these hunters 30,000 years ago, your first memories might have been of the first great gathering you attended. You were so young at the time that your own memories of the gathering are just vague impressions of sound and images—lots of people, talking, moving, and singing, and the flickering light of many fires. You don't even remember the long journey to and from the gathering. But even years later, people in your family talk of that gathering. As they travel or work, they discuss what they remember learning at the gathering. As they snuggle together under furs to keep warm during the long winter nights, they reminisce about how they felt during the gathering's nighttime events. They try to rekindle those feelings during the smaller gatherings with local families that are held every few months.

You're not yet fully grown when your family starts preparing for the journey to the next great gathering. You know that this time you will have many strong memories, and you're also sure that attending the gathering will change your life. As you help with the preparations, you think about the way the last gathering changed the lives of people in your family. Your favorite uncle's future was decided there. A little while ago he left the family to live with a woman he had met at the gathering when they were both children. They had liked each other, and it was planned that they would meet again when they became old enough to marry. When they met, they still liked each other, so now they've joined with some of her relatives to make up a group that has gone off to explore the lands to the south and east. The grandparents arrange a party at the gathering so that young people meet each other and perhaps find their future mate. The idea of meeting strange children makes you feel shy, but making friends and meeting mates is one of the things that gatherings are for.

The last gathering had a big effect on your great-grandmother's life. She died right afterward, on the journey home. All elders go to the gatherings, even if their families have to carry them there. The kinship between elders helps bind families together and their memories ensure that the families survive. When elders meet at a gathering, they talk about the old times with the help of the spirit talkers, who can hear the voices of the spirits of dead ancestors. This makes the elders' memories richer and clearer. The world is changing all the time, but time is a circle. The elders' memories help their families prepare for the future when the old times come back.

A few years ago, your favorite cousin, who was born the same summer as you, started hearing voices that no one else can hear. Some family members are excited about this, saying that it's a great honor to have a spirit talker in the family. But your grandfather has told them not to encourage her. He says that at her age she should be listening to what her *living* ancestors are saying and not bother with the dead ones. Your cousin says she can't help hearing the voices, but she won't tell anyone what they say, not even you. At the next gathering your grandmother will take her to meet the leader of the spirit talkers to see what he says about the voices.

Another change caused by the last gathering was that your grandmother started making you eat a share of the stuff from inside an animal's stomach. A woman at the gathering told everyone that it may prevent the sickness that causes bad skin and aching joints. She said that she used to not eat from the stomach because she didn't like the smell, but when she started holding her nose and eating it anyway, her skin cleared, and she was able to move without pain. She said it had worked for other people too, and children were healthier if they ate it. You didn't like eating it at first but now you're used to it.

Your family hunts bison, horses, and the other animals that live on the grasslands, but they specialize in hunting mammoths. Mammoths are so big and dangerous that it's only safe to go after one if you're a specialist. To kill one, you need to be able to launch a spear with enough force to penetrate its thick hide. You need a spear thrower[2] and you also need a lot of help. It usually takes a hit with several spears to kill a large mammoth. They have to be launched all at once or the wounded mammoth will attack the hunters. Men from several families must work together. Families usually camp together near the paths the animals take to get water, ideally in a place sheltered from the wind and easily defended from lions and bears. The lookouts know when a large mammoth is nearby because, even if you can't hear them, you can feel their heavy feet shaking the ground. When one comes near, it's important to keep out of its way, but trackers follow it and signal the hunters with smoke. Small mammoths are easiest to kill, but you need to kill a large one if you want valuable materials like ivory, tough hide, and large bones. It's foolish to even think of trying to tackle a large mammoth until enough experienced hunters have arrived and they have agreed on a strategy. When a mammoth is killed, there's so much meat that it's shared with all the nearby families, and even families from further away come by to pick up some meat if they need it. Meat doesn't go bad if it's kept frozen, and it can be kept cold even in the summer if it's buried.

Some hunters prefer to work with close relatives, but hunters from different families can hunt together, as long as they feel they can trust one another. Your father and grandfather got angry with your favorite cousin's father because he kept behaving in an untrustworthy way. After that, he took his wife and daughter away to live with some other families, but they also got angry with him and he was killed in a fight. After that, his widow and your favorite cousin came to live with your grandparents.

Hunting is only a small part of what mammoth specialists do, and anyone growing up in a family of mammoth specialists has a lot to learn. All the different parts of the mammoth carcass can be used, but to be most useful, they need to be treated. The skin from some parts of the mammoth can be made flexible enough to be used to make boots and clothing. Other parts can be made stiff enough to be used for the walls and ceilings of shelters. Your family even knows how mammoth skin can be used to line the hull of a boat. Some of the bones of the mammoth make good building material, but others don't and can be burned as fuel. It's hard to get mammoth bones to burn, and that's another thing than mammoth specialists know how to do. The cinders of burnt mammoth bones can be broken up into gravel and used for the floors of shelters. Some mammoth muscles have a covering that can be stripped and twisted to make strong threads and ropes used in the making of clothing, shelters, and hunting nets. The ivory from the tusks can be carved into useful objects like needles, fishhooks, spear-throwers, combs, and moon calendars, as well as beautiful gifts.

You're better at close work like sewing and carving than the other children. You're even better than many of the adults. Your eyes can see tiny details even when the light is dim, and your hands stay warm and nimble when you're working, even outside in the cold. Other children get frustrated with close work, so they do other things, like taking messages to other families and helping with hunting. Messengers need a good memory, and they must know how to read the sky so they don't get lost. You sometimes take messages. You're learning to read the sky and how to use a moon calendar. Grandad thinks everyone should have these important basic skills.

Mammoths provide much more useful material than the specialist families need, but you trade with other families. One of your grandmothers grew up in a family that travels around shores fishing and harvesting shellfish. She met your grandfather when he was showing them how to improve their boats. Ever since their marriage, the mammoth families and the fisher families get on very well, and some of the fishers go to the great gatherings. Fishing families can get salt from their relatives that live near the seacoast. They also have far more shells and fish bones than they can use, and they pick up and work flint cobbles that have washed into the river. People use flint or a similar kind of stone to make tools that need to be hard and sharp like knives, dart and spear points, chisels for working wood and ivory, and hide scrapers.

As part of the preparation for the gathering, you have been carving beads from pieces of mammoth tusk. Making beads is good practice for working with ivory, and it was decided at the last gathering that children learning to carve ivory would make beads and bring them to the next gathering to decorate a gown for the Leader of the spirit talkers. Only in the last year have your carving skills become good enough to make round beads with holes going down the center. You've made eight of them. As the families prepare for the long journey, your beads are packed, along with the ivory and skins that will be traded. *(To be continued.)*

Why do we think some human ancestors lived like this 30,000 years ago?

The families in this story had a culture that was adapted to the habitat in which they lived, but this was a changing habitat, so their culture had to be adapted to dealing with change. They had to keep learning, and they also had to remember the past, because they never knew what the environment would throw at them next.

The last ice age left many scars on the Earth's surface and plenty of evidence just below the surface. Both can give us an idea of what it was like to live during this crucial period of our history.

Clues to a lost world and its people

It was about 120,000 years ago when the alignment of the Earth and Sun resulted in less of the Sun's warmth being absorbed by the planet. The ice sheets that covered the poles and mountains gradually grew as, summer after summer, less of the winter snow thawed. As the area covered by white shiny ice grew larger, more of the sun's energy was reflected from the surface. This made it even colder. With more and more water trapped as ice, sea levels fell. And because water evaporates less readily into colder air, the weather became drier and the sky less cloudy.[3]

Evidence of the dramatic temperature fluctuations that started 80,000 years ago comes from the composition of the chemicals that were laid down as mud on the bottom of oceans and lakes and as ice in the glaciers (see Figure 6.2). The fluctuations were likely due to changes in the way heat was moved around the planet by the upper atmosphere winds and ocean currents. These rivers of air and water became unstable, changing position and shifting direction. As a result, some areas suddenly got warmer while other areas became colder. Then the changes would reverse.[4]

This chapter's story is set in the northern part of the planet, on the edge of a huge expanse of grassland that spread south from the ice sheet that covered northern Europe and stretched eastward into what is now Siberia. At times, the grassland spread across Asia and into North America. Because of the low sea level during the ice age, there was dry land between eastern Siberia and what is now Alaska.

To a casual observer, these grasslands would have looked like the grassy habitats we have today (known as "steppe," "prairie," or "veldt"), but they were different. The grass grew in a layer of soil that lay on top of permanently frozen ground. Scientists have called this ice age habitat the "mammoth steppe."[5] Despite the cold and long winter nights, the mammoth steppe was highly productive during the spring and summer, providing food for many herds of grazing mammals. Giant bison, horses, reindeer, woolly mammoths, and woolly rhinoceroses grazed the treeless plains and were hunted by bears, wolves, and large cats. One of the reasons scientists have been able to learn so much about the mammoth steppe is that some of the animals that grazed the plains died in conditions that allowed bits of them (and sometimes their whole body) to be frozen into the ground. The occasional unearthing of the frozen and mummified body parts of extinct ice age creatures has given biologists the opportunity to learn about both the animals and the environment they lived in.[6] It's been possible to analyze the soil trapped in their hooves, examine the contents of their stomach, and even pick from between their teeth the bits of grass they were munching just before they died.

Families of humans hunted these mammals, and because many of them lived in areas that remained sparsely populated after the ice age ended, evidence of their existence remained undisturbed for thousands of years. In the low temperatures, their physical remains and other remnants of their lives were better preserved than those of people who lived further south.[7] In

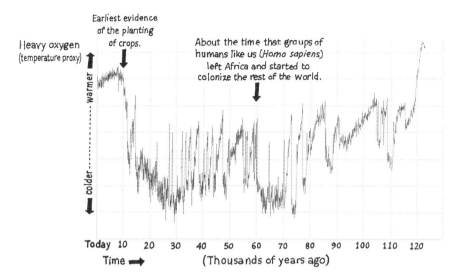

Figure 6.2 This graph plots the level of Oxygen-18 in layers of ice that built up in Greenland over the last 122,000 years. In colder temperatures, less of this heavy isotope falls as snow, so its levels are an indicator (or proxy) of the air temperature. By extracting a "core" of ice going deep down into the ice pack, researchers can track changes in temperature going back tens of thousands of years. Greenland is one of the coldest places on Earth today, and the amounts of Oxygen-18 detected in the deeper (earlier) layers of ice show that it was even colder during the ice age. They show that temperatures have been quite stable over the last 10,000 years but fluctuated dramatically during parts of the ice age. Some of these fluctuations have been linked to climate events in other parts of the world and suggest that temperatures all over the world were unstable for thousands of years. Further support is provided by the observation that people only began to farm in the last 10,000 years. These results suggest that, before then, the climate would have been too unstable to allow farming.

hundreds of sites across Europe and northern Asia, evidence has been found of the humans that lived on the edges of the mammoth steppe. They were there for tens of thousands of years (between about 12,000 and 45,000 years ago). The evidence is like a series of poorly composed snapshots taken at random. Many of them show extraordinary scenes in startling detail, but it's hard to link them up to create a clear scene. Paintings and carvings on the walls of caves depict the animals of the steppe—horses, bison and mammoths, lions, and bears. Models of animals carved in bone, ivory, and antler have been found, as well as several carved female figures (see Figures 6.1 and 6.3). Many of the female figures appear to be pregnant, with large breasts and buttocks. Were pregnant females considered especially beautiful? Perhaps these figures were dolls for children to play with. Some of the cave paintings and carvings are seen today as works of art, but there are also less lofty depictions—outlines

Figure 6.3 This 11.1 cm (4.4 inch) tall figure carved in limestone is believed to have been made about 30,000 years ago. It's known as the "Venus of Willendorf," named after the Austrian village where it was found.

of people's hands and many strange little drawings, some of which could be what we today would consider pornographic[8] (see Figures 6.4 and 6.5).

Mammoth steppe people seem to have traded tools and materials over long distances. Shells have been found in human settlements that had been transported over 500 km. Their stone tools included blades, awls, chisels, and spearpoints. They had needles, fishhooks, and little flutes made of ivory, bone, and antler. A stone spearpoint and ivory shaft have been found embedded in the shoulder bone of a mammoth, proof that they really were able to kill these massive beasts and didn't just scavenge bone and ivory from animals that were already dead. Pieces of clay found in ancient settlements in what is now the Czech Republic were found to have patterns likely to be impressions of nets. The people living here may have woven nets to capture small mammals, birds, and fish.[9]

Bodies have been found of several dozen ancient humans who lived and died or near the mammoth steppe. The earliest are Neanderthal, but by

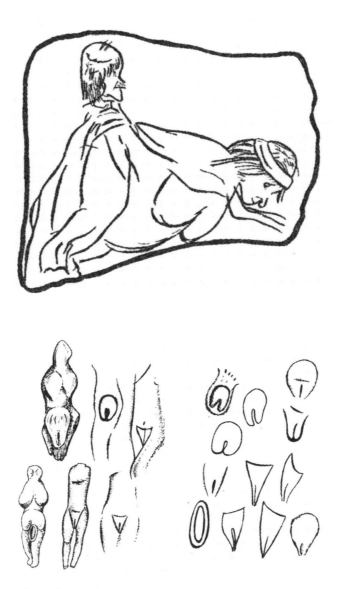

Figures 6.4 and 6.5 The peoples who hunted on and near the mammoth steppe left many drawings on the walls of caves. By no means were all of them drawn by great artists and some might be what we would today call pornography. The larger sketch (above) may be of a couple copulating, but it's hard to tell. Some of the tiny sketches (below right) look like they could be of hoofprints. But such images are sometimes seen in contexts (below right) which suggest that they could represent female genetalia.

Source: R. Dale Guthrie, *The Nature of Paleolithic Art* (Chicago: University of Chicago Press, 2005). Reprinted with publisher's permission.

45,000 years ago, populations of *Homo sapiens* were living there too. These people had developed the means to use deadly force against the large animals they hunted, and perhaps toward individuals who stepped out of line as well. But there's no evidence that groups went to war against each other. Archaeologists have become expert at recognizing the signs of death by violence. They have dug up and inspected many bodies from sites of battles and massacres. But the battle-ravaged skeletons and mass graves have all been from more recent times. Wars, aggression, and genocide may not have been a feature of the lives of people on the mammoth steppe or any of the more ancient peoples.[10] Their tools don't include weapons of war, like shields, which, unlike spears, have no function in hunting. And their artwork has no depictions of battles or victory celebrations.

In the cold climate of northern Eurasia, even very old bones are often preserved well enough to yield samples of DNA that can be analyzed. By comparing the DNA and other chemicals found in the bones, it's possible to get an idea of the relationships between people who lived in different times and places, and to make guesses about how they moved around their vast hunting ground. Evidence of their presence has been uncovered as far north as the Arctic Ocean in Siberia. There was no wood on the treeless plains, so they built shelters with massive bones and covered them with animal skins. Like people living in the Arctic in more recent times, they might have burnt greasy bones to cook their food and keep warm.[11]

How had these ancestors advanced so quickly?

The humans hunting the animals of the mammoth steppe had figured out how to make a living in a habitat very different from the African grasslands where *Homo sapiens* had evolved and lived for tens of thousands of years. Not only had their culture made them adapted to a very different environment; it had also become extraordinarily complex. What had happened to speed up their rate of cultural evolution? And why were they thriving while other kinds of humans were dying out? Ever since the 19th century, when Neanderthal bones were found in Europe, scholars have been wondering how our own ancestors managed to replace these much stronger-looking humans. The evidence found since hasn't provided a definitive answer to this question. No evidence has been found to suggest that *Homo sapiens* from Africa were intellectually superior to Neanderthals, at least not at the beginning of the ice age.

But the evidence of the temperature fluctuations does help to explain why surviving was difficult. Our ancestors might also have been close to extinction when the long period of climate chaos began. Groups of *Homo sapiens* had begun leaving Africa over 100,000 years ago. Archaeologists have found evidence of their settlements in areas east of the Mediterranean. But the descendants of these early colonists either died out or scurried back to Africa when the climate got difficult. The genetic similarity of all humans alive today suggests that during at least one point in our history, the population of *Homo sapiens* was very low, perhaps only a few thousand people. Genetic evidence suggests that everyone alive today is descended from humans who were living in Africa 60,000 years ago, although it also tells us that most of us are also descended from Neanderthals and a fair few of us from Denisovans as well. The genes of these other human populations entered our lineage through matings that occurred after some *Homo sapiens* left Africa around 60,000 years ago.[12]

So, why did *Homo sapiens* populations survive while Neanderthal and Denisovan populations died out? It had long been suspected that the heavy-boned Neanderthals weren't very bright, but in the late 20th century some scientists suggested that the brains of both the Neanderthals and early *Homo sapiens* might have been disadvantaged in some specific ways. They proposed that these earlier forms of human lacked imagination, language, and abstract reasoning. Some *Homo sapiens*, they argued, might have gained these abilities between 40,000 and 50,000 years ago—a baby could have been born with a genetic mutation that changed the way its brain developed. This random change in a DNA sequence was transformative, they explained, because it brought about a "cognitive revolution." This baby, which they said was probably born into an African *Homo sapiens* population, developed a brain that was able to think in the same way as we do.[13] It grew up to be successful and had many descendants who inherited the mutation. The mutants were able to work together in a way that was impossible for those with the old-style brain. They could think and communicate in abstract symbols and this allowed them to invent new tools and dream up new solutions to their problems.

It's possible that a genetic change could have somehow reconfigured the human brain, but we tell a different story in this book because we don't think the cognitive revolution idea is well supported by the evidence as it stands today. If humans today are benefiting from DNA sequences that Neanderthals lacked, it should be possible to identify those sequences by comparing the DNA of humans and Neanderthals. Scientists have been working to find such sequences since 2014, when the first complete genome of a Neanderthal woman was reported. Comparing Neanderthal DNA with that of *Homo sapiens* has revealed several small differences, and a few of these are in parts of the genome thought to influence brain development. But, so far, no evidence has been found to support the suggestion that a key genetic change happened in the last 100,000 years that redesigned our ancestors' brains and dramatically changed the way they thought.

The cognitive revolution story is also undermined by archaeological evidence. Archaeologists investigating African sites, such as Blombus Cave, near Cape Town in South Africa, have found remnants of more sophisticated stone tools and other evidence to suggest that the people living there between 50,000 and 100,000 years ago had brains just like ours. These people etched patterns on rocks and pieces of ostrich egg, ground up colored stones to make paint, and carefully broke holes in tiny shells, perhaps so that the shells could be strung as beads on a bracelet or necklace (see Figure 6.6).[14] It's compelling evidence that people had a brain capable of creativity and ingenuity much earlier than 50,000 years ago. But the evidence is patchy and doesn't give the impression that a cognitive revolution had taken place. People seemed to be creative and imaginative for a while but didn't progress further and may have even gone back to their old less sophisticated ways. If a genetic change was providing humans with upgraded mental abilities, why wouldn't they be using the abilities all the time?

Developing stone age social networks

The scientists who developed the cognitive revolution idea in the late 20th century lived in a time when computer users were more concerned about kilobytes of RAM than megabytes per second. They were exchanging emails with their colleagues and text messages with their friends, but no one had Wi-Fi and people still used fax machines to send documents. When scientists of this technological era puzzled over the evolution of the "computer" inside the human skull, they thought about hardware upgrades. They hadn't yet experienced the way that better network access can vastly increase what computers can do.

We and many other people studying human evolution now believe that it was better network access that allowed our ancestors to think in more sophisticated ways and survive the climate chaos. No matter how intelligent their members, small groups of humans can seldom find solutions fast enough when changing environments are continually bringing new problems. Families were more likely to survive if they formed and maintained connections so they could share information and ideas and help each other out in times of trouble.

Maintaining a large social network isn't easy for peoples who make their living by hunting and gathering. They usually live in small groups and move

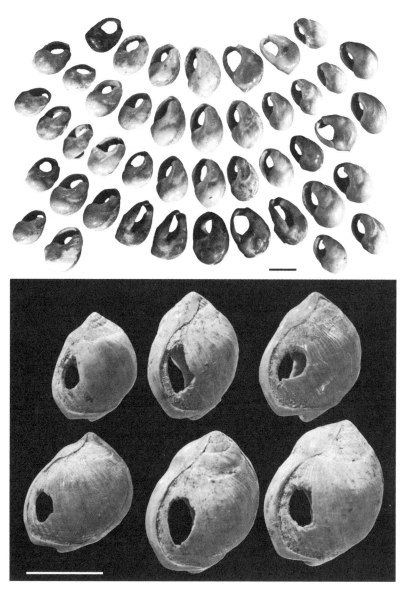

Figure 6.6 These snail shells found in 75,000-year-old sediments on the floor of Blombos Cave in South Africa suggest that the people who lived in this cave enjoyed creating decorations. The shells are only half a centimeter across, too small to have been collected as food. They seem to have been carefully selected to be of a similar size and pierced in the same way. They may have been used as beads and strung on a length of cord. Traces of ochre on the shells suggest that their creator had colored them red.

Photo credit: Image courtesy of Professor Christopher Henshilwood.

on as food gets scarce. Few habitats produce enough food to support large numbers of foragers, but there are times and places where food is abundant for a while. For example, on the shores of bays where fish gather to spawn, exhausted fish can sometimes just be collected from the water. If foraging families knew about such a place, they would probably try to come back every fish-spawning season. This meant that the families met up regularly and had the chance to share and trade both ideas and the small number of goods they could carry with them. They could build on each other's ideas in ways that benefited all of them. While food was easy to find, they had spare time to do things like make beads for jewelry. Parts of the southern coast of Africa, near Blombos Cave, may have been places with seasonal abundances of food and this may be why evidence of more complex culture has been found there.[15]

But if gatherings depended on the environment delivering a seasonal bonanza, then cultural advances it triggered would be hard to sustain. Environments change and eventually fish start to spawn somewhere else. The disappointed families would go their separate ways and, with the reason for meeting gone, they would lose their opportunity to share. This would make their lives duller and also more precarious, because they no longer had access to the extra knowledge that gave them extra strength. When the temperature fluctuations began, social connections became more vital, but also harder to maintain. Droughts, floods, and landslides cut off the routes people traveled and damaged the food supplies they relied on. Climatic shocks would have led to culture being knocked back to the most basic levels. Skilled and knowledgeable people died, and their skills and knowledge died with them. Small isolated groups become both culturally and genetically depleted because, without contact with people outside their family, there was very little choice of people to mate with and exchange ideas with.[16]

So what was it that allowed our ancestors to evolve their culture fast enough to adapt to the climate fluctuations and then go on to do things like invent music and create subterranean art galleries? We think that some families came to realize the importance of keeping in touch and helping each other out. They started to develop social tools to create a group of connected families and encourage other families to join. The groups with the most effective social tools managed to stay connected and to survive. Their descendants inherited this culture of connectedness, which included both the desire to keep connected and the customs and habits that ensured these desires were met. Our ancestors belonged to these groups.

Evolving a culture of connectedness didn't just allow our ancestors to survive. It transformed their lives. Once families started to meet and communicate regularly, they began to explore the potential of their large brain. This very hungry brain made them good at learning and sharing information, but maintaining it was barely cost-effective when there wasn't much to learn and share. Tapping into a larger social network meant that the computers inside their skulls had access to more data and more programs that they could use to process the data.

Most of the social tools our ancestors used to stay connected were in their minds. They were the beliefs, customs, habits, memories, and so on that people shared. But members of a culture often also share objects that symbolize their beliefs, represent their memories, or just demonstrate their feelings. These objects are social tools as well. And there are some objects that provide practical help in keeping people connected. Hundreds of pieces of bone, antler, and ivory have been found that might have been performing this function for the mammoth steppe people. These objects have been carefully etched with little lines or dots and many are worn and shiny, suggesting they'd been carried around and handled often. Fingers had been run along the etched lines again and again. Some of the lines had colored earth, such as ochre, rubbed into them, perhaps to mark them.

In the 1960s, a science journalist called Alexander Marshack, who was working on a book about the origins and development of scientific thinking, became fascinated by these objects. Some of the things that had been made by ancient Europeans are today considered to be great works of art, and they're displayed in major museums. But many more of them are displayed or just stored in small museums near the towns and villages where they were found. Artists considered the patterns of lines and notches rather mundane, but Marshack was intrigued and traveled around making a meticulous study of them (see Figures 6.7 and 6.8). He formed the idea that each mark might represent a day, and the length of the mark might represent the phase of the moon on that day.[17] If he was right, these objects were moon calendars displaying the days of an entire three-month season. Not all the patterns exactly matched the phases of the moon, but producing fine marks with points made of stone couldn't be easy. He reckoned that even a rough representation would be useful for a people who had not invented writing and may not have had a system of counting. The marks would give groups a way to agree on the day of their next meeting months, even years, ahead. Then the calendar would make it easier for each group to keep track of days so they would arrive at the meeting on time.

Not everyone agrees with Marshack's interpretation of the markings, but if he's right and these objects are moon calendars, then they are early social networking devices. Our ice age ancestors may have invented many means of staying in touch, such as a system of signals that could be communicated by

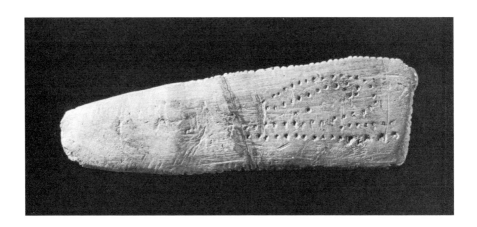

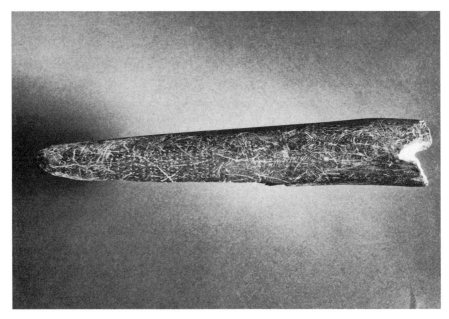

Figures 6.7 and 6.8 Two of the many photographs taken by Alexander Marshack of the etched creations of people who hunted the animals that lived on the mammoth steppe. Objects like these, made of many different materials, have been found all over Europe. Marshack believed that one of the purposes of these objects was to keep track of days so that people could organize meetings.

Photo credit: Gift of Elaine F. Marshack, 2005. Courtesy of the Peabody Museum of Archaeology and Ethnology, Harvard University, PM 2005.16.2.262.1 (Fig. 6.7)

Photo credit: Gift of Elaine F. Marshack, 2005. Courtesy of the Peabody Museum of Archaeology and Ethnology, Harvard University, PM 2005.16.2.318.38 (Fig. 6.8)

smoke, flag, drum, coded whistles, or courier. But these would leave no trace. The scratched depictions of phases of the moon have survived and they could be tangible evidence of the importance our ancestors placed on staying connected and getting together.

Over tens of thousands of years, your ancestors traveled to many gathering places. The remains they left behind hint at what went on at these meetings.

Part two

You travel to the gathering place with the slow group that includes the old and injured people, small children, and mothers with babies. The older children have to help carry supplies and babies. Teams of hunters and messengers run ahead to prepare the stopping places and get food ready. The journey has to be planned carefully with the nearby families that travel with you. It's the only way to make sure that everyone gets to the gathering on the right day and in reasonable shape. Timing is critical. No one can afford to wait around for other families to arrive. They'd soon run out of food. It's the middle of summer so the nights are short, and you can travel a long way each day. Even so, the route must be as short as possible while still passing through areas where water containers can be filled up and wood can be collected. You're travelling south, so each day gets a bit shorter and the land is getting wetter. You see more and more shrubs, and eventually trees start to appear.

You set off walking beside your favorite cousin, each of you weighed down by a mammoth-skin blanket. Yesterday she was excited and looking forward to the journey because she would at last meet the spirit talkers and tell them about the voices she hears. But this morning she's quiet and you don't press her to speak. You've often tried to imagine what it would be like to hear voices. Are they talking to her now? Is that why she's quiet?

Finally, she says, "I can talk to you because it's too difficult on my own and I need to be strong."

"What do you need to be strong about?" you ask cautiously.

"We'll have to use this journey to say good-bye and get used to the idea of being apart," she says in a matter-of-fact way. But her tone changes and she starts to cry as she says, "I will have to stay with the spirit talkers, so I won't be coming back with you."

You are stunned. You never expected this, and part of you doubts her. How could she know this? But tears still to come to your eyes to at the thought of saying good-bye. "Does your mother know?" you ask, "and Grandma and Grandad?"

"No!" she says, "And they mustn't know. They will ask questions and may try to stop it. Something is going to happen that everyone will think is bad. I need to be ready to help make good come of it."

"But if it's bad, shouldn't they try to stop it?"

"No one can stop it and I don't know what it is. I just have to be ready to make good come of it."

For the rest of the journey, you and your cousin spend as much time as possible together, trudging along side-by-side, carrying your heavy loads. At night you curl up next to each other. You talk and talk, trying to remind each other of every good and bad thing that has ever happened since she and her mother came to live with you. Your cousin's memories are so clear and detailed. You wonder if the spirits are helping her remember.

As you get closer to the gathering place, you see the forest that you've heard about. You see more trees in one day than you remember seeing in your whole life. The gathering is in a place where trees have been cut down to create clearings. It's a perfect place. There's a river nearby, no shortage of wood for fires, and the trees provide a shield from the wind. A low-lying section is marked off as a latrine area and each family has been allocated an area with some poles, logs, and kindling piled in the center. Some of the poles will be used to build a shelter with the skins you've been carrying, and the logs will be burned for cooking. It's already crowded when you arrive, but you soon find your family's area and it doesn't take long to build the shelter and set out the skins and the ivory the family have brought to trade. Strange adults are wandering around looking for old friends and relatives. Your parents recognize one of them and start talking. The young children play noisily, pleased that the journey is over. But the older children, who're having their first real experience of a great gathering, are quiet. It's hard to be surrounded by people you don't know, especially when some of them look so strange. One guy, who's wearing a hat with reindeer antlers, looks at you and waves. You wave back. It might just be their clothes that make them look so different, but they seem to move differently too. And they sound strange. You feel embarrassed to talk in case other people will think that you look and sound strange.

Your grandmother tells you to unpack your beads while she takes your cousin to meet the spirit talkers. You give your cousin a hug, and your grandmother says, "Don't worry, she'll be fine" and takes her hand. You watch as they disappear into the trees. Then you find the pouch holding your eight round beads. Your brother, who's only a toddler, starts to scream. His forehead is bleeding, and he seems to have fallen over. It's only a scrape but he's tired and cranky. Your mother is still talking so you hold and cuddle him until stops crying, starts sucking his thumb, and drops off to sleep.

When your grandmother comes back, she's on her own and very happy. She says that the Leader and his twin sister invited your cousin to help them prepare for the singing practice. Then you and your aunt set off with your grandmother to take the beads to the place where the Leader's robe is being made. Your grandmother, who seems to know everyone, introduces you to the three women who are sewing beads onto the robe. A crumpled mass of soft leather lies between them. It's already covered with beads and there's a pile of more beads on the ground waiting to be sewn on, but most of these beads are the box-shaped kind. Your handful seems very small, but the women look at them carefully and admire their roundness. They ask you to pick the one you like best and point out where on the gown you want it to go. Your grandmother helps you spread the gown out to get a sense of its shape and you point to the front so that your bead will be over the Leader's heart. One of the women takes your beads and, in no time, stitches your chosen one into the row so that it nestles comfortably with all the other beads.

People are still arriving but there's already plenty to see. Your grandmother wants to watch the men practicing for the spear-throwing competition. They're using different designs of ivory spear-thrower and the results of the competition will help people decide which design is best for them. You're more interested in carving, so you walk around looking at the figures and jewelry that people are showing off. When you admire a flute made of a hollow bird bone, the man who carved it tells the story of its making, how he chose and prepared the bone. His accent is different, but you can understand what he's saying. When you tell him that you carve mammoth ivory, he gives you a piece of reindeer antler and some tools and lets you feel how carving antler is different from carving ivory.

After a while, the guy wearing the reindeer antler hat comes over and asks if you would also like to taste reindeer meat. You follow him to where a large group of reindeer-hunting people are gathered around a fire. There's a strong smell of meat being cooked, and you realize how hungry you are. A woman, who turns out to be the sister of the flute carver and the mother of guy in the reindeer antler hat, gives you a skewer of thinly sliced reindeer tongue. It's delicious. Their different way of saying some words ends up making it easier rather than harder to talk to them. They laugh and tease you for saying things in a funny way, so you insist that you pronounce them the right way and that they're the ones that sound funny. It's sort of stupid but everyone has a laugh.

Because of the trees, it becomes quite dark when the sun gets low it the sky. Orange light twinkles through their trunks, casting long shadows and making everything seem magical. People start heading to the meeting area for singing practice. You see your family and look for your cousin but she's not with them. The meeting area has a raised platform in the middle. The clearing is large enough, but only if everyone presses close together. Someone starts to sing one of the songs you sing at home and everyone joins in. There are so many voices with different accents and tones: the rasping voices of very old people, the piping voices of the little children, and words being quietly spoken by the people who can't sing in tune. You've been told so many times about the magic of the gathering. There are so many different people, but you all make one voice—a voice louder than you've ever heard before. When the chorus comes, you find a hand to hold and do the steps together, all moving like one body.

This morning you felt that all these people were strangers. Now you're pressed close to them and they feel like members of your family.

As that song ends, someone starts a song that you love even more, and as you sing, you hear some strange new voices, unworldly voices that make the melody but use no words. It's the spirit talkers and the crowd parts so they can get to the platform. Three of them are playing flutes and two are carrying drums. Their faces are rubbed with ochre and seem to gleam with red, warm life. As they go up the steps of the platform, you see that one of them is smaller and realize that it's your cousin, wearing the bangles and cloak of a spirit talker! You want to shout out to her but stop yourself. It doesn't feel like she's your cousin anymore. She's a spirit talker and belongs to everyone.

The Leader isn't with them, but his twin sister is leading. After the song ends, she welcomes everyone, saying that all the families have now safely arrived. Then the other spirit talkers on the stage speak up, talking in the voices of spirits, who are also saying hello to the audience. The audience calls out messages to their ancestors who have recently crossed over and their spirits shout back. Some of the exchanges are funny and people laugh. You keep watching your cousin to see if she gives a message from the spirits. When she does, you can't understand what she has said but some of the elders in the audience gasp in amazement and start to smile and nod. The woman leader smiles also and tells the audience that they have been joined by a new young talker with such a good ear that she can hear the voices of ancestors from long ago who spoke a language that is now almost forgotten. "This will be another unforgettable gathering," she says.

After that, there's more singing accompanied by the flutes and the drums. You would be happy to keep singing all night, but after two more songs the practice is over, and the spirit talkers form themselves into a line, step down from the platform, and file out. As soon as they're gone, people start talking and the buzz of conversation gets louder as everyone walks back to their shelters. You ask the reindeer hat boy and his sister if they will be at the party tomorrow. They say they will, so you agree to see them there and run back to be with your family.

As you expect, there's a throng of people there, including your favorite uncle. His wife is there too, and she's expecting a baby. People are asking them about the lands where they now live to the east and south. They're in a valley on the other side of some mountains that used to be impassable, but because the ice is melting, they were able to find a way to cross them. They want to find out if any of the elders or spirits know about these lands. The hunting is good and there are lots of trees and strange plants. They plan to take back ivory and ivory tools to trade with people they met there. These people eat a lot of plants and speak a different language, but children from the two peoples are playing together and can already understand each other.

The main topic of conversation is, of course, your cousin and how she is now a spirit talker. You get the impression that your grandfather isn't very pleased about it, but her mother and your grandmother are bursting with pride. You feel sick with worry. What is the bad thing that's going to happen? Will she be able to make some good come from it?

You see that your little brother is almost ready to fall asleep, so you pick him up and take him into the shelter and curl up with him. You miss your cousin so much.

When you wake, the sun is high in the sky and you can hear your mother explaining to your little brother that the party is only for older children and he will go to the spear-throwing competition with his father. The party! Perhaps the spirit talkers will let your cousin go to the party.

Your cousin isn't at the party, but otherwise it's better than you expected. The grandmothers bring treats for the party so that the children can taste food from different places. You like the dried fruit, but your favorite is the chunks of strange chewy stuff that comes from the home of a kind of bee. You spend most of the time with the children you met from the reindeer families. They tell you all about the ways of reindeer and ask questions about mammoths.

The Elder Talk starts right after the party. Everyone goes to the meeting area at the beginning to show their respect to the elders, but once it gets started, the older children are expected to take the younger ones away and keep them safe and quiet until it's over. The adults need to discuss questions and problems with the help of the oldest elders and the spirits of dead ancestors. Some of the elders are so old that they must be carried onto the platform. Your grandparents aren't old enough to be up on the platform, but they're standing close to it.

The spirit talkers arrive wearing no ochre and don't seem in any way magical. You catch your cousin's eye, and she gives you a quick smile. Then the Leader strides in. Just seeing him makes you catch your breath. He's magnificent. As he moves, the sea of beads on his gown click against one another producing a husssshing sound, as if they are all cheering for him. You feel so proud that your bead is close to his heart. Even before he steps onto the platform, he announces the first item on the agenda: the growth of marshy areas and how it's changing the reindeer migration route. The young children stare at the cloak in fascination at first, but soon get bored. Most of the children creep away while they're still discussing the first item. You decide to stay. You understand the problem about the growing marshes. The world changes as time travels around the circle. The past is remembered by the elders and that helps people to know what to do. No one living today remembers a time like this. Ground has become mud and marsh, and forests are growing in places where there used to be only grass. This means there's more wood and water, which is good, but it is also causing the animals to find new migration routes. They're often not where the hunters expect them to be.

The elders discuss what can be done. Some think it would be possible to track animals better if an old signaling system is revived and improved. The details of this are hard to follow so you just watch the Leader. You see how vital he is to the meeting. The spirits give him energy, patience, and wisdom. He can tell when someone has something to say, even if they don't know themselves. If a person has trouble finding words to explain their point, he helps them by making suggestions or repeating parts of what they have already said. If they drift off topic, he gently brings them back. If people disagree, he finds the common ground in the points each side are making. The other spirit talkers are there, but the Leader is doing almost everything. The ancestors in the spirit world are all speaking through him.

Once the elders have gone through the most important topics, many of the parents, including your mom and dad, leave to check on the children and start cooking. You shift your focus to your cousin. She looks well enough but it's hard to tell. As the afternoon wears on, the Leader starts to tire, and other spirit talkers play a bigger role. When your cousin talks, it doesn't seem to be her voice. It's not just that the accent and tone are different. She talks with far more confidence, like a grown-up. As the smell of meat cooking becomes stronger, the pace of the meeting increases. Everyone seems more inclined to reach agreement without much argument or discussion. Finally, it's over. The spirit talkers walk out slowly, surrounding and protecting the Leader. He's exhausted and looks like an old man. He leans on his sister.

You leave the meeting area to find that people are already starting to eat. Different families are chatting and sharing the food they've brought. A lot of questions were resolved during the Elder Talk and the adults believe they can now go away and make more detailed plans. The elders are exhausted, and the children are worn out from playing hide-and-seek and other games among the trees. After eating, most people decide to have a nap while they wait for sunset and the closing ceremony. You lie down but can't sleep for worrying. The gathering is nearly over, and you almost manage to convince yourself that your cousin is wrong and that there will be no "bad thing."

The first part of the closing ceremony is like singing practice. People start singing and the spirit talkers come in, their faces red with ochre. But this time, the Leader is in front. He's a young man again, full of energy. As soon as he arrives, adding his voice to the singing, the words of the old songs seem to have more meaning—meanings you had never thought of before. Everyone does the movements along with the Leader. As he moves, the beads and bangles he wears find a new voice. They seem to laugh with the joy that everyone is feeling. They glow and flicker in the light of the flames. You've never enjoyed singing the old songs so much and you feel that from now on they will always mean more to you.

At last comes the song that everyone has been waiting for. It starts slowly but its tempo gradually increases with the pounding of the drums and the cheering of the beads and bangles. Some of the spirit talkers sing a different melody that seems to dance on top of the song everyone else is singing. As the pace quickens, the crowd moves together like a rapidly beating heart, breathing with one breath. You can hear the Leader's voice mingling with the sounds of the flutes.

Then, suddenly, you feel rather than hear a shout of surprise and pain. You stop singing, feeling confused and overcome with dizziness. Other people must feel the same because no one is singing. In the grey dawn light, you see that the Leader is lying on the platform, his twin sister is kneeling beside him with her head on his chest. Blood is spurting out of the Leader's neck and soaking into his beaded robe and his sister's hair.

The silence is complete as everyone watches the blood stop spurting. Then you hear the leader's voice as clear as can be. Everyone's eyes turn from his body on the platform and look at the source of the voice. It's coming from the lips of your cousin who is standing at the far corner of the platform. It seems to be saying, "I'm all right. I'm OK." And then your cousin walks forward, and her lips say, "I understand now that the spirits wanted me to cross over to be with them. My sister was the instrument of their wishes. Don't punish her for doing what she had to do." There's a pause and then the voice continues, "Now that I've crossed, I will work in the spirit world to help gather the knowledge of dead elders and pass it to you through the spirit talkers. My death may seem like a bad thing to you now but in the long run it will be good. The spirits believe this will help us through the difficult times that are coming."

There's another pause. When the lips speak again it's with your cousin's voice. She just says, "I'm so tired," and then sits down on the platform with her hands covering her face. You run to her and put your arms around her. She sleeps with you in your family's shelter that night, but the whole family now knows that this is the last time she will be with you.

When you wake up, a hole has already been dug in the forest floor. Everyone watches as the Leader's body, still wrapped in his blood-soaked beaded cloak, is lowered into it. His sister stands alone, her hair matted with her brother's blood. But then your cousin leaves your side and goes to stand with her. You watch as the beads you made are covered with earth. The only sound is the falling earth. Everyone is silent, even the little children who are far too young to understand what's going on. Then you suddenly realize that all of you are like little children. None of you understand what's going on or why the Leader had to die—not the elders and not the spirit talkers. The important thing is that you are all together sharing this memory and this feeling. Perhaps it will make sense when you have all crossed over and are together in the spirit world.

On the long journey back from the gathering you think about how this gathering has changed your life and changed you. You will talk about these experiences for years to come, but right now you just want to think about ordinary things. Simple things, like trying to keep your little brother amused, seem precious. Mostly you walk beside your grandfather because you sense that he is missing your cousin as much as you are. As you begin to enter familiar territory again, he says to you, "I told your cousin not to listen to those voices because I wanted her life to be easy and safe. I wanted her to get married and have children. Now she's a servant of the spirits and we don't know what will happen to her. But I suppose we should be proud that she's able to serve in this way."

Then he looks at you and smiles. "I'm so glad that there are other ways to serve," he says. "And I was very pleased to see you spending time with that family of reindeer hunters. We don't know enough about reindeer. They're coming further north nowadays so we're seeing more of them. Perhaps when you're a bit older we could arrange for you to meet up with that family again?"

A sudden death 30,000 years ago, with beads

This story is, of course, fiction, but we've tried to make it consistent with evidence found at an archaeological site on the outskirts of Vladimir, a city 120 miles northeast of Moscow. In 1955, workers in a clay pit found a human skeleton that seemed to be very old. Archaeologists from the local university were called in and began to investigate. The site, which has been given the name "Sunghir" (see Figure 6.9), has now been examined by a wide range of scientists from many countries. As new analytical techniques are developed, more is learned about this extraordinary place. Samples from the site have been sent to laboratories all over the world.[18] As the evidence has piled up, the Sunghir site has provided a more and more detailed and intriguing snapshot of the life on the mammoth steppe.

Carbon-14 dating suggests that people were in Sunghir between about 25,000 and 34,000 years ago. The layer of earth containing the evidence of human occupation (what archaeologists call the "cultural layer") is about a meter thick at Sunghir. It contains tools and evidence of toolmaking. There are remains of the hearths of people's fires and of the animals that they had killed and brought to the site. These include large animals like mammoths, reindeer, horses, and bison, and smaller animals like hare, squirrels, lemmings, and foxes. Pollen and plant remains in the cultural layer suggest that the area was forested when people were there—a forest of mostly pine, spruce, and birch trees. The presence of trees suggests that people were there during

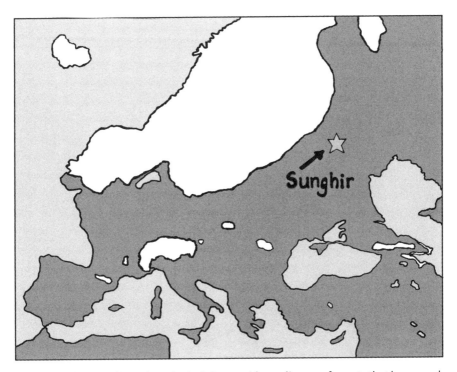

Figure 6.9 The Sunghir archaeological site provides a glimpse of events that happened around 30,000 years ago. At this time, the position of coastlines was different due to the lower sea level. Thick sheets of ice (in white) covered high elevations and a large area of northwest Eurasia. Much less snow fell to the east of this ice sheet and grass grew here, creating the grasslands known as "mammoth steppe." When temperatures rose for a while, the permafrost sometimes thawed in an area and this allowed trees to grow.

a warm period. For most of the last ice age, trees couldn't have grown so far north. Evidence from ocean sediments suggests that a relatively warm period did occur here around 30,000 years ago. The temperature was colder than today but slightly warmer than average for the last ice age.

The cultural layer at Sunghir is full of interesting clues about life on the edge of the mammoth steppe, but the site is most famous for two graves that archaeologists found just underneath the cultural layer. One of the graves held the skeleton of a man whose age at death is estimated to be between 35 and 45 (see Figure 6.10). Even though they had been lying in the ground for about 30,000 years, the contents of the grave were extremely well preserved. The man's skull and the bones of his neck and shoulders are stained red with iron oxide, suggesting that the upper part of his body had been rubbed with clay or ground-up stone containing the mineral known as ochre. On his forehead were 12 pierced fox canine teeth, and he wore a painted stone pendant

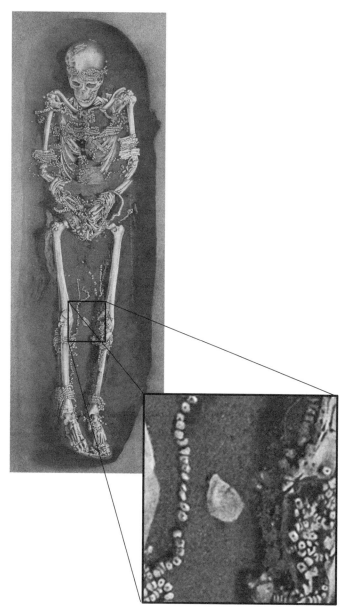

Figure 6.10 The grave of a man aged between 35 and 45 found at the Sunghir site in western Russia. This drawing is based on the many detailed photographs taken as archaeologists uncovered his remains and the object buried with him. A sharpened flake of stone found between his knees may have been the blade used to sever the artery in his neck.

Image credit: Drawn by K.N. Nikahristo and supplied by the Science Archive of The Institute of Archaeology of the Russian Academy of Science.

on his chest and 25 mammoth ivory arm bands above his elbows. The grave also contained about 3,000 tiny beads carved from mammoth tusks. Most of them are box-shaped, but some of them are smaller and round. The position of the beads suggests that they had been in strings that had perhaps been sewn onto a robe. The robe would have rotted away within a few years of the man's burial, but when the grave was uncovered, the earth was holding the beads in position around his skeleton. Examination of his skeleton revealed that his death was due to a wound that seemed to have been expertly inflicted. It was a sudden stabbing incision of his lower neck, deep enough to damage his backbone, cut the artery going to his brain, and cause immediate death. A sharpened stone point found in the grave may have been used to kill him.

Our story about the mammoth steppe people is also consistent with another body of evidence—the observations made by social scientists of how human groups use beliefs, customs, rituals, and charismatic leaders to build and maintain connectedness. On the strength of this connectedness, joint projects, such as a signaling system, can be envisaged and carried out. The completion of a successful joint project gives further strength to the connectedness. People of many cultures believe that spirits exist and have the custom of remembering, honoring, and sometimes talking to their dead ancestors. Such beliefs and customs can help maintain connections between people by reminding them of the ancestors they share. They also help to preserve memories of events from the past as well as the knowledge of the ancestors that may be useful in the future.

When members of a group gather, they seldom just talk, exchange information, and plan joint projects. Social connections made casually tend to be untrustworthy, and information gained through them can be useless or even harmful.[19] It's hard to think of an example of a meeting between humans that doesn't involve some ceremony, even if it's only the shaking of hands, the sharing of tea or coffee, or the wearing of ties and high heels. Customs like this force people to be alike and do things together. This builds memories of shared experiences and feelings, creating conditions in which people can form connections of good quality. When people take part in events that they find meaningful, their bodies and brains change. Studies suggest that this can generate physiological effects similar to those that cause mothers to bond with their young.[20]

Our ancestors who lived 30,000 years ago knew nothing about oxytocin or endorphins, but they must have noticed the way people respond to stories, ceremonies, music, and singing together. Group activities that were most effective and popular were treasured and became more elaborate over time. Extra flourishes were probably tried out and became permanent if they proved popular. As well as building commitment to the group, complex rituals provide a test of commitment. Group members who forgot the words to an anthem or failed to show proper respect to a sacred object were treated with suspicion. Children were most likely to survive if they were born into a culture with rich traditions connected a wide and strong social network.[21]

The beliefs, customs, and rituals of a group can generate behaviors that are impossible for outsiders to understand. The grave of the man buried at Sunghir is mysterious to us. The 3,000 beads it contains would have taken thousands of hours to carve. Seen from the perspective of today's materialistic cultures, this suggests that the man in the grave was very wealthy—that he might have been a powerful and greedy despot who was overthrown and executed by his people. But why would his murderers have buried the body of a hated dictator with valuable beads? It seems more likely that both the beads and the man were valuable to the people of the mammoth steppe because they helped them maintain the connections that were vital to their survival.

The other famous grave found at Sunghir contains the bodies of two boys buried head to head (see Figure 6.11). One is around 12 years old and the other is aged around 10. Their upper bodies had also been rubbed in ochre, and they too were buried in clothing decorated with mammoth ivory beads. As well as about 5,000 beads around his skeleton, the older boy had more than 40 fox canine teeth on his head and more than 250 around his middle. Two animal figures carved in ivory were with his body and beside it was a human thigh bone that had been polished, its ends cut off and filled with ochre. The younger boy's clothing was decorated with about 5,400 ivory beads, two ivory disks carved with a pattern, and two pieces of deer antler than had been pierced. Lying on either side of his body are 16 spears made of mammoth ivory varying in length from a quarter of a meter to nearly two and a half meters. There's no evidence of how the boys died.

The boys were buried together, but probably their burial occurred at a different time than that of the man. The similarity of the graves suggests their occupants were from the same cultural group. Analysis of DNA from the three skeletons and the thigh bone confirms this. The human remains are all from males. They weren't closely related but their DNA is similar enough to suggest they are from the same population, and it is a population large enough not to be inbred. Analysis of the chemical composition of the bones reveals that the thigh bone came from a man who had grown up in a different location from the other three.

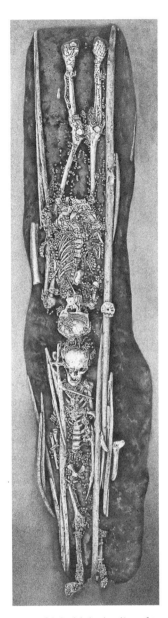

Figure 6.11 A second grave at Sunghir held the bodies of two boys laid head to head. The smaller one is estimated to be about 10 years of age, and the larger one about 12. This grave also contained thousands of ivory beads and many other objects.

Image credit: Drawn by K.N. Nikahristo and supplied by the Science Archive of The Institute of Archaeology of the Russian Academy of Science.

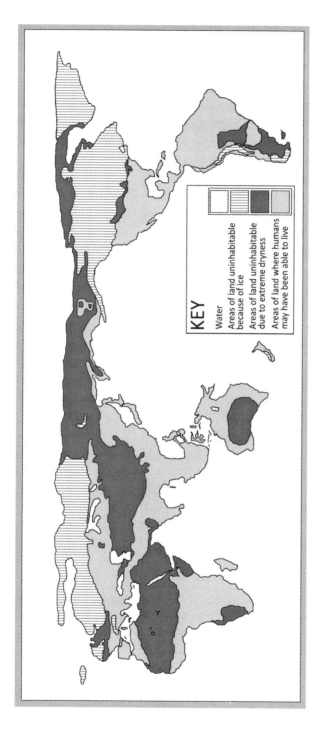

Figure 6.12 As well as having a very unstable climate during the last ice age, large areas of land were uninhabitable by humans. These areas were either too dry for almost anything to live or they were covered in thick sheets of ice. Uninhabitable areas separated people living in southern Africa from those living in Eurasia and prevented people migrating to habitable areas of the Americas. Only after the world had become warmer could people begin to explore the Americas.

This map depicts the world about 20,000 years ago, when ice coverage was still near to its peak. It's a simplified version of a map drawn in 2001 based on hundreds of reports analyzing data from all over the world. The more complicated map and citations of the many people who provided data on which the map was based can be seen at https://intarch.ac.uk/journal/issue11/rayadams_index.html

This evidence supports the idea that the people who visited and died at Sunghir were connected to a large and strong social network that covered a large area. Their rich cultural traditions helped the population maintain feelings of community over long distances. They could rely on each other in an uncertain environment and work together to maintain and adapt a large body of knowledge and know-how. The DNA evidence also shows that some of the ancestors of many people living today were part of this population.[22]

When culture became "cultural"

We've talked a lot about culture in this book, using the very broad definition of "culture" favored by biologists who study culture in animals. They define culture as simply the body of information and behaviors that members of a population share, and as they study animals more closely, they're finding evidence that more and more animals are sharing information. English-speaking Westerners often use a narrower definition of culture. According to this definition, culture is a human thing that is found in art galleries and concert halls. It's something that adds magic to our lives and gives us a glimpse of a world (or worlds) beyond our everyday experience. For them, culture is poetry, myth, music, art, and drama. It turns eating into a banquet, fire into a spectacle, and noise into a beat that a stadium full of people can rock to.

Culture defined in this way is a small but important part of culture in the broader sense. And it has a very practical purpose. The magic it adds are the feelings that allow us to trust and share with others, even people we don't know. These feelings are conjured by a vast kit of social tools our ancestors invented during the last ice age. Equipped with these tools, our ancestors' cultures were transformed, and they were able to survive the climate chaos.

Even though some parts of the world experienced a warm period about 30,000 years ago, the ice age was far from over (see Figure 6.12). It was after about 20,000 years ago that the northern part of the world started to become more habitable. Temperatures rose and there was more rainfall. Eventually the climate stabilized. Then our ancestors could begin to really explore the potential of having a large brain that is connected to a large social network. They would go on to invent many more social tools.

7
Building Today's World

This chapter is about the penultimate period of our evolutionary history—the period that is just ending now. It began about 20,000 years ago, when the world started to warm up again and today's physical world started to emerge—a world with less ice, more rain, higher sea levels, and reasonably stable vegetation zones.

During this period, the foundations of today's cultural world were also built. Cultures became more complex and, in some parts of the world, complexity increased at an accelerating rate. The humans that lived during this time left behind a great deal of evidence of what their lives were like, including stories about some of the events and people that, they believed, had shaped their lives. An awful lot of stories have been told about this time, some based on evidence and some based on legend. Many of them are now changing as new evidence is discovered and old evidence is analyzed in new ways.

The story we're telling to illustrate this chapter is set roughly in the middle of this period—about 10,000 years ago, when the Earth's climate had been stable for over a thousand years and a lot of the world's land area was covered in forest. It's likely that many of your ancestors who lived 10,000 years ago made their home near the edge of a forest. They lived in family groups. Some families would have moved several times a year and built temporary shelters. Others may have been more settled.

Part one

If you had been born as one of these ancestors, you would have spent the first part of your life with your mother, perhaps tied to her back during the day and lying beside her at night. By the time her next child is born, you're able to run around with the other children.[1] Your first memories are of the days you spend with the children and the evenings you spend around the fire with your mother, aunts, and grandmother. You see your uncle and grandfather sometimes, but mostly they're away guarding your land to stop the fire demons

Figure 7.1 The broken skull of one of 61 skeletons found beneath that sands of the Nile Valley at an archaeological site known as "Jebel Sahaba." This is believed to be the site of a massacre that occurred around 13,000 years ago.
Photo credit: Marta Mirazon Lahr.

burning it. Your father died fighting the fire demons and now your mother is married to his brother.

For a lot of the year, you and the other children spend the daytime in the safe areas near camp, collecting firewood and looking for things to eat. When you're little, you just follow around after the older children watching what they do, but as you grow you learn more and more. Some kinds of food can be eaten right away but other kinds must be taken back to camp to be cooked or prepared in some other way. Learning to spot food takes time because different areas have different kinds of food and each area changes in the different seasons of the year. In the spring there are nests containing eggs or baby birds and ducklings. In the autumn fruit and nuts are ready to be picked. The oldest child, who is your cousin, has a bag made of animal skin for carrying the food that is to be taken back to camp. Your cousin is very good at using a sling to shoot stones, and if one of you spots an animal, he tries to kill it. When he does kill something, all the kids go with him back to camp and watch your mother, aunt, or grandmother skin and gut it. One day, your cousin finishes making himself a new sling and gives you his old one. You practice with it so much that the other children laugh at you and say that your flying stones are scaring away the animals.

In the winter, and sometimes in the hot dry part of the summer, there isn't much food to be found. But you don't go hungry because of the acorn stores. In the autumn, acorns drop from the oak trees and everyone works to gather them up before the acorn-eating animals get to them. Even the men sometimes come to help. Then you carry them to the dry rock crevices where they can be safely stored. Children who are small enough to squeeze into the crevices pile the acorns against the back wall. Then the older ones hand them gobs of wet clay brought up from the stream to build a wall in front of the acorns. Once the clay is dry, the wall is hard and looks just like the back of a crevice. No one will know the acorns are there except the people who put them there. Many animals don't eat acorns because they taste nasty and make them sick. The only reason your people can eat acorns is that your grandmother's great-great-great-grandparents figured out how to make them safe to eat. Warm acorn toasties are the best thing for cheering up a cold dark winter night! After your people learned how to make acorns edible, they started looking after oak trees. If a young oak tree is being crowded by the seedlings of other trees you clear them away to give the oak more soil and sunlight. You also have the custom of saving the largest acorns and planting them in places were old trees have fallen down or just outside the forest. Many strong healthy oak trees now live in the forest and the grasslands just outside.

It's a sad day when your cousin goes off to be with the men fighting the fire demons. He says he's excited that he will soon be able to use his sling to shoot stones at the enemies. He says he will hit them right between the eyes so that they die instantly. But he's also sad to be leaving. The children have never seen a demon because the men keep them away from the places where people camp and search for food. Now your cousin is going to be one of those fighting men, and he may die like your father did. He promises to come back often and says he'll tell you more about what the demons are like. According to the stories your grandmother tells, the demons are just ordinary when they're children, before they learn to set fires. She sometimes talks about a time long ago, even before her own grandmother was a child. In those days, she says, there was no fighting because the demons lived somewhere else and hadn't started setting fires in the forest. It seems hard to believe that people could be friends with anything that destroys the land and trees that provide food.

After your oldest cousin goes off to be with the men, his younger brother wants to take his place, carrying the bag and telling the younger children what to do. But you and most of the other children like your older sister better and usually agree with her about what to do. As a result, she is the one who carries the bag, and you find and bring back plenty of food. This makes your cousin angry, but he can't do anything about it. Then, one day your sister is suddenly told that she must get ready to leave the camp and go to live with a new family. Your stepfather sometimes brings other fighters with him when he comes to visit. On one of these visits, a man saw your sister and decided that he wanted to marry her. Your uncle agreed and now they say she must go to live with his mother and his first wife. Your sister is miserable about going and your mother and uncle have an argument about it. Your mother cries and starts shouting at him. Your uncle slaps her hard on the face and she has to give in. The camp feels gloomy for the next couple of days. Only your cousin is happy. He says that, once your sister is gone, he will carry the bag and all the children will have to do as he says.

Your sister and mother set off to go to her new family. You already feel terrible and your cousin makes it worse by being extra bossy, saying that you can't have your own way now your big sister is gone. You try to stand up to him but he's bigger and keeps slapping the side of your head. You're so angry and miserable that you just sit by yourself, crying and thinking. That's when you decide to run away. You go back to camp and pick up a bed skin, a blade, and your sling. You put them in a bag and set off in the same direction that your mother and sister took. They've left a trail that's easy to follow and you can go faster than them because they're carrying more, including the baby. You catch up with them just as it's starting to get dark and they're settling down for the night. As soon as you see them you start to cry, and they cry too. You share some of the food they've brought. Your mother says she understands why you've come but begs you to go back because it'll be harder for everyone if you come with them. You agree to go back and be strong until your mother returns. Then the four of you cuddle together under your skins to go to sleep.

As they continue on their journey to your sister's new family, you begin to retrace your steps back to the camp. You take your time and the sun is already high in the sky when you get close to the camp and realize that something's wrong. There's a strong smell of smoke. You go on cautiously and then freeze when you see that a line of flames and smoke runs across your path between you and the camp. It seems to stretch a long way in both directions. From what you can see, everything on the other side of the line is scorched. The camp must already be burned! You stay motionless as you try to think what would be best to do. Then you take off. You run parallel to the line of the fire, going deeper into woods.

Finally, you come to the end of the line of fire and see two men and a boy beating down the flames. You don't know them but think they must be people from another family trying to put out the fire. Even so, you approach them cautiously. They can't hear you because of the crackling of the fire—and thank goodness for that! Once you're close enough, you realize that they're demons. You've many times heard fighters imitating the strange way demons talk, laughing at how they mispronounce words. That's exactly how this man and boy sound when you hear them shouting to each other! You crouch down into some undergrowth until they've moved off and their backs are to you. Then you run.

For a long time, you just try to put as much distance as possible between you and the demons, not thinking about where you are or where you're going. When it starts to get dark, you find a stream to drink from and unpack your sleeping skin. It's too dark to look for something to eat, but you're not hungry anyway. You wonder why you're not crying and why you feel so calm. It must be because you're completely alone except for the forest spirits. There's no person to hear you cry and the spirits are telling you that you can (you must!) rely on yourself. You curl up in your sleeping skin and listen to the forest. It sounds just as it has your whole life. As you lie there, you make plans. A few times you seem to be carrying out those plans but then discover that you're still wrapped in your sleeping skin. It may be dreaming, or it may be messages from the forest spirits.

By the time the sky starts to get light, you've decided that you must walk toward the sunrise because the edge of the forest is probably in this direction. Yesterday, when you were so afraid, you wanted to go deep into the forest to hide from the demons. But to survive and find people you need to go to the edges of the forest. This is where most of the food is and this is where people's camps will be. You set off, moving slowly enough to make very little noise, listening for the sound of voices and looking for signs of people. You start out heading toward the sunrise but end up going more and more to the right because that direction has an "edge of forest" look to it. The trees are further apart, and light is dappling the forest floor. And there's so much food around! Blackberry bushes are growing in the best-lit areas and some of the berries are ripe. As you pick them, you notice that birds and deer have been eating the blackberries but none of them have been picked by people. You can hear water rushing in the stream nearby and have a drink. You also hear plenty of animal noises—scratches, grunts, and calls. But there are no human voices. Why are there no people in this part of the forest? You see a bush with some ripe hazelnuts. You pick as many as you can easily reach and put them in your bag for later. Squirrels chase each other around the tree branches. You surprise a few deer and you see lots of evidence that pigs have been digging in the earth. This is a wonderful place for a camp! You find some stones and take out your sling and make a few practice shots. You put a handful of the stones in your bag, with one ready in the pouch of your sling.

You set off again and keep walking until sunset without coming across any people or getting to the edge of the forest. Before wrapping yourself in your bed skin, you crack and eat about half the nuts. Lying in the dark you wonder if you can be happy living in a place where there's plenty of food but no people. You decide that it would be too lonely without people and ask the forest spirits to guide you to some people soon.

You haven't gone far the next day when you come across some more blackberry bushes. These have empty calyxes, a sure sign that the berries had been picked by human hands. You walk even more quietly and finally hear voices. With your heart pounding, you climb a tree and sit in its branches listening. As the sound becomes more distinct, you can tell that they're speaking like demons, and also that the voices belong to children. You catch glimpses of them and see that it's a party of just children, no grown-ups. Your grandmother says that when demons are children they're just like normal people. These children certainly look normal. They're doing the same things as you were doing with your sister, little brother, and cousins just a few days ago. It seems like another lifetime now.

Your fear keeps you sitting in the tree, but once they've gone you climb down. Part of you wants to run, but the rest of you just doesn't have the heart. Instead, you sink to the ground and sit there. A wave of sadness and self-pity crashes over you and you feel that you would rather be set on fire by a pack of demons than be alone for another night. *(To be continued.)*

Why do we think our ancestors lived like this 10,000 years ago?

Ten thousand years ago, temperatures on the Earth were similar to those we experience today and no longer fluctuating wildly. This had allowed communities of plants, animals, and microbes to settle down together to start building complex ecologies. A vast diversity of habitats had emerged, and human groups were exploring them, learning about the food and useful materials they could supply. Oak trees of many different kinds are found in many parts of the world and many groups of people collected and stored acorns. In their different regions, people needed different knowledge, tools, and skills. And they also evolved different social tools.

When the ice caps and glaciers started to retreat about 20,000 years ago, new lands were exposed. Moisture was released, allowing rainfall to increase, and this brought new life to large areas of almost barren desert. Permanently frozen soil began to thaw, and trees replaced grass on many of the plains and hillsides. Forests returned and the mammoth steppe slowly disappeared, along with the giant woolly creatures that lived there. The human populations which possessed social tools that allowed them to adapt to rapidly changing climates thrived. Their numbers grew and they expanded into areas that had once been uninhabitable. It's not known exactly when humans first arrived in the Americas,[2] but the colonization of western North America by people from northeastern Asia was well underway by about 15,000 thousand years ago, and within a thousand years the descendants of these colonists were living all over North and South America.

The high-altitude winds and ocean currents that distributed warmth around the planet continued to be unstable for several thousand years after the warming began, so the temperature fluctuations of the ice age took a while to subside. North Africa, western Eurasia, North America, and possibly other areas experienced a very rapid rise in temperature about 14,700 years ago, but then it dropped again, fluctuated a bit, and then, about 12,900 years ago, it plunged. In some areas, the average annual temperature dropped by as much as 6 degrees Celsius (43°F) and several hundred years of ice age conditions followed. This last cold period, which scientists call the "Younger Dryas" (see Figure 7.2), ended about 11,600 years ago, and after that temperatures rose rapidly, climates stabilized, and today's vegetation zones became established.[3]

Figure 7.2 The cold period known as the "Younger Dryas" got its name from this tiny Alpine flower, with the scientific name, *Dryas octopetala*. By the end of the 19th century, scientists were already aware that they could learn about conditions in the past by digging to reveal the layers of earth and plant material that had been deposited at the time.

The peaty earth of some bogs in Europe has built up over tens of thousands of years. Each year, pollen from the plants growing nearby blew into bogs and became trapped and preserved in their layers. As a result, bogs carry a record of how local habitats changed as temperatures fluctuated during the last ice age.

Dryas octopetala was common in many parts of Europe during the coldest parts of the last ice age, so their pollen was common in the layers that formed during these times. Then, as temperatures began to warm about 15,000 years ago, warmer climate plants, shrubs, and trees began to grow there. Then, in the layers dated between about 12,900 and 11,700 years ago, the Dryas pollen was back, indicating that the area was once again experiencing a time of bitter cold. Scientists studying ancient climates began referring to this cold period as the "Younger Dryas," and the name stuck.

People were in tribes and tribes were sometimes in conflict

The social tools that evolved during the ice age, such as widely shared beliefs and rituals, had made it possible for our ancestors to connect with larger numbers of people and maintain strong social networks over long distances. The strength of these networks wasn't based on people rationally deciding that it would be a good idea if everyone pulled together. The strongest networks were those whose members participated in group events that made strangers feel as close as if they had known each other for years. The emotions generated during these events made people want to trust and be worthy of trust, and to maintain shared rules and customs. As a result, by the time temperatures started to stabilize, our ancestors' cultures included a collection of social tools that fostered peace, mutual aid, and the sharing of information between many families. These tools were passed down the generations along with the genes for building a brain that could learn to work with these tools and be influenced by them. As a result, families, even if they lived far apart, felt closely connected and, as the ice age ended, they started working together to take advantages of the new opportunities the emerging habitats presented.

But in more stable conditions, the social tools didn't work in the same way as they had during the ice age. While the climate was in chaos, families had a common enemy—the ever-changing environment. Individuals and isolated families could seldom figure out ways of coping with the latest threats. Everyone was at risk and most families would probably need help at some point, so it made sense to have the tradition of giving strangers the benefit of the doubt unless there was good reason not to.[4] People had little to lose by being generous, and they had a lot to gain by receiving generosity from others. When the people from many families put their ideas and resources together, their chances improved.

In a more stable climate, the challenges people faced were not so sudden and random. Habitats developed that could support more people. Populations grew in these places, and it became unnecessary to travel long distances to maintain a useful social network. Once habitats were pretty much the same year after year, it made sense for people to travel less and invest time in learning how to more efficiently exploit the land around them. By experimenting with the plants growing in their habitat, foragers could find out which were edible and nutritious, and which could be made edible and nutritious by processing. For example, many learned not to pick bean pods when they're green but to wait till they're ripe and their seeds have become more nutritious. The ripe seeds are highly toxic, but if they're soaked and then boiled, they become safe to eat. Experiments in food processing were time-consuming and risky, but they were a worthwhile investment for foragers who expected to be living in the same stable habitats for many generations. It was also worthwhile to experiment with ways of modifying habitats to increase the number of useful plants growing there or to attract more animals to hunt. Over time, people began making more and more major modifications to their habitats. What we call "farming" wasn't really invented. It evolved gradually.[5]

By 10,000 years ago, our ancestors had developed quite a bit of knowledge about the places where they found their food. The more expertise a family possessed, the less vulnerable its members were. The less vulnerable they were, less they needed to seek out and exchange knowledge with others. More expert families could obtain more food and raise more children. They passed on their expertise to their youngsters, who grew up and moved into nearby territories, often competing with the people who were already there. Families who were generous with their expertise could be taken advantage of, so it made sense to place limits on feelings of fellowship. Those who decided to treat their most valuable discoveries as family secrets were more successful. Perhaps at first they only shared these secrets with the families that were related to them by blood or marriage, but if they were successful, this would soon amount to quite a large number of families. Networks of culture-sharing families became what we think of as "tribes." As time passed, tribal boundaries tended to change as families formed new connections or when groups within tribes became separated. Over the centuries, tribal networks grew, shrank, split, and merged so that each group possessed a patchwork culture made up of beliefs, habits, customs, expertise, and social tools inherited from their various antecedents. The mix kept changing and evolving as each generation added new knowledge and ideas that they hoped would make them better adapted to current conditions. People also decided, often unconsciously, what old knowledge and ideas weren't worth remembering.

Archaeologists and historians have found quite a bit of evidence of group-on-group conflict during the last 10,000 years. The conflicts took many forms and occurred in many different places, but scholars have noted common features in the behavior of combatants, and we still observe them in people today.[6] The observations gave support to the apemen competing in the savanna stories and the idea that it might be in the "nature" of humans to fight for territory and hate those who are different. They suggested that warlike behavior might be somehow programmed by our genes and generated automatically by our human brain. If this were the case, the peaceful coexistence of different human groups would be impossible. A deeper view of history gives peacekeepers and diplomats more reason to hope. Far from being part of our ancient biology,[7] going to war seems to have begun rather recently in our evolutionary history. The archaeological evidence suggests that warfare was rare before about 8,000 years ago.[8]

The earliest archaeological evidence of group-on-group violence was found beneath the sand of the valley of the River Nile at a site called Jebel Sahaba (see Figure 7.1). Sixty-one skeletons of men, woman, and children were found buried there, along with bones from the body parts of other humans. Nearly half of the skeletons showed clear signs of violence, cutmarks on the bones, or bits of weapon still embedded. Some bones showed evidence of injuries that had healed, suggesting that they had survived earlier violence. The Jebel Sahaba site has been dated to about 13,000 years ago. This is so close to the date estimated for the sudden drop in temperature at the beginning of the Younger Dryas that it seems unlikely to be a coincidence. During the preceding warm and moist period, the land had been more productive, allowing the human populations to grow. The sudden shift back to ice age conditions killed the forests. Productive lands became deserts and the glaciers grew once more. It's easy to imagine how conflict might have broken out as groups of starving refugees met and started competing for limited supplies of food and water.

It's in the nature of hungry animals to compete aggressively for limited amounts of food. Humans aren't special in this regard. The way we are special is that even though we experience the feelings that generate this aggression, we can, and do, try to ignore these feelings and curb our aggression. Individuals often choose not to satisfy their own hunger so that a limited supply of food can be shared among their group. It's not individual greed that turns our aggression into warfare. It's the opposite. We live in groups that share resources, so fighting for resources usually ends up being a battle between groups rather than a fight between individuals.[9]

Hatred and intolerance of people who are different does seem to "come naturally" to humans, but lots of behaviors come naturally to our flexible brain. Once there were tribes competing for space in a richer and more hospitable environment, "tribalism" evolved, but it evolved culturally, not genetically. Each tribe told stories favorable to themselves that exaggerated the differences between people of different tribes. The exaggerations ("stereotypes") became part of everyone's folklore. Each tribe had a set of social tools for keeping the peace within their own group by distributing resources, resolving conflicts, and maintaining trust. But these rules and customs didn't apply to outsiders. Group members could recognize who was an insider, who was an outsider, and who was an enemy by the way they spoke, behaved, and looked.[10]

Tribes may have been more successful if they encouraged their members to see outsiders as strange, crazy, or not really human. From their perspective, members of other tribes did have odd customs and beliefs and didn't understand "normal" rules of conduct. Going one step further and "demonizing" members of enemy tribes may have been a useful social tool. If competition between tribes became outright conflict, the fighters who saw their enemies as demons rather than people like themselves would likely be more brutal and merciless. Tribes won more battles if their fighters went to war equipped with strong feelings of fear and hatred of the enemy. By winning battles, tribes got more territory and their numbers could expand. The children of the victorious tribes inherited the culture that helped them win the victory, including the practice of dehumanizing the enemy.

Our ancestors were the more successful members of each generation and were likely to have been part of more successful tribes. We have "inherited" from them the tendency to dehumanize people who look and behave in ways that seem abnormal to us, but there is no reason to believe that it is programmed in our genes. Our genetic inheritance provides us with a brain that is capable of feeling emotions like anger and fear. This brain is also good at remembering and analyzing information, and determining what is familiar and what is novel. But it is our cultural heritage that tells us how to categorize the information we analyze and what emotions we should attach to different categories. Our brain can't help seeing the differences between people, but the idea that some of these differences make certain people less-that-human is culturally inherited.

Culture, nature, and the age-old problem of raising young

People living today have mixed views of our forager past, and these are often expressed in terms of what is "natural" and what is "cultural." Many of the most positive views of forager life see it as more "natural" and therefore better for both humans and their environment. It's sometimes claimed that foragers saw themselves as part of their environment and ate the kinds of food that our body evolved to eat. Those with a more negative view of foraging point to all the evidence showing that our foraging ancestors weren't careful stewards of the environment and often severely damaged the habitats they exploited. Those who view foraging life more skeptically tend to celebrate the way that modern culture has made life easier. Given the hardships they endured, the argument goes, it's hardly surprising that our foraging ancestors seldom thought about environmental stewardship. Modern culture has helped to alleviate a lot of hardship, brutality, and unfairness that is inevitably part of a "natural" lifestyle.

A closer look at the lives of our ancestors forces us to abandon both sides of the argument. The whole "natural" versus "cultural" split is meaningless. Having culture is part of our "nature" as humans. Tens of thousands of generations of our foraging ancestors relied on culturally adapting to their environment. For example, there is no "natural" human diet. As with other animals, the human diet must include certain nutrients. But our ancestors culturally evolved ways of getting those nutrients from the plants, animals, fungi, and bacteria they found in a very wide range of environments. Amino acids from salmon or beans are no better for us than amino acids from beef or chicken. It's true that some environmental chemicals harm us, but our ancestors learned what was safe to eat and culturally evolved ways of detoxifying many foods. In the process of exploiting the habitats they lived in, our ancestors changed them. They also created new habitats.

What makes the natural versus cultural idea even more silly is that our beliefs about what "feels natural" mostly come from our culture—as do our ideas about what is brutal, disgusting, unfair, and abnormal. To achieve a better understanding of the complicated and rapidly changing cultural world that we've inherited, we need to stand back from these feelings and ideas and try to stop having them color our thoughts. It's more useful to look at how cultures formed and changed over the last 20,000 years.

Cultural change is "evolutionary" in the sense that each new generation builds on the culture they inherited from their elders. Kids often imagine that they're being revolutionary, but that's because the cultural "home" they've inherited is so much a part of their lives that they can't really grasp how big it is or understand its structure. What each new generation does is renovate the cultural home. Their changes may be radical—they may strip out the carpets and totally replace the kitchen. But they can't tear the home down and start from scratch. No single generation can create an entire culture, and we can't survive as human beings without some sort of cultural roof over our heads. We need things like language, knowledge of how to survive, and beliefs about the purpose of life.

Stories told about the past often depict cultural shifts as being triggered by key events (e.g., battles, discoveries, inventions, disasters) and key individuals (e.g., generals, scholars, religious leaders, tyrants) who were usually men but were sometimes gods and goddesses and occasionally women. The actions of individuals did play a role in shaping historical events, and such events can influence culture. But culture belongs to a population—a networked population with a complicated environment and a complicated history. Culture therefore changes by a complicated process. Examining the last 20,000 years with an evolutionary lens yields a picture that has fewer heroes and villains. Leonardo da Vinci might have been brilliant, but he didn't change the world. He was part of a world that was changing.

Sometimes cultural change is rapid and sometimes it happens more slowly. As a rule, the more people that are connected to a culture's network of users, and the greater their diversity, the more interesting and vibrant the culture can be. Large populations can share a bigger body of cultural information, and this information tends to change faster in such populations. That's why culture is changing so fast in today's huge modern societies. But just because people are always renovating the cultural house that they received from their forebears doesn't make it easy to predict or influence how the renovations will proceed. Attempts by outsiders to persuade people to discard elements from their cultural legacy can make these heirlooms seem much more precious. And some elements may be much more vital to the stability of a population's cultural edifice than anyone suspects, especially outsiders.

Some elements of culture, such as our technical know-how and the social tools that ease our interactions, have obvious practical uses. Others seem frivolous, arbitrary, and even harmful. Why do people make such an effort to decorate their bodies with clothing, paint, and jewelry? And why did so many cultures evolve the idea that it's OK for men to beat their wives? On closer examination, some cultural elements that seem bizarre do make sense. For example, a tribe may develop a ceremony to celebrate boys reaching manhood that involves cutting skin on their face to create scars. This seems an unnecessarily painful and risky ritual, but scarification ceremonies are not uncommon.[11] It's because scars can be a social tool—a sign of tribal membership. A man whose face shows the correct pattern of scars can count on the help and trust of similarly marked distant tribesmen. A less painful ceremony that simply awards young men with a maturity badge is likely to be less effective. This isn't just because a maturity badge is more easily stolen or faked. It's also because receiving a badge isn't a memorable shared experience that demands strong commitment and generates powerful emotions.

Over the last 20,000 years, warfare, natural disasters, and migration caused some cultural networks to split apart, shrink, and become isolated, while others grew, overlapped, and sometimes merged. This jostling around of tribes carried on for thousands of years in the post-ice-age world. During this time innovations were very slow to develop and spread, but gradually the pace of cultural change quickened, and people's lives became more complex. What historians sometimes refer to as the "rise of civilizations" involved the forging of links between different cultural networks. This started slowly and had many setbacks, but gradually new ways of doing things became established. This was thanks mostly to the invention of social tools that allowed people who belonged to different tribes to interact peacefully and profitably despite the cultural differences between them. More about this later.

Our ancestors were discovering many new ways of obtaining and using resources from their environment, but the main problem of our species—raising our difficult, large-brained offspring—continued to be tackled in pretty much the same way. Small teams of people—"families"—worked together to raise a new generation of members. There were some useful innovations, such as new kinds of foods to give older babies who are only just getting teeth but already need more nourishment than can be provided by their mother's milk. Several cultures worked out how to extract milk from other mammals. This may have begun by about 8,000 years ago. Milk would have been very helpful for increasing the survival chances of older infants and toddlers, but more evolution (both genetic and cultural) was necessary before older children and adults would gain much benefit from milk.[12]

The fact that humans organize family life in such a wide range of ways is evidence of how difficult our ancestors found the problem of raising children. If they had been able to find a set of beliefs, customs, and rules that was really effective, families would be organized in more or less the same way in all human populations. But the fact is, there are no "traditional family values." All cultures have beliefs about what families should be like and which family members should provide what kind of help to mothers. But their beliefs are all different, and every culture is aware that families can't live in perpetual harmony or meticulously follow "the rules."

The idea of marriage is widespread, but there's no consensus on what "marriage" means other than that, for most cultures, it identifies a "father"[13] of a child. What the father is expected to do varies. In a few cultures, such as the Mosuo people, who farm in Yunnan and Sichuan provinces of China, fathers may bring their children small gifts, but they play little or no role in their upbringing. Men's responsibility is to their mother's and grandmother's family, and any work they do helps the children of their sisters and female cousins. But it's the women who own the farms, and they do most of the work raising crops. Some cultures expect men to assume much more responsibility. Maasai men, who herd cattle in parts of the Rift Valley in the East African countries of Kenya and Tanzania, may not do much hands-on baby care, but they're expected to be in control of their family's welfare and also their behavior. The father is considered the owner of everything the family possesses and can wield great power over his wife (or wives) and children.[14]

Most families are somewhere in between the Mosuo and the Maasai. But not all. For example, when anthropologists first contacted people who live by hunting, gathering, and farming in tropical and subtropical South America, they found a considerable number of tribes who believed that children could have multiple fathers. Any man who had sex with a woman just before or during her pregnancy believed that they had contributed to the creation of her child. They were expected to help her by providing her with food during her pregnancy and after she gave birth. These peoples believe it best for children to have multiple fathers so that mothers get enough support. Anthropologists studying these people have found evidence that they were right. Children with multiple fathers were more likely to survive than those with only one recognized dad![15]

Our ancestors may not have discovered an ideal way of organizing a team to raise young humans, but they still managed to do it. They raised children in conditions that most of us today can't imagine surviving. The children they raised learned their family's ways and this influenced how they raised their own children. If you're a member of a modern society, your family may not be playing a very large role in your life right now. You may not even have a family. But families—whatever they're like—continue to be important, even in modern times, but they were generally much more important before modern times.

It is, however, hard for many people living today to comprehend how much more important "the family" was in the past. Modern societies have many institutions—from hospitals to universities—that help mothers produce and raise the next generation. Until recently, it was just families. Only your most recent ancestors would understand the idea of going out to work or school. Most worked and were educated within a family—sometimes moving between different families as they grew older. If they were injured or sick, they were supported and nursed by members of their family. If they were mistreated by someone from another family, their family might try to gain compensation or revenge. If they were accused of mistreating someone else, their family would often defend them and negotiate on their behalf. Wealth belonged to the family and would, family members hoped, be passed down the generations. For most families this "wealth" consisted of their home, the skills and tools they used to make a living, and their connections with other families—their "good name within the community." Sometimes quarrels within families led to individuals being "disowned," and these individuals had no connection to society unless they could find another family to be part of. The fact that your ancestors managed to raise successful children meant they spent at least part of their lives as members in good standing of families who had useful skills and worked reasonably well together.

It certainly would have helped if families were happy and their members loved each other, but all families really had to do was function as well as, or better than, the families they were competing with. You may ask why we are suddenly talking about families competing when we've just described close connections between families and claimed that they were and are bound up in cooperative cultural networks. It's because, even if families cooperate, it's inevitable that they are also competing for reproductive success. The winners of this competition were the families that managed to raise more children. If families were cooperating, the winners achieved their success within the bounds of rules agreed upon by the families. They didn't make it a habit to cheat or steal from one another or commit adultery. They mostly helped one another in times of need. And they arranged marriages between their children. They cooperated, but, at the same time, families were still competing.

When families managed to raise more children, future generations contained more of their descendants. Both their genes and their family practices would be inherited by more members of the next generation. Child survival was partly down to luck, but hard work and parenting skill certainly made a difference. To raise a larger-than-average number of children, families had to get a larger-than-average share of the food and the other resources available in the environment. And they had to bring up their children to help in this endeavor—to be a credit and not a burden to the family. Ideally, children would grow up to be adults who could maintain and even advance the family's position in the community and/or tribe.

In chapter 5 we pointed out the obvious fact that the men and women who lived and died without producing surviving children aren't our ancestors. The same logic applies to families. There may have been families whose members decided not to raise children. They understood the association between mating and pregnancy and could have easily worked out how to satisfy their sexual urges in ways that didn't involve the risk of pregnancy. When times were hard, they could have judged it better not to bring children into the world, as their lives would include a great deal of suffering and hard work. This was a perfectly human (and humane) choice to make. But our ancestors didn't belong to these families. The beliefs and customs of families like this weren't passed on to the next generation. Our ancestors behaved in ways that led to reproductive success. They may have belonged to families who believed preventing pregnancy was wrong because the decision about who should have a chance at life belonged to a higher power. They may have believed that the effort they put into raising children was worth it because of the way children and grandchildren enriched their lives—providing the pleasure of their company and security in their old age. Or they might have simply believed it was their duty to have children. It doesn't matter what they believed as long as it made them produce children even when times were very hard. It's thanks to those beliefs that we exist.

The sorts of beliefs and customs about family life that were passed on seem to vary depending on how families made their living. Anthropologists observing life in premodern populations have found that families who grow crops or herd livestock tend to be more rigid and hierarchical than nomadic foragers. The children of farmers and herders are given work to do—harder and more complicated work as they grow older. They're likely to be punished if judged to be disobedient or lazy. Members of these families are taught that it's their duty to put the needs and honor of their family above their own needs and desires. They learn to respect and obey their elders. An age-based hierarchy reduces arguments and uncertainty, and elders usually do know more. If children survive, they graduate to new levels in the family hierarchy. Child mortality was high in these families, mostly because of disease rather than severe treatment and lack of care. But the children who died tended to be the ones who were physically weaker or had poorer judgment.

Foragers tend to be more easy-going and indulgent of their children. Some may argue that this parenting style is more "natural," but the farmers and herders managed to raise more children. Their ways were passed on to those children and, over the centuries, farmers and herders claimed more and more of the Earth's habitats.

The rules of family life didn't need to be fair and the things they taught their children didn't need to be true. All they needed to do was encourage family members to behave in ways that produced surviving children—and grandchildren. For example, a family needed to raise girls who grew up to be mothers. Girls didn't just need to learn mothering skills; by helping to care for younger family members, they needed to grow up to be willing to endure pregnancy and childbirth. Families were most likely to be successful if they believed (and taught their girls) that having children isn't just a duty but that it's the most fulfilling thing a woman can do. A mother who taught her daughters that they had a choice about whether or not to get married or allowed them to foster ambitions to do things that interfered with childbearing would likely have had far fewer grandchildren than mothers who "knew" that good girls just want to be mothers.

But it took a great deal of work to give a baby the chance of reaching adulthood, so giving birth to too many babies was also a problem. Families that had customs for regulating sexual behavior were, potentially, able to produce babies at a close to optimal rate. In many families, marriage played an important role in regulating births. As we mentioned in chapter 5, we believe that many human groups had evolved the custom of arranging marriages for their children by 100,000 years ago. As well as helping to bind families and establish the family connections of newborns, marriage could serve as a birth control measure. By making it a rule that young women must not have sex until they're married, families could be better prepared to care for the babies they would produce. One way of making it likely that this rule would be obeyed was to control the information children received about sex. For example, adolescent girls could be brought up to believe that sex is vulgar and very painful for women, but that they must submit to their husband's carnal urges in order to have children. After she married, a woman might discover that sex isn't so bad after all. But she might still believe her unmarried daughters should be kept ignorant and frightened about what they will have to suffer on their marriage beds. Such beliefs may generate anxiety in young women, but a little anxiety is not as bad as being tempted to have sex before marriage and then endure a risky pregnancy and labor only to give birth to a baby with little chance of survival. Some cultures evolved the custom of performing operations on the sexual organs of girls that many people believe reduces their chances of enjoying sex.[16]

After she marries, a woman is expected to have sex and produce babies, but producing them too quickly is inefficient. It exhausts mothers and their helpers and reduces infant survival. Breastfeeding can help to delay a woman's next pregnancy by triggering the hormone system that inhibits ovulation. But once the baby is old enough to need food as well as breast milk, this system is unreliable. Increasing child survival may be the reason behind some curious "old wives' tales." For example, believing that having sex taints a mother's milk can discourage couples from having sex when their baby is small and needs the milk. And once their toddler is ready for weaning, it may encourage them to have sex frequently in hopes that it will make the mother's milk taste bad so that weaning will be easier. Beliefs don't have to be true to serve a practical purpose. A belief that optimizes the spacing between pregnancies increases the chances of a family having plenty of healthy young members.

Anthropologists have found a great deal of cultural variation in how men and women treat one another. In some premodern cultures, women controlled the family's wealth. If a man dared to raise a hand to his wife, he risked being sent back to live with his mother. It may be tempting for people living today to see these cultures as "enlightened" or "advanced," but there's a more practical explanation. Populations with these beliefs tended to live in habitats where woman and children could make a living without much help from men. Men may have not had much power in these families, but they also didn't do much work. The more that families needed men to work for them—the more they needed protection from predators and invaders and the more their survival depended on activities that required strength, such as the handling of large animals, the more power men had over "their" women and children.[17]

Our ancestors inherited beliefs and traditions about parenting, gender roles, and sexual behavior that many people today see as silly, unfair, and downright evil. While we (the authors of this book) have sympathy with this view, we also think it's important to recognize why these beliefs and traditions persist. They exist because, for many generations, the families that held such beliefs were the winners of the Darwinian competition. When times were really tough, they were the families who managed to produce babies who would survive to adulthood and go on to have babies themselves. The beliefs and traditions that brought this success continue to exist because humans have a brain that evolved to be part of a population that shares cultural information. We unconsciously absorb elements from the culture that surrounds us. Changing minds takes time. Each new generation continues to "inherit" a great deal of the culture of the previous generation. Cultural change can't be instantaneous.

The fate of the forager child who ran away illustrates the complex relationship between the culture of tribes and the way families make their living.

Part two

You sit on the ground perfectly still for several minutes, listening to the forest and trying to gain strength from the spirits. Then you hear the faint crack of a branch breaking. You look up, expecting, and half hoping, it to be the demon children. You see instead that it's a deer. Silently, you rise to your feet reaching for your sling. The deer continues to chew the mouthful of leaves that it has plucked. You have a clear shot at the side of its head. If your stone hits hard enough, you might stun it for a few seconds. As you release your stone, the deer turns its head toward you so that the stone hits square between its eyes. You hear a bark and two other deer standing near it leap away, crashing through branches. The deer you hit is falling, its legs crumpling beneath it. You run to it. It's already starting to come around when you get there, so you throw yourself onto its shoulders to prevent it getting up. It starts wriggling trying to throw you off.

"I've got a rock," you hear a girl's voice say. "Watch out, I'm going to try to hit its head." The demon children are all around and some of them are holding the deer's legs. There's no way it can escape now. Finally, the girl manages to land a blow on its head. You stay in place, lying across the deer's shoulders until its body starts to twitch and then finally, it stops breathing. You take a deep breath and start to laugh. So do all the other children. Finally, the girl, who is the oldest of the children, says, "Wow, this is amazing! How did you catch it? Was it sick?"

"It wasn't sick. I shot it with a stone using my sling and that stunned it," you say, trying your best to imitate their demon way of speaking.

"What are you going to do with it?" the oldest girl asks.

You don't answer because you have no idea what to do with it.

She seems to think that you can't decide and says, "I think you should share it with us since we helped kill it. Will your family allow that? We didn't know there was another family nearby."

"There is no other family nearby," you say.

All the children look at you puzzled, and then one of the younger ones asks, "You're on your own?"

Something about the wonder and pity in his voice touches you and you start to cry. And then you can't stop crying. You hide your crumpled up miserable face behind your hands. The oldest girl puts her arm around you and then all the children get close to you. You feel damp little hands stroking and patting your tummy and legs. This makes you cry more.

Finally, the oldest girl, who reminds you so much of your big sister, says, "Well then, it looks like you'll have to share the deer with us. Let's take it back to camp."

"That's fine with me," you say. You take your hands away from your face and smile at them to show you that you're finished crying. You, the oldest girl, and some of the other children manage to lift the deer between you and start to struggle along. The girl sends a boy running ahead to get help. He soon comes back with two men who, when they see the size of the carcass, run over and tell the children to put it down. "You kids killed this with a stone?," they ask. "Well done." Then they pick it up onto their shoulders and set off. They don't seem to notice that one of the children is a stranger. The oldest girl takes your hand and you all follow them.

The camp, when you get there, looks pretty much like the camps that your family makes, with shelters made of branches and skins. It's bigger though. The men carry your deer to a tree where there's already a deer hanging by its hind legs, and they string your deer up next to it. The youngest children are excited about the deer and the men tease them by pretending to take credit for killing it. The children scream in indignation. The oldest girl is still holding your hand. "Sit here," she says, pointing to the ground. "Don't look so frightened. No one's going to hurt you."

You sit and watch as she goes further into the camp, finds a woman, and starts talking to her. She points you out to her and you try to give a smile. It feels wonderful to be around people again, even strangers—even strangers who talk like fire demons. These people seem normal and, if they're different, it's in good ways. You haven't been around men much because the men in your family are mostly away fighting. When they're around, they're grumpy with the children. They just want to eat and be with the women. In this camp, the men seem friendly and there's much more food around. You can only remember two times in your whole life that your father and uncle brought a deer back. In this camp, there are two deer hanging at the same time! There are so many deer living near this camp that children can bring one back!

The girl comes back. "My mother says you can stay with us if you want," she says. "I'll be leaving soon to get married and you can help us look after our baby. I hope you're good at seeing snakes. A snake bit my little brother last year and he died."

You assure her that you're good at seeing snakes and follow her to where the woman is sitting. Her baby is a boy and he's just learning to walk. You kneel in front of him and hold his hands so that he can stand up. You smile at him and make encouraging noises; he smiles back and touches his wet mouth to your cheek and laughs when you put your mouth against his cheek and blow, making a farting noise. All babies love that.

Your life with a new family begins. It's not so very different from your old life. You spend part of it in the camp, mostly with the baby. Part of the day you leave the baby with his mother (your new mother) and go with your adopted sister and the other children to collect firewood and food. You watch the children and try to copy the way they do things. But you're used to foraging in areas where there's less food, so you often spot things they don't see. And none of them is as good as you at shooting stones with a sling. Killing the deer was a fluke—a gift from the forest spirits—but you manage to kill quite a few small pigs, hares, and squirrels. Once you see a snake and kill it and your sister is very pleased. She says it may be the one that bit her brother. Snake is good to eat but so full of little bones! Your new family also sits around the fire in the evenings telling stories, and some of the stories are similar to the ones you know. And for the first time you hear men telling stories. For a while, you hardly say anything, except to the baby. But you listen hard to what everyone says. When you do start to talk, you pronounce your words just like everyone else in your new family.

It doesn't take you long to find out that you're living with a family of "fire demons"— at least that's what your old family would call them. You hear the men making plans to go and burn a place, as if setting fire to the forest is just a normal thing to do. The place they're going to burn is quite far away so some of the men and older boys will be away for a few days. You're relieved that your adopted sister won't be going with them, and after they're gone you quietly ask her why they're going to burn the forest. She's surprised you don't know and tells you that forests need to be burned every few years to clean out the dead stuff so that more food will grow and there will be more animals to hunt. Men manage the fire, she says so that it gets rid of dried out, diseased, and useless plants. This allows the good plants to get more sunshine and water.

You discover another surprising thing about your new family when the oak tree by the camp begins to drop its acorns and you start to collect them. You hear the children laughing and look up to see that they're laughing at you. You look around confused, your face turning red.

"It's because you're picking up pig food," you sister says gently. "You know that people can't eat those, don't you?" she asks.

You nod and put them back on the ground. They call acorns "pig food"! There are many magnificent oak trees around the camp, and it almost makes you feel sick not to pick up their acorns. You wonder if these trees had been planted by your ancestors. They might have lived here once but were driven away by the people they thought of as "fire demons." Your cousin and uncle may right now be trying to kill the relatives of the people you now live with. It's horrible to think about this so you try not to. You just know that the people in your new family are not demons. They're people and they're kind.

As the time approaches for your new sister to marry, you think a lot about your old sister and the terrible day that she left to get married. Your new sister is happy about getting married and going to live on the grasslands. Her husband-to-be isn't old, and she's known him all her life because he's the son of her favorite aunt. She says that her aunt is very clever and hard-working and she's looking forward to having her as a mother-in-law. When the time for the wedding finally arrives, quite a large party leaves the forest for the two-day journey into the grasslands. Some of the time you carry your little brother and some of the time you carry the young pig that you killed, which will be part of the wedding feast.

When it's time to stop for the night, you stay up with your adopted father to keep the first watch. You have never spent a night outside the cover of trees and feel you can't get enough of looking at the wide sky covered with stars. By the light of the stars you can see the shapes of huge animals slowly moving around in the valley. "Those are cattle," your father says. "They grow large dangerous horns. I've seen a cow fight off a pack of wolves going after her calf. She killed one and chased the rest away."

"Do your brother-in-law's people hunt those animals?," you ask.

"Of course," he says. "But they don't hunt them so much as fight them off. Some of the grasses make big fat seeds. The women and children harvest them because at the end of summer they become dry and can be kept all winter. When there's not much other food, we can grind up the seeds to make powder, called "flour," and add it to water to make a dough that can be cooked on flat rocks heated in the fire."

This is pretty similar to what your old family does with acorns, but you don't say anything because he might laugh at you for eating pig food. Instead you ask, "How do the men prevent those huge animals eating the grass?"

"Not all the grass," he laughs. "Only the fat seeded grass. That only grows in special places and my sister thinks she's found ways of getting more of these kinds of grasses to grow. My brother-in-law is kept busy preventing cattle, goats, and other grazing animals eating and trampling the grass that she grows in special plots. The cattle are difficult but he's finding the smaller grazers quite easy to herd. They use their hunting dogs to help."

You notice some movement in the distance and point it out. It's two people walking, probably men. As they get closer your father stands up and waves to them. They shout back. The younger one is the man that your sister is going to marry. As you meet by the fire, you can see that he's a tall and handsome young man. But you can also tell that he's been crying. Your father tells you to go and wake people up while he adds wood to the fire.

When people are gathered, your sister's future husband starts to tell his story. His younger brother has died, he says, and their mother now thinks she might have to give up all her plans for growing and harvesting the fat grass seeds. Last summer she had noticed tiny pods on a few of the seed heads and after the seeds were harvested, she saw that some were darker than the rest (see Figure 7.3). It was easy to pick these dark seeds out and throw them away. But this summer most of the seed heads had the little pods on them and many of the seeds were black.

"I offered to eat a few of the black seeds to see if they were OK," he says, his voice breaking with the effort of not crying. "But my little brother said he would do it because I shouldn't take the chance of getting sick with my wedding day coming."

He pauses as more tears flow down his cheeks.

"He just got stomach upset at first," he continues, "but then we noticed that he was no longer himself. It was like an evil spirit had entered his body. It got more violent and he shouted and cried in agony as it made his body stiffen and shake. Then he became unconscious and his body started to die before our eyes, starting with his fingers and toes."

"It was the worse death I've ever seen," the older man says.

Your sister's wedding turns out to be very different from the one that everyone had been looking forward to. But she tells everyone that a marriage forged in sadness can be even stronger than one that starts with a celebration. She seems to be looking forward to the challenges ahead, and this makes everyone feel more hopeful. You feel so much admiration for her.

People eat the wedding feast with the feeling that they must build up their strength for winter. Many baskets of the fat grass seeds had been harvested, and your family was planning to take several back to the forest to make flour to help see them through the winter. But the grain is speckled with the black seeds and everyone is now afraid to eat it. You travel back empty-handed with grim thoughts of how hungry this winter will be. The route home takes you past many large beautiful oak trees. You look on the ground beneath them, but their acorns have all been eaten by the pigs, deer, mice, and other animals.

Once back among the trees, the family sets up camp near a small marshy lake, where birds spend the winter, and which is sheltered from the wind by a high and rocky ridge. The following afternoon, you go out with the older children and men to catch birds with a net. The first sweep catches quite a few, but it frightens the rest

Figure 7.3 Spores of the fungus *Claviceps purpurea* infect the tiny flowers of grass or cereal plants in the spring and produce a pod that shrinks and turns black as the seeds develop. Apart from its dark color, this pod can look like a cereal grain. But it's full of fungal toxins that can be fatal, in even small amounts.
Photo credit: Dominique Jacquin.

away, so you give up for the day. As you walk across a large flat rock by the side of the lake you notice that it has a number of bowl-shaped indentations, just like those in the rock where your old family used to grind up acorns. You look up and, on the ridge, you see crevices where acorns could be stored.

 Winter wears on and people get thinner and thinner. Like most babies, your little brother hasn't learned that hunger just has to be endured. His crying and whimpering are annoying, but at least they show he isn't sick. The youngest baby in the family is silent and everyone expects her to die any day.

Finally, you decide that, "pig food" or not, it's time to explore the crevices. You pick up a bag and climb to the lowest crevice and squeeze inside carrying a sharp rock. The back wall of the crevice does feel like it's made of clay and a few blows with the rock make a hole in it big enough for your hand to fit through. A rush of satisfaction and joy passes over you as your hand feels the acorns hidden behind. But you also feel sadness or perhaps guilt. These acorns must have been the food store of people who had been driven away by the men they thought of as "fire demons."

It's almost dark when you get back to camp with your bag full of acorns. People are only a little bit interested when you tell them that you've found a place where acorns are stored. They watch curiously when you crack and peel them, and the two toddlers are told not to try to eat the "nasty-tasting nuts." Perhaps because the acorns have been lying in the crevice for many years their shells are dry, and they peel easily. By the time you're ready to sleep, nearly half the acorns are shelled.

The next morning you look at the store of collecting bags and choose a thin-skinned one that water can ooze though. You take it and the shelled acorns down to the stone by the lake with the bowl-shaped indentations. You find a rounded stone to use as a grinder, put a handful of acorn nuts into the largest of the indentations, and pound them into a powder. You transfer this to the bag and start on the next handful.

When you're about half finished, your adopted mother comes down carrying the baby and watches you work. She says, "I see what you're trying to do. You think you can make a flour from the acorns."

"Yes," you say, "and I think I can wash away the bitterness and make the flour good to eat. The people in my old family did this every winter."

To show her, you fill the bag with water and swirl the flour around with your hand. Then you tie the bag to a branch with its bottom in the water so that the flour can soak. Your wet hands are so cold you can barely move your fingers and you put them on your face and neck to warm.

"Take the baby," your mother says, "I'll do some grinding."

The two of you work together taking turns holding your brother and grinding the acorns. When you've finished, the bag is almost half full of flour and you give it a good rinse. "We must let it settle," you tell her, "and pour off the bitter water and rinse it again. After several rinses the water that comes off won't taste bitter and the paste will be good to eat."

You go back to the others sitting by fire. A single thin duck was caught in a trap and it's cooking. That's not much to share with the whole family, but it's better than nothing. You go back to the lake several times in the afternoon to rinse the flour, and that night, after your brother has gone to sleep, you and your mother work to crack and peel the rest of the acorns. Hunger seems to have made the family listless and no one bothers to ask what you're doing.

The next morning you give the flour one more rinse and taste the water. There's no trace of the acorny bitterness, so you squeeze as much water as you can out of the bag and take the paste up to show the rest of the family. You and your mother tell them about how you made it. Everyone is now very interested, but they're not convinced that it's safe to eat. The description of the death of the grassland boy has everyone very rattled.

Even though it would be better cooked, you eat several gobs of the paste raw and cold to show them that it's safe. You love the familiar taste and the full feeling in your stomach. Then the whole family goes down to the lakeside to watch you and your mother grind acorns and rinse the flour. That night, seeing that you have suffered no ill-effects, everyone has a taste of the paste and tries cooking it on hot rocks as if it were flour dough. Your mother gives the baby his first acorn toastie and he eats it greedily. The next morning a few people say their stomach feels a little strange, but the others tease them, saying that their stomach has just forgotten what it feels like to have food in it. Everyone wants to go up to explore the crevices.

That winter, the family eats their way through the stored acorns. The babies and toddlers all survive, and your mother's younger sister even becomes pregnant. The following autumn, the whole family works together to gather acorns and you show the children how to build a wall of clay to protect a store of acorns in the back of a crevice. You're now almost too big to squeeze into the crevices yourself. Stories are told around many campfires about the child who discovered the acorns and started to grind and wash them. After talking it over with your mother you decide to say that the spirit of the grassland boy came to you in a dream and told you about the acorns. This means you don't have to mention that you once lived in a different family. But, more important, it shows that the boy's spirit survived his horrible death and that good has come of it. The story makes his family on the grassland very happy.

You spend your life far away from the border where the fight continues between your old people and the people they think of as "fire demons." You never see your old family again and don't talk about them, even with your children, because you don't want to confuse anyone. When you're old and your children have children of their own, you begin to feel that the knowledge you brought your new people changed them in ways you don't like. They have become more like your old people, with men spending a lot of time away. Once they started collecting and eating acorns, there was less food for the pigs and deer of the forest. This meant there was less game for the hunters to kill and less meat to eat. Of course, having the acorns meant that less meat was needed, but that wasn't the point, as far as the men were concerned.

It wouldn't have been a problem, except that it made the men feel unnecessary and unappreciated. They still hunted and did the burning to keep the forest clear of dead stuff, but many of the young men began to prefer living on the grasslands, in the company of each other and their dogs. They would often bring a sheep or goat carcass but seemed to be more interested in collecting and wearing fancy furs, foxtails, and wolf fangs. Instead of hunting for meat, they spend a lot of time killing the carnivores that prey on the grazing animals. They say that this will pay off in the long run because there will be more grazers for people to eat. Time will tell.

Lessons from the evolution of life

It's sometimes suggested that the genes of humans are no longer evolving because our cultures protect us from natural selection. It's easy to see why people say this. Thanks to culture, we don't die from a lot of things that would likely kill us if we were trying to survive on our own in the wild. But the only way that a culture could stop our genes evolving would be to ensure that everyone in the population has exactly the same number of surviving children. (As far as we know, no cultural has attempted to do this.) Whenever some people have more children, whatever the reason, their genes are being selected. The next generation will contain more of their genes, so the genes of the population are evolving.

Our cultures help to create the environments that we live in, so they cause different characteristics (and genes) to be selected. For example, as people evolved ways of producing food through agriculture, they often ended up eating a much more restricted diet than their foraging ancestors. In some cases, their diet contained large amounts of substances that foragers would seldom if ever encounter. Some people had genes that allowed them to survive pretty well on this new diet. They had more descendants, so these genes became more common in the population. The evolution of genes was being driven by the evolution of culture. The human evolutionary story during the last 20,000 years is mainly about our cultures evolving—evolving at an ever-increasing rate—and often driving the evolution of our genes.[18]

On the face of it, it seems as though cultural evolution can have little in common with the evolution of genes. There's no mind or intelligence driving the evolution of genes. Random errors in the copying of information create variation among members of a population. Those variants are selected by the blind cruelty of natural selection. The organisms lucky enough to have a combination of variants that makes them well suited to the environment are most likely to survive and have offspring.

Cultural change does have intelligence behind it—human intelligence. People often work hard to increase their knowledge and figure out ways of improving things. Our ancestors were looking for new kinds of food and developing new ways of dealing with the environment and each other. They often thought hard about whether a new way of doing something was a genuine improvement or whether it was best to stick with the tried and proven way. While the evolution of genes is a random natural process, the evolution of culture seems, on the face of it, to be more like "intelligent design."

But if you go a bit deeper and examine how cultures *actually* change, you find a process that looks rather similar to the evolution of life and not at all like intelligent design. We humans can't "design" our culture because none of us are intelligent enough. Our cultural "home" existed long before we were born and will continue long after we're dead. It has rooms we've never explored and corridors we don't even know about. We may think that a new appliance or larger windows will improve our cultural home, but it's more hoping than thinking. The new appliance may catch on fire. Knocking out part of the wall to let in more light might cause part of the roof to collapse. We can't know how any change will affect culture even in the short term—and certainly not in the long term. The woman in the story who worked hard growing and harvesting cereals couldn't know that her plots were at risk of being infected by a toxin-producing fungus. And it's not just ignorance that limits our intelligence. Our emotions do too. Memories of the horrible death of a much-loved boy can have a larger effect on the direction of cultural change than information gained in years of quiet research and experimentation.

Ten thousand years ago, it's likely that many experiments in growing cereals ended in tragedy as crops were attacked by insects, parasites, and microbes, or wiped out in droughts or floods. In some cases, cereal-growing experiments might have had the unexpected effect of encouraging hunters to become herders instead. The experience of protecting cereal plots from grazing animals might have made them realize that they can control these animals, especially the smaller ones, like goats and sheep. It may have given them the idea that it would be best to kill only the males and work with their dogs to protect the females and their young from predators.

Cultures can evolve quickly because humans are always coming up with new ideas, and this means there's plenty of variation to select from and potentially pass on to the next generation. Which of the many cultural variants are adopted and which are abandoned isn't entirely random, but neither is it always the result of an informed or rational decision. And since the ways in which the adoption of a new variant will affect culture are unpredictable, anyone who imagines that we humans can comprehensively engineer our cultures is deluded. (Some of the most tragic periods in recent human history began with people thinking that they knew how to engineer their society to make it better.) People's ideas and efforts can and do influence cultural change, but the full effect of these influences can only be appreciated long after the changes have taken place. This doesn't mean that we shouldn't collectively or individually try to improve our cultures, but we need to understand that such initiatives are in the nature of experiments. They are as likely to fail as to succeed.

In short, cultural evolution is a lot more like the evolution of life than it is like intelligent design by a supernatural agent. Some of us like to think that recently developed social tools like science and democratically elected governments are increasing our ability to build a better world. But our abilities will grow faster if we recognize the limitations of these tools. And if we want to understand how our cultures evolve, it makes sense to look at what biologists have learned about how life evolves. There are some lessons to be learned from this evolution. Three of them are highlighted here.

Lesson one: Don't think of evolution as "progress"—but there is a trend

When people talk about "evolution" in everyday life, they usually mean "progress." CEOs proudly talk about the "evolution" of their business from a two-guys-in-a-garage start-up to a billion-dollar corporation. When people think that things have changed for the worse, they don't see the change as evolution. No one ever talks about the "evolution" of a city from a prosperous industrial hub to a wasteland of derelict factories, empty houses, and toxic soil. But it doesn't make sense to allow our ideas about what is better to excessively color our analysis of cultural change.

Led by Darwin, biologists have shown again and again that the evolution of living organisms shouldn't normally be thought of as progressive. Earthworms are no less adapted to their environment than the songbirds that eat them. And there is no reason to think that human beings are the most advanced or "highly evolved" animal. It's true that living things evolve to become better adapted to their environment, but this isn't "progress." Living things are part of an intricate ecological balance that is always changing. An adaptation that makes one kind of organism more successful may alter the balance in a way that reduces the effectiveness of the adaptation. Or it may favor an entirely different kind of adaptation. Because environments are always changing, there can be no progress toward being perfectly adapted.

If we want to understand cultural change, we must try to stand back from the culture that has shaped the way we think. This requires humility and a lot of self-examination. We may have very strong ideas about how things have changed (or should change) to make the world a better place. These ideas may be evidence-based and currently very popular, but the effect of any change will be complex. Just because many people would like to make their small part of the world a better place doesn't mean there will be "progress" toward a "better" world. All we know is that there will always be change. Events will take place and change will happen. In many cases, more can be gained from making the best of a situation than trying to bring radical change.

Even though we shouldn't regard it as "progress," we can see an obvious trend in both biological evolution and the evolution of human cultures. Over time, life on Earth and human cultures have both increased in complexity. As well as a huge range of single-celled organisms, life on Earth now includes many multicelled animals, plants, and fungi. And from a culture not much more complex than apes, human cultures have evolved to be massively complex. The increasing complexity in both life and culture is partly due to changing environments. For example, animals couldn't survive on the Earth until the atmosphere contained enough oxygen to support the metabolism of large organisms that move around. And human cultures couldn't become massively complex until the ice age ended and the climate became more stable.

The other important factor that allowed increased complexity to evolve was the appearance of what we might call "networking tools." Multicellular organisms aren't just a collection of cells that happen to live together; they're a network of cells that are constantly communicating and working together. The tools that make this possible evolved genetically. Random errors in DNA copying eventually produced genes that created substances that allowed simple cells to connect and live and reproduce together as parts of more complex cells. In time, other genes appeared that produced substances which caused groups of cells to communicate and work together more effectively. These networked groups of cells eventually evolved into animals, plants, and multicelled fungi.[19]

In the case of cultural evolution, more complexity became possible with the invention of new social tools that allowed people to become connected in new ways.

For many centuries, different tribes explored and learned about the habitats they lived in, but competition between the tribes prevented more complex cultures from forming. Information flow between different tribes was slow and haphazard. In our story, a tribe gains the knowledge of how to make acorns edible because a family adopts a child whose home had been destroyed. If human evolution was just a matter of natural selection on genes, such an adoption would make no sense, because unrelated children don't carry the genes of the people who are caring for them. Once you recognize the role that culture plays in human evolution, it's possible to see why the adoption of children is common in so many cultures. A cultural group benefits from the introduction of strong and healthy new members, especially since they may also bring knowledge and ideas. This isn't true in the city-like hives of bees and colonies of ants and termites, because it's the genes of these insects that create the "social tools" that allow individuals to communicate and work together. Members of social insect colonies are almost always closely related genetically. In the case of humans, the social tools are inherited culturally. Human groups are bound together because they're culturally related, and members don't need to be close biological kin. An adopted child can be brought up in the ways of its new family and become a useful member of the tribe.[20] Over the millennia, our ancestors culturally evolved social tools that created ways for families and tribes to interact safely and profitably. Allied tribes exchanged knowledge, ideas, and members, allowing their cultures to adapt more rapidly and become more complex.

As farming expertise grew and people began to build permanent settlements, new social rules were needed to recognize the effort invested in producing homes and crops, and to deal with increased population density. Once they began to live in settlements, it was a lot harder for our ancestors to move away from annoying neighbors. Rules and means of enforcing rules gradually emerged. The historical rulers who are most familiar to us today (Julius Cesar, Genghis Khan, etc.) are less known for their popularity than their greed, cruelty, and strong-arm tactics. A reputation for being "bad" probably enhanced a leader's effectiveness as a social tool. In some social environments, rulers are most useful if people fear them. With the help of an armed militia, rulers can keep the peace between families and protect them from lawless outsiders. If their enforcement of rules helped people maintain a functioning network, everyone within the network had the chance of a better life. Large projects that benefited everyone could be completed, like the building of a system of reservoirs and channels to irrigate land and allow more food to be grown. This gave families more options, and they could specialize. Only some of the families needed to farm. Others could develop skills and tools that allowed them to trade their labor for food. Some could make things, some could be miners, some could organize transport or trade, some could keep records, and so on.[21]

People could trade much more widely when, about 5,000 years ago, some populations developed ways of symbolizing the value of goods as "money." Money is a wonderful social tool, but it's only effective if everyone involved in trade believes the coins of the realm really are worth something and are willing to obey the rules that maintain this belief. It took a strong ruler with plenty of power and treasure to convince people that it was safe to participate in the earliest economies.

Economic benefits and skillful enforcement help keep networks together, but they function most smoothly when everyone involved has the same basic beliefs about how the world works and what life is for. This is difficult when networks include people from tribes who honor different ancestors, eat different food, enjoy different music, and have different ideas about the unseen worlds beyond everyday life. At some point a new kind of social tool evolved in some population that at least partially solved this problem. Stories were told and rituals performed that claimed to be more fundamental than those of the different tribes. People were urged to abandon the beliefs and world-view of the family they had grown up in. If they accepted new stories, learned to take part in the rituals, and obeyed the rules, they could be members of an overarching larger and more powerful tribe. This super-tribe may have honored a pantheon of gods that included all the small tribal ones. Or it may have acknowledged only one all-powerful deity. Rulers were often seen as having some sort of mandate from heaven. People may have been encouraged to see rulers as gods or the servants of the gods (or the one God).

The new overarching belief systems ("religions") added an emotional "secret sauce" that made it palatable to live and work with strangers from different families who had a different cultural background. Some systems taught that worshipers could be reborn and share a heavenly father who protects, punishes, and rewards his children like a good earthly father does. Cultural differences among members could be tolerated, even appreciated, because, through their shared religion, they were now recognized as having kinship based on important beliefs that they all shared, including beliefs about how people should treat one another. This allowed large numbers of families to trust one another, cooperate, and live in peace—at least for a time.

Over the years, rulers, realms, and religions came and went. All were vulnerable to plague, pestilence, famine, natural disasters, discontent among the citizenry, and attacks from marauding armies of outsiders. But there was a general trend toward increasing complexity.

Lesson two: Cultural curiosities and "runaway selection"

After Charles Darwin published his ideas about evolution by natural selection in 1859, he faced a lot of questions. Among the most difficult were about biological curiosities like the peacock's tail. Surely peacocks would be better adapted to their environment if they didn't have such huge gaudy tails. The explanation Darwin proposed in his book published in 1871[22] was "sexual selection." The evolution of the peacock's bright blue color and extravagant tail was, he said, driven by the mating preferences of pea hens. The evolution started sensibly enough—dowdy-looking males are likely to be sick or hungry, which suggests they aren't very well suited to the environment. Females who avoided them and chose to mate with a snazzier-looking male had chicks that were better adapted and more likely to survive. But this set up a positive feedback loop that was likely to run away. Males would look snazzier and snazzier until, after many generations, they were producing exaggerated characteristics like the peacock's tail.

Ronald Fisher, the early 20th-century statistician, explained the "runaway" mechanism in more detail. It happens because the chicks of a choosy female don't just inherit the qualities that made their dad healthy. They also inherit the precise flashy characteristics that made their mother like him. AND they inherit *her* tendency to be attracted by those characteristics. If their mother chose the male with the biggest and most elaborate tail, not only will her sons tend to have big tails, but her daughters will tend to be keen on big tails. When the daughters are grown up and ready to choose their mate, the males strutting their stuff in front of them will have, on average, bigger tails. The same will be true for the next generation, and the next, and so it will go on.[23]

Culture also provides some pretty spectacular curiosities—the pyramids of Egypt for example. Why did people living in the Nile Valley about 4,500 years ago spend so much time building these massive tombs for the pharaoh who ruled them? The answer may be a similar runaway mechanism. Positive feedback loops are also created as cultures evolve, particularly in the evolution of social tools. For example, community projects can be useful social tools because, when people work well together, they develop the habit of trusting one another. This creates a social environment in which small squabbles can be resolved before they develop into costly and time-consuming disputes. The completion of some community projects can also bring practical benefits. People might get together to build a bridge or dig a well. But the planning of obviously useful projects can cause rather than heal divisions in a community. For example, arguments may arise over where a well should be dug so that all families benefit equally.

Less controversy is generated by projects to provide benefits that transcend the practical values of everyday life. This may be why the people of the Nile Valley started building tombs that would allow their pharaoh to organize things in the afterlife. Building the tomb together and sharing a sense of pride in its magnificence would have generated community feelings, and these would have helped to maintain peace and prosperity, as well as give people a lot of hope that the afterlife would also be good. It wouldn't have taken much persuasion to convince people that the lifestyle they enjoyed, and hoped to continue to enjoy, depended on the building of tombs. Their children inherited this belief and so were ready to work on a project to build an even bigger tomb for the pharaoh's son. As long as the peace and prosperity lasted, the projects got more ambitious.[24]

Today's world is littered with the remains of great projects carried out by past communities—pyramids, temples, statues, stone circles, and so on. Part of the attraction of visiting these magnificent monuments is to try to imagine what was going through the minds of the people who built them. We (the authors of this book) can offer no insight into the details of what they were thinking, but we think it very likely that the cultural beliefs that inspired these monuments evolved through a runaway process. They are the cultural equivalent of a peacock's tail or the tall antlers of a stag. Like them, they are signs of the health and vigor of a community. Producing them helped to maintain this health and vigor. But the potential problems of cultural runaways are obvious. They divert people's efforts away from finding food and raising children. Being in the grip of a cultural evolutionary runaway can make a population vulnerable, and we see many examples of this in today's cultures, as we shall discuss in the next chapter.

Lesson three: It's all about the survival of copies

For decades people who told stories about human evolution described Neanderthals as an "extinct species of human." These stories had to be revised when it became possible to determine the DNA sequence of genetic material from their ancient bones. Buried within the genomes of most people living today are DNA sequences so similar to those found in the ancient Neanderthal bones that scientists had to conclude that they must be copies of Neanderthal genes. The best explanation for the presence of these sequences is that the people who have them inherited them from Neanderthal ancestors. There may be no Neanderthal populations left, but copies of most of the genes that were once carried by Neanderthals continue to survive scattered among modern human genes of all Eurasians.

Copies of cultural information can travel much more widely and rapidly than copies of genetic information. They can also change faster. But it's still all about copying. If one population invents something useful and someone from another population copies it, the idea can take on a life of its own. It can be picked up, tried, passed on to others, and changed many times as it travels from population to population. For example, humans all over the world coordinate their actions using the same way of organizing time, with sixty seconds to the minute, sixty minutes to the hour, and so on. Historians studying ancient civilizations have suggested that this way of dividing time was developed by people who lived over 5,000 years ago in the valleys of the Tigris and Euphrates Rivers that run through what is now Iraq. It is, of course, possible that these people picked up the idea from people living somewhere else.

The most systematic studies of the way that cultural information is copied, modified, and spread are done by researchers known as "historical linguists,"[25] who compare the words of a people's language and the way they pronounce those words. By the age of five a child has learned over 2,000 words, but the word sounds that children make aren't precise copies of the ones made by their parents. Children tend to copy the way other children speak, so pronunciation changes over the generations. In this chapter's story, the "fire demon" people and the acorn-eating people could understand one another because they were descended from a single population that had lived several generations earlier. They had inherited mostly the same vocabulary, but pronunciation had changed in both tribes. If a dispute between factions of a population gets so bad that the sides stop talking to one another, it will only take a few generations for pronunciation to change enough for combatants to identify the enemy by the sound of their voice. After about 20 generations, they'll barely be able to understand each another.

But, even after thousands of years of separation, linguists will be able to detect similarities in vocabulary and grammar that reveal the languages to be "related"—descended from the same "parent" language that was spoken in the past. Historical linguists have become experts in identifying the shared ancestry of languages, and they have developed techniques for estimating how long ago the common parent of two related languages was spoken. They make minute comparisons of the pronunciation used by a wide sample of current speakers. This also allows them to develop plausible reconstructions of parent languages not spoken for thousands of years. In the case of some ancient parent languages, such as the spoken (or "vulgar") version of Latin, from which the "Romance" languages (French, Spanish, Italian, etc.) evolved, documents exist where the parent language is written down. Such ancient documents have both confirmed and helped to extend the techniques used to reconstruct ancient languages.

Of the more than 7,000 languages spoken today, a few dozen are known as "isolates," unrelated to any language so far discovered. But most of us speak a language that belongs to one of the main families. The Chinese languages are part of "Sino-Tibetan" family, which has more than 450 members. Arabic is one of the 350 plus languages of the "Afro-Asiatic" family. And over half the world speaks at least one of the 430 or so languages belonging to the "Indo-European" family, which includes the languages spoken from Iceland to Sri Lanka, from Vladivostok to Spain. It also includes English, which is spoken by over a billion people as a second language.

Rewriting history as new evidence comes in

For years people have looked with awe at the mysterious stone circles and standing stones dotted around European landscapes (see Figure 7.4). No doubt, many of the visitors to these sites have tried to imagine their ancestors digging gigantic holes, dragging huge stones into position, and carefully standing them up where they would still be proudly standing thousands of years later. Future visitors to these sites shouldn't bother to try to imagine their ancestors erecting these stones, because their ancestors probably weren't there. We now know that the men who built the stone circles have very few descendants living today.

Now that scientists can extract and analyze DNA fragments clinging to the insides of ancient bones, it's possible to learn much more about the thousands of skeletons that have been found in archaeological digs all over the world. Geneticists can't "decode" the DNA of these long dead people and give us a picture of what they were like—at least not yet. But they can discover how these ancient people were related to each other and to us. As this new evidence builds up, some of it is bound to raise questions about the accuracy of stories that people have long believed about their ancestry. Developing stories that are consistent with the new evidence may turn out to be politically fraught, but if people want stories of their ancestors to be based on evidence, histories will have to be adjusted as new evidence is presented. Ideally, new stories will be developed slowly, by scholars who take account of evidence from many sources and keep in mind how cultures evolve.

The techniques for extracting and analyzing DNA from ancient bones were mostly developed in Europe, and people of European descent are the first to have some of their beliefs about their ancestry severely shaken up.[26] For instance, there will have to be a major rewrite of the chapter of European history often referred to as the "Bronze Age," which started a little over 5,000 years ago and lasted for a little more than 2,000 years. This was a time when knowledge of how to smelt metals was developing and spreading around Eurasia. In the Bronze Age, the best tools and weapons were made of bronze. The major

Figure 7.4 One of the authors (Pete Richerson) standing between two of the 100+ stones that were erected more than 5,000 years ago in what is now the English village of Avebury. It was once assumed that the stones were placed here by men who were the ancestors of today's Europeans. Studies of ancient DNA have made it possible to test assumptions like this. It's now believed that the men who built the many stone circles of Europe have very few descendants living today.
Photo credit: Lesley Newson.

shake-up of Bronze Age history seems to have nothing to do with the smelting of metals, but rather with the way people used the land and moved around. Early ideas for how the history of this age should be rewritten are already stirring controversy,[27] and the story of this controversy provides an example of how evidence and lessons from evolution can help in the development of historical stories.

DNA has now been analyzed from the remains of hundreds of people who lived in different times and places in Europe. The findings from the bones of people who lived earlier than 5,000 years ago were, in some respects, not surprising. They suggest that the population of hunter-gatherers living in Europe increased as the world warmed and the ice retreated. Then, around 9,000 years ago, people from a different population began to migrate into Europe. They were the descendants of families that had developed cereal farming in the region east of the Mediterranean. This population gradually moved across Europe as their population grew and they cultivated more and more land. Before long, there was mixing between the two populations. By 5,000 years ago, the chromosomes of people who lived in Europe carried a mixture of the DNA sequences characteristic of two populations: the hunter-gatherers who had long been in Europe, and the more recently arrived Near Eastern farmers.

What *was* surprising about these results is that the DNA of these people looked very different from that of today's Europeans. It's not that their genes are very different. The *instructions* coded in their DNA are pretty much the same, but there are tiny differences in parts of the DNA that don't affect the information carried in the genetic instructions. It's as if the same instructions are being spoken by a different voice. The chromosomes of today's Europeans have sections of DNA "spoken in the voice" of the ancient hunter-gatherers, and they also have sections carried by the "voice" of the Near Eastern farmers. But in almost every person of European descent, there are also large sections of DNA spoken in a third "voice." This suggests that a substantial chunk of their genetic material is inherited from a *third* population—a population that didn't live in Europe 5,000 years ago. DNA carrying the voice of this new population only appears in the bones of people who died less than 5,000 years ago, but within about a thousand years, copies of this population's DNA were a big part of the chromosomes of almost everyone living in Europe.[28] The only explanation for this is that masses of new people moved into Europe starting around 5,000 years ago and admixed with and substantially replaced a lot of the people who were already living there.

Archaeologists who had spent their careers digging up evidence and developing stories about Europe's past had found nothing to make them suspect that such a huge population turnover had taken place. They had observed that the building of stone circles had stopped abruptly and that there were changes in farming and burial practices, and in the shape and decoration on the broken clay pots they dug up. But such cultural shifts happen all the time and could have simply been due to changes in fashion and ideas. The physical stuff the archaeologists had found gave little sign that there had been a massive migration. But historical linguists were finding plenty of evidence of the people who had moved into Europe.[29]

As mentioned above, Indo-European is the largest language family. More than half the world's population speaks an Indo-European language, and this includes most people of European descent. It now seems likely that the people who began migrating to Europe 5,000 years ago were speakers of the ancient language that historical linguists call "Proto-Indo-European," or "PIE." PIE is the ancestor of all Indo-European languages (see Figure 7.5). The people

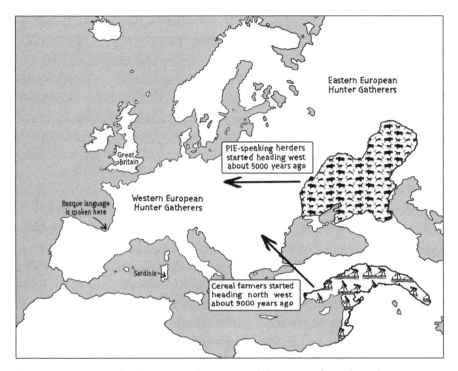

Figure 7.5 Most people of European descent speak languages from the Indo-European language family. These languages evolved from the "Proto-Indo-European" (or "PIE") language, which was probably spoken by the herders who migrated into Europe from the East.

Genetic studies show that the ancestors of Europeans are from three different populations: (1) Western European hunter-gathers, who had lived in Europe since before the end of the last ice age; (2) cereal farmers from the Near East who started migrating into Europe about 9,000 years ago and had made it across Europe to Great Britain by about 6,000 years ago; and (3) PIE-speaking herders who started migrating into Europe about 5,000 years ago and made it to Great Britain about 500 years later. The herders are partly descended from the Eastern European hunter-gatherers who are related to the mammoth steppe hunters such as those whose bodies were found at Sunghir.

It appears that when the herders spread across Europe, few of them made it to the Italian island of Sardinia. People from Sardinian families have DNA very similar to the DNA found in remains of people who died before the herders came to Europe. And it may be that the Basque language, which isn't an Indo-European language, is descended from one of the languages spoken in Europe before the PIE speakers arrived.

who spoke it didn't develop a way of writing their language down, but those who spoke some of the ancient PIE "daughter languages," such as Latin and Sanskrit, left lots of scrolls and tablets carrying samples of their language. By comparing the PIE daughter languages, the 300-plus languages that are spoken today, and the earlier PIE daughter languages that were written down, linguists were able to reconstruct parts of the PIE vocabulary. If most of the daughter languages had roughly the same word for something, they reckoned that they had inherited that word from their common ancestor language. Eventually, the linguists' knowledge of PIE had grown to the point that they could almost hold conversations in it. They couldn't hold real conversations, because a vocabulary from thousands of years ago lacks words for most of the things people are interested in today. Even so, the historical linguists reckoned they knew PIE well enough to gain an insight into the lives of the people who spoke it.

Many archaeologists and historians thought it most likely that PIE was spoken by early cereal farmers who migrated into Europe from the Near East. The historical linguists disagreed. They pointed out that PIE speakers didn't have the vocabulary of a Near Eastern farmer. They didn't have words for *palm tree*, *olive*, or *monkey*, which would surely be familiar to people living in the Near East. On the other hand, PIE did have words for *birch tree*, *salmon*, and *beaver*, which people in the Near East would never see. Also, the PIE vocabulary would have been more useful to livestock breeders than cereal growers. There were words for *bull*, *cow*, *ram*, *ewe*, *lamb*, *pig*, and *piglet*. Words in their vocabulary suggest that they didn't just raise animals in order to eat them. They could talk about *wool* and *shearing*. They probably milked some of their animals, because they had words for *sour milk* and *curds* and *whey*. And they had a word for *saddle*, which suggests they rode horses.

The DNA evidence resolved the argument over who had spoken PIE. The historical linguists were right. As well as inheriting a large chunk of their DNA from the people who moved into Europe 5,000 years ago, most Europeans also inherited their language.[30] Further investigation has revealed that the historical linguists were also right about where the PIE-speaking immigrants had come from and how they had lived. They were from an area that is now part of Ukraine and southwestern Russia, stretching from north of the Black Sea to north of the Caspian Sea. It was a grassland or steppe landscape, but people living here would have seen birch trees, and also salmon and beaver. They were herders and breeders of livestock, who sheared and milked their animals and rode horses. They're often referred to as "The Yamnaya," but this isn't what they called themselves. The name was given to them by the Russian archaeologists who first investigated sites where they had lived. *Yama* is the Russian word for "pit," and it refers to the way their dead were buried—in a pit covered by a burial mound. Analysis of DNA from Yamnaya skeletons indicates that some of their ancestors were from the same population as the very ancient skeletons buried during the ice age at Sunghir and other places in Russia. It is mostly through them that modern Europeans are related to these ancient ice age mammoth steppe hunters.

The DNA evidence can't tell us why there was a big population turnover in Bronze Age Europe, but it does provide one more clue. It's possible to see that the DNA from the new population was mostly coming from the male side. Not long after their arrival, new generations of Europeans were mostly being fathered by immigrant men and their male descendants. The immigrant males had largely replaced the males from the population that had long lived in Europe.

It's perhaps not surprising that among the first stories being proposed to explain this population turnover were tales of violence and bloodshed. It's possible to imagine hordes of eastern invaders sitting astride their horses wielding bronze swords or stone axes. The peaceful cereal growers may have performed ceremonies in their stone circles attempting to persuade their gods to help, but it was to no avail. The men were massacred, and the women were captured and raped by the murderous intruders.[31] The problem with this story is a lack of evidence. Archaeologists are good at recognizing signs of violent death in the human remains they find. If the arrival of the immigrants had brought genocide, archaeologists should have seen evidence of more violent deaths at the time of their arrival.

Whether the takeover was violent or relatively peaceful, it happened rapidly, because the immigrants arrived at a time when the people already living in Europe were laid low and vulnerable. Archaeologists have found plenty of evidence that cereal growing was not going well in many places. Their crops seem to have produced good yields when they first arrived in an area, because their population increased for a while. But then the communities started to decline. Perhaps their crops started failing due to depletion of the soils, a change in the climate, or plant diseases and pests.[32] Another source of vulnerability may have been epidemic disease. DNA from the bacteria that causes plague has been found associated with skeletons of young adults who died around the time the immigrants arrived.[33]

The people living in Europe may have also had beliefs and cultural practices that made them vulnerable. The social tools that were helping to reduce conflict during the hard times may have entered a "runaway" spiral that ended up preventing them from trying to develop practical solutions to the problems they faced. If people were responding to hard times by performing more complex ceremonies and building larger and more elaborate stone circles, they would have had less time to help find ways of providing food for children.

The male newcomers may have been better providers. If they had expertise in the breeding and handling of livestock, they could have taken over the depleted farmlands and used them as pastures. Land that was rested and grazed would have eventually become fertile again. If the culture of these immigrants encouraged them to be responsible hard-working fathers, many women may have chosen to have them as husbands rather than men of their own culture. They may have seen that if they had children with these men, their children would be more likely to survive. Perhaps the children of the herders received milk from their cows and were wrapped in snuggly woolen blankets made from the fleeces of their sheep.

It is, of course, impossible to know what happened, but we shouldn't simply assume a story in which women are no more than pawns in conflicts between men. The genetic evidence shows that women from the population that had been living in Europe had had children with the immigrant men, and these children survived to be the ancestors of today's Europeans. Women had to work hard to raise these children, and it's perfectly possible that they realized that their work would be easier and more rewarding with the help of the PIE-speaking newcomers. Their children learned about herding animals, but their mothers also probably taught them a bit about growing crops. And these children, with their mixed cultural inheritance, grew up speaking PIE.

South-central Asia is also populated with people who speak languages descended from PIE. Are these people also partly descended from immigrants who came from Yamnaya populations? More research will shed light on this, and also on the ancestries of peoples living in other parts of the world. Some of the evidence being collected will raise questions about the accuracy of long-accepted legends and inspire new stories. If scholars want to be guided by evolutionary theory as they argue over how history should be rewritten, they must keep in mind that the foundation of all stories of human evolution isn't one of warfare. It's a story about people having children and children surviving.

From an endangered species to a very successful one

When the world began to warm up 20,000 years ago, humans were very rare animals. There were probably fewer than a million members of our species on the whole planet, scattered around the habitable regions of Eurasia, Africa, and Australia. After 10,000 years of warming, much more of the Earth's land surface was productive enough to support humans. People had colonized

North and South America. But there still may have been no more than two million humans on the whole planet.

Gradually, our ancestors discovered new and better ways of making a living. New knowledge became part of the culture of populations and was passed down the generations. The evolution of new social tools increased the complexity of networks and the connectedness of populations. Exchanging cultural information became easier and the speed of cultural change accelerated. This made our ancestors more successful in biological terms—they could extract resources from the environment more effectively, and this allowed the size of the human population to increase. Larger populations needed new social tools, and these tools increased connectedness and complexity even more. This further accelerated the speed of cultural change. By 4,000 years ago, the middle of the Bronze Age, humans were no longer rare. The population was probably reaching 100 million about 3,000 years ago, as the Zhou dynasty began rule over people in the Yellow River valley in China. By 200 years ago, there were over a billion of us (see Figure 7.6).[34]

Not all populations were able to develop the same level of connectedness. In the fertile valleys of large rivers, populations could reach a high density, making it relatively easy for peoples to connect. Families living in less productive habitats were relatively isolated, so their cultures remained less complex. Connecting up was also difficult in mountainous areas like the

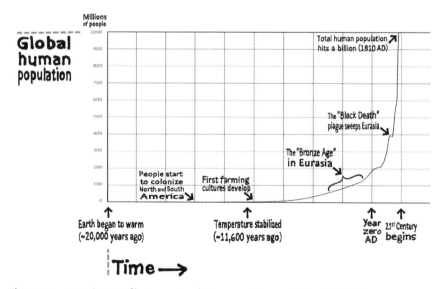

Figure 7.6 An estimate of human population growth over the last 20,000 years.

Melanesian island of New Guinea or the Balkans of Europe. The geography provided each tribe with defensible boundaries, so they tended to fight and compete for resources rather than working together to find ways of making their habitat more productive. In general, populations evolved the social tools that worked best in the short term to allow families to survive in the environment where they lived.

8

Another Transformation

Modern Times

Human populations adapt their behavior as conditions change. That's what's useful about culture. It evolves. But what we're going through now is more than just normal cultural evolution. Human life is being transformed again, as it was during the last ice age. Thanks to the ice age transformation, our ancestors evolved a much more complex culture and gained the capacity to adapt to sudden and dramatic changes in the climate. The transformation we're experiencing now is just as profound, or perhaps even more profound.

In chapter 6, we argued that the ice age transformation occurred because our ancestors developed new social tools that allowed them to create networks of many connected families that were spread out over a wide area. After the ice age ended, this new capacity to connect allowed networks to expand as populations grew and moved around, although networks would also contract if conditions changed and connections were broken. Even so, over thousands of years, the trend was toward increasing connectedness and cultural complexity. In the last couple of thousand years, the increase happened at a faster and faster rate.

In a way, today's world can be seen as the latest stage in the process described in chapter 7. The human network has become global, and people are inventing many new ways to connect. But it's more than this. Human networks aren't just growing—they're changing shape as well. In this final chapter we'll explain what we mean by "changing shape" and why this change in our information networks is transforming the way humans think about our lives, ourselves, and each other. European populations began the transformation a few centuries ago. Their descendants now live in many parts of the world. They include those who call themselves "Westerners," and also the people descended from those who lived in Eastern Europe. People of European descent continue to experience dramatic and rapid change, suggesting that, even for those who began the transformation first, it's far from over.

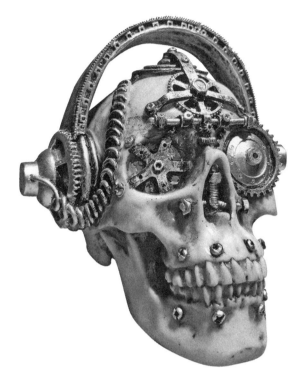

Figure 8.1 Steampunk skull cyborg sculpture, made of polyresin and hand painted, is available from Atlantic Collectibles.
Picture credit: Lesley Newson

Not long after the transformation started in Europe, similar changes began in other parts of the world. Today almost all populations are affected. Let's call this transformation "modernization." What we think of as "modern" changes all the time, so modernization is a good word for this process.[1] An account of the lives of two women born in the middle of the 20th century illustrates the speed and direction of cultural change that Westerners have experienced.

Part one

Even though The War had ended several years before they were born, Ruth and Rebecca remember that grown-ups talked all the time about what it was like during The War. Their father, who had actually fought in The War, wouldn't tell them anything about the fighting. He just said that he was lucky and that being in the navy had taught him how to be a plumber.

Ruth and Rebecca were twins. They weren't identical. Ruth had dark brown hair that was straight, while Becca's was lighter and slightly curly. But their mother's mother (whom they called Grandma) often told them that they were identical in all the important ways—they were both healthy, clever, and pretty. Their dad's mother (whom they called Nanny) talked about them getting a little brother. They remember thinking that they were getting a brother, but then their mother said there was no brother because she was happy with her two girls.

Their father's plumbing business was doing well, and just before the twins started school, they moved into a brand-new house on the edge of town. It was only a 10-minute walk from the school they would go to. The new house was so far from their old neighborhood that they couldn't see their grandparents every day. This made the twins and their grandparents sad, but it also meant they didn't have to dress up in uncomfortable clothes and spend a long time in church with their grandparents every Sunday. Instead they went to a church with a Sunday school for the children, and that was quite fun. The twins also liked going to regular school. They didn't think the schoolwork was hard, and because they had each other, they never felt lonely or left out. Even though they often argued, they were best friends.

The family got a television when the girls were six. They only watched it the evening with the whole family sitting together because their parents believed people should be busy during the day. Their mother was always doing housework or helping with their dad's plumbing business by answering the phone, opening letters, and sending out bills. Their dad said that girls ought to be given jobs to do to help around the house, but their mother said the best way they could help was to keep out of her way. So the twins kept themselves busy by playing. After school, they mostly played in the neighborhood with the other kids if the weather was good, skipping, playing ball games or hide and seek—all the usual things. On weekends and during the holidays, they had more time so they could have adventures. Every year, a new street of houses was built in their neighborhood, but the workmen were only there during the week, so on weekends the kids could climb around in the partially built houses, imagining what the rooms would be like and seeing how the building progressed week by week. They knew that playing on building sites was dangerous and that if one of them had an accident or caused any damage they would be in trouble, so they were careful, and the older kids shouted at the young ones if they saw them doing anything silly or dangerous.

For their ninth birthday, the twins got bikes, which allowed them to go further away from home. They sometimes went to town to do errands or get books from the library, but on sunny days they usually rode out into the country. Once they came across some loose dairy cows trotting down the road and helped the farmer turn them around and shoo them back into their pasture. After that they often visited the farm and spent a happy afternoon helping him move cattle around, feed calves, stack hay bales, and do other farm chores. One day, they watched a bull "servicing" a cow, and the farmer told them it was so she would have a calf. They would have liked to know more about this, but they knew that grown-ups didn't want to talk about it. They didn't tell their parents about any of their adventures.

When the weather was too bad for them to play outside, they read books about kids catching robbers, smugglers, and spies. They longed to have adventures like these. Their neighborhood didn't have any obvious criminals, but their farmer friend often complained about the people who lived near him in a big old house he called "the Beecham Place." The house was hidden behind trees and surrounded by a tall fence. Becca thought it likely that these people were criminals. One day, instead of going to the farm, they decided to investigate the Beecham Place. The fence was too high to climb, but they found a spot where animals had dug an opening under the fence. It only took a bit more digging to make a hole big enough to wriggle through.

As they crept through the trees, they saw more and more of a huge and beautiful building. It looked like a palace, with windows taller than a man. Some of the upstairs windows were doors onto a balcony. There were colorful flower beds all around the house and vines growing up the walls. The twins crawled under a bush, which allowed them to get a good view without (they hoped) being seen themselves. At first, they just stared at the house, pointing things out to each other, like, "Wouldn't it be wonderful sit in the window of the tower on a stormy day?" and "Do you think there might be secret passages or ghosts?" They couldn't wait to get a look at the people who lived there, but there was no one around. No one entered or left the house for what seemed like ages. Becca decided that the criminals must be out, which meant it was OK to creep up to the house and look in the windows. Ruth refused, saying they must stay hidden. So the girls continued to sit there until they got bored and went home. Becca often talked about going back and doing some more spying, but Ruth refused. Becca didn't want to go on her own.

Shortly after the twins became teenagers, everything changed. They no longer had the same energy and ambition, and they became worried that boys would call them "tomboys." They also began to realize how important it was to learn about the clothes and makeup they should wear. Becca got her period first, but Ruth was only a few months behind. They were both unhappy a lot of the time during their teenage years, partly because, instead of supporting each other, they argued, insulted, and taunted one another. The feelings of trust and friendship they had had as children were gone. They had different interests and made different friends at school.

School work continued to be easy for them though, and their teachers urged them to think about careers. Their parents weren't so sure about girls having careers, but Becca convinced them that a business administration course would give her practical skills so that, once she was married, she could help in her husband's business like their mother did. The family couldn't afford to support both girls at university, but Ruth wanted to stay home and be near her current boyfriend. So Becca went to university and Ruth took some business classes at the local college. Their mother kept track of every penny they spent on Becca while she was at university so they could give the same amount to Ruth later (with interest), when they could afford it. (*To be continued.*)

Growing up in a changed and changing culture

By the middle of the 20th century, the cultural transformation was well underway in European populations, including the populations mostly made up of the Europeans who had migrated to the Americas, Australia, and other parts of the world. Modernization had also begun in populations in the Far East. Ruth and Rebecca grew up in a time of rapidly accelerating cultural change.

Ever since the ice age, humans had grown up and lived as part of a family, a tribe, and a local community. But starting in the 1700s, more and more Europeans saw themselves as being part of a "nation." This growing "nationalism" increased the scale of conflicts between groups. From 1914 to 1918, most of the European nations were caught up in a terrible war. Millions of men put on a uniform and fought for their country. It's estimated that about 20 million soldiers and another 20 million civilians died in this war. From 1939 to 1945 another massive war was fought, which they called the Second World War, or World War II. It involved many more nations and fighting took place over a much larger portion of the world. The death toll was nearly twice that of the earlier war and far more civilians died than soldiers. Survivors experienced the war in many different ways, but the shared experiences of wartime destruction and the rebuilding that followed had a profound effect on the culture of all the populations touched by the war. The parents of Ruth and Rebecca's generation feared that their children would also have to fight, and perhaps die, in a war.

As the importance of nation had grown, the importance of family had gradually diminished. By the end of this second "world war," the families that made up Western populations were smaller and more dispersed. Relatives often lived far apart, and many people saw a "typical family" as a household containing just a mother, father, and their children. This tiny "nuclear family" didn't spend much time together. Adults went out to work and children went to school. It was still common for mothers to stay home and work as "homemakers," but "the home" was seen to be a place where people slept and relaxed rather than a place of work. Family members left the home to be with non-family members whom they worked, played, and learned with. People no longer felt the same obligation to their wider kin. They might have been willing to help their relatives out in times of trouble, but if some family members were very prosperous, they were only expected to share their good fortune with their own children.

As people's obligation to their kin diminished, their obligation to their state increased. They were expected to take part in discussions about how they would be governed and to elect a government representative. They had to pay taxes to help support the government, and if the government declared war, young men were expected to fight. Government departments increasingly took over family obligations. If the black sheep of the family got into trouble, it would be up to the government to punish him, and the family might not feel any obligation to help out his long-suffering wife and children. They would have to apply to the government for help.

Over time, religious organizations also changed. As family members became more dispersed and interacted with a wider range of people, they were exposed to different religious beliefs and often started to question what they were taught as children. The idea that belief should be based on evidence rather than faith is useful in modernizing populations because it can help people find common ground on which to build shared beliefs, even if they come from different family and cultural backgrounds. It doesn't always help, however, because it's impossible to absorb even a tiny fraction of the information available on a complex issue. Usually, people just tend to go along with their friends or match their opinions to someone they regard as an expert. People from different backgrounds often put their trust in different experts and different sources of evidence. And many religious people believe that they're consulting evidence when they read their sacred texts.

The shrinking and dispersal of "the family" coevolved with the development of new ways of providing the services that had previously been the responsibility of families. Over time, this brought many improvements. Thanks to modern medicine, we're much healthier than we would be if we only had "old family remedies." Trained educators can provide a much wider range of knowledge and skills than we could ever get by observing our family elders. We have a more reliable, varied, and convenient food supply, and we live in more hygienic environments.

The transformation has made some aspects of childrearing much easier, but raising children is more than just a matter of feeding and protecting them. Humans evolved to grow up and live as part of a family, and by diminishing the family, the cultural transformation has changed childhood. And it keeps on changing. Since the transformation began, each generation of children has grown up in a world that is different from the world their parents grew up in. This is important, given that the experiences we have when we're young help to shape our emotional and neurological development. No one knows what kinds of experiences might help children develop into adults who will thrive in the modern world. Before the transformation, the details of family life varied enormously, but the basics were the same. Premodern families were teams made up of people of different ages. Children were expected to be part of the family team, and once they were able, they were expected to contribute and provide care for other family members who needed it. Youngsters who didn't fit into the team had a poor chance of survival.

By the middle of the 20th century, Western families were no longer functioning as teams. Children of Ruth and Rebecca's generation were sometimes given household chores and expected to care for their younger siblings, but they were freed from the need to help find, produce, and process food. They no longer labored. Some of their time was spent in adult-supervised education and games, but the rest was playtime, which they were expected to organize themselves. They spent time with children from other families and organized play activities, often using social tools such as games that allowed them to compete without doing much harm to each other. They consumed knowledge and entertainment provided by adults and used their imagination to turn it into something that seemed to make sense. Time and experience brought a better understanding, and many children of this generation reached adulthood reasonably prepared for the changing world they would live in. Judging by their social media posts, many people who grew up in the middle of the 20th century believe their childhood experiences were much better than the experiences of children today.

As they approached adulthood, this "postwar" generation of Western children was the first to be faced with a range of choices about what their future would be like. They were taught that their future depended on their own performance and the decisions they made. They didn't have to be limited by their family background. Family background continued to matter, as it does today. But young people were increasingly able pursue a kind of life their parents wouldn't have dreamed of. The widening of choice was experienced most profoundly by the female half of the population. Their mothers had mostly believed that it was "normal" and "natural" to want to be mothers and to devote a large part of their efforts to making a home for their husband and family. Since the 1960s, more and more girls have been encouraged to believe that they should be considering other options.

Part two

Becca's time at university was expensive for her family because she chose to go to a university in a large city. She loved city life, and even before she'd graduated, she had a job lined up in an advertising agency. The creative side of advertising fascinated her, she told them in the job interview, but she was more interested in managing and growing a business—getting new clients, finding creative talent, and nurturing it. At first, her salary was barely large enough to cover her food and pay the rent on a tiny shared apartment. But her years as a student had taught her to be frugal.

Back at home, Ruth's expenses were lower—she still lived with their parents—but her chances of getting a good job were also lower. She was hired by a dentist in the town to do his office administration and some reception work. At the Christmas party held at the dentist's home, she met his son Paul and they started dating. He'd been to university and was now studying to be an accountant. He said that he hated the idea of being an accountant but didn't know what else to do. It rather annoyed Ruth that he didn't appreciate the opportunities he was being given. But he was sweet and funny, and he obviously loved her. Ruth couldn't help loving him back. His father joked to Ruth that he was sure that she could make something out of him. Ruth told him that she would keep working on him. She turned down his first marriage proposal because, at 23, she just wasn't ready to make up her mind.

Their grandad's heart attack was totally unexpected. Becca came back to attend the funeral, and she and Ruth held hands during the service, feeling closer than they had for years. Afterward, they and their parents talked about Grandma. She had become very forgetful and couldn't cope on her own, so Ruth and Becca worked together to clear out Becca's old bedroom so that Grandma could move in. Having Grandma live with them was a nightmare from day one. She kept forgetting that Grandad was dead and often demanded to be allowed to go home to him. Their mother had trouble keeping an eye on her and still manage to do all the shopping, cooking, cleaning, and paperwork for the plumbing business. A couple of times, Grandma escaped and had to be brought back in a police car. The doctor gave her what he called tranquillizers to make her calmer. But they made her so dopey that she was unsteady on her feet and needed help getting washed, dressed, and going to the toilet. Ruth had to do the shopping and spent many evenings doing the work for her dad's business that her mother couldn't get done because she was too busy with Grandma. However hard she worked, it was never enough. The house smelled bad.

Becca heard all about Grandma's problems during her long phone calls home. She felt guilty not being there to help but she didn't even have a bed at home anymore. And her own work kept her very busy. The agency had given her more responsibilities and raised her salary. The woman she shared her apartment with was engaged and planning to move out soon. Becca wondered if she should look for another person to share the apartment with or see if she could afford the rent on her own. She also wondered if she would ever find a boyfriend. Men at work were always flirting with her, but she was used to men wanting to sleep with her. She'd had sex with a couple of guys at university but always regretted it afterward. (What would her parents think if they knew she wasn't a virgin?!) She had male friends, but they were all attached, mostly to her girlfriends. She knew that women today didn't need to get married and have a family, but she had always imagined that one day she would. She was 25 already!

When Ruth phoned Becca to tell her about Grandma's fall, she didn't mention that Paul had proposed to her again and that she had said "yes." She told her at Grandma's funeral two weeks later. Grandma had broken her hip and, while in the hospital, she'd caught the flu, which soon turned to pneumonia. She had died in her sleep. Everyone agreed that her death was a blessing, both for her and the family. Paul and his family were at the funeral, and Becca got to meet them all. She decided that Paul wasn't as boring as she had feared. She saw his loveable side and decided he would be a perfect husband for Ruth.

The other "blessing" from Grandma's death was the inheritance. She and Grandad had put aside quite a bit of money and their mother inherited half of it. She could now afford to compensate Ruth for staying home while Becca went to university. Ruth suggested to Paul that they use the money to buy a business they could run together. He thought that was a great idea and his parents said they would give the couple a similar lump sum as a wedding present. Ruth and Paul set about researching ideas for businesses and the additional financing they would need. They ended up investing in a McDonald's franchise. Their town didn't have a McDonald's, and these new food places seemed to make money everywhere they opened. Shortly after their wedding, their restaurant-cum-hamburger stand was nearly built, and they were interviewing for staff. Paul finally discovered what he was good at. People loved working for him, and he and Ruth encouraged the employees to feel they were part of a family. The business did even better than they had hoped, and they could raise salaries and pay bonuses to reward loyalty.

Becca might have felt jealous of her sister if she hadn't fallen in love at last. She'd known David for a while but had dismissed him at first. She saw him as just another good-looking young guy who wanted to make money taking photographs. He was wittier and more charming than most, but she had found that irritating. Photographers need to be judged by the pictures they take, not by their personality. It was seeing David's pictures that changed her mind. And seeing him work made her realize that his personal charm was part of his artistry. He used it to animate the faces of his models and capture the kind of expressions that made images that attracted the eye. His magic worked with male models, children, and animals, but it worked especially well with women. He made it look effortless, but that was because he was a good craftsman. His skill at lighting could make a plate of spinach look sexy.

When Becca began to talk to him about his work and career plans, her admiration was informed and genuine, not flirtatious. She wanted to know him better professionally because she was getting tired of working at the advertising agency and thought it was time to start her own business, marketing and managing talented photographers like David. They discussed this over several long and rather boozy dinners.

Looking back a few years later, Becca decided that their relationship turned into a romance and later a marriage because David couldn't trust a woman he wasn't sleeping with. He wanted there to be love between them, so he turned his charm on Becca and she fell in love. Once she was his wife, he felt free to need her. With *her* skill and talent, she could create a business around *his* skill and talent, leaving him free to do what he wanted. Becca and David were much better business partners than marriage partners.

Ruth and Paul's son, Thomas, was born 18 months after their marriage. He was the first grandchild of both their parents, which gave him a special status. The restaurant staff loved him too. This time Becca *was* jealous of her sister and readily admitted it. She and David had been married for eight months, but both agreed that it would be several years before they could even think about being parents. Ruth became pregnant again when Thomas was two and a half. A few months into the pregnancy, her doctor recommended that she have a new kind of test called an "ultrasound scan." He explained that the baby's heart sounds were unusual and, given her family history, this probably meant she was carrying twins and he was hearing two heartbeats. A scan would tell them for sure, because it would make it possible to see the baby or babies in her womb. Ruth and Paul left Thomas with his Grandma and drove for several hours to the university hospital to get an advanced peak at what turned out to be two babies. It was magical to see their tiny twitching bodies and little hearts beating.

Then, 11 weeks and two days later, Ruth went into labor and gave birth to two dead babies—two little girls, just like her and Becca. She was inconsolable. She felt guilty that she hadn't been able to keep the babies alive, and even more guilty that her grief was getting in the way of her love for Thomas. She knew that she should be grateful that she had a beautiful, healthy little boy, but she could only think about the two dead babies that she could do nothing for. She tried to be the sort of mother he needed, but she was sure that he could tell that she was empty inside. Paul and her mother were wonderful. Her mother confided that, when she and Becca were three, she'd had a very difficult labor and the baby—a boy—had died. Everyone had expected her to just put it behind her and get pregnant again. It was because she couldn't put it behind her that Ruth and Becca had no younger siblings, even though Nanny had very much wanted a grandson. One of the reasons they'd moved to their new house was to get away from Nanny's nagging.

Ruth was grateful that her friends and family understood and were willing to give her time. She started to feel herself again three months later when Becca came to spend the weekend with them. It was listening to Becca's problems rather than talking about her own that helped Ruth most. She wasn't surprised to hear that David had "discovered" himself unable to be faithful to one woman, but she was puzzled at Becca's attitude. Becca clearly cared about David, but they didn't have what Ruth would call a "real marriage." Becca showed no sign of being hurt by his infidelities. She said she knew David wouldn't leave her because he needed her. The life he enjoyed was organized by her. The public profile, travel to exotic locations, parties, and the cash flow were all her creation. He admired her organizational talent just as she admired his artistic talent. Drinking and cocaine had become a big part of his life, and that worried Becca. But, so far, they weren't affecting his health or his work. The worst part of being married to David was, she said, their sex life. His most serious affair had been with a woman who had introduced him to bondage and discipline. This woman was no longer around, but she had left David with a taste for sex of what he called the "non-vanilla" variety. Becca didn't want to go into the details of what this meant, but she assured Ruth that she was OK and had only tried cocaine a few times and didn't like it. Then she burst into tears and said that she was desperate to feel a gentle hand on her body—to feel cuddled and cared for. Ruth and Becca hugged one another and cried for a full five minutes.

When Ruth said good-bye to Becca on Monday morning, she felt much stronger and started to get back to her normal routine of work and caring for her son. She had planned for Tom to start going to nursery school five mornings a week. This was partly to give her time to concentrate on her newborn twins, but also because people were now saying that children who attend preschool are better prepared for real school. She decided that she could use her free mornings looking at ways to expand their business. The local council had just announced a scheme to improve the downtown shopping district. They were looking for businesses to invest. Should they open a new McDonald's restaurant downtown? She tried to think of a new restaurant as being like a new baby.

Spending the weekend with Ruth had made Becca see her life in a new way. Describing her troubles made her realize what a trap she was in with David. They were joint owners of the agency that she had set up and developed. The agency paid them both good salaries and she was saving money from hers. But this income was trivial compared to what the business itself was now worth. The problem was that the value of the business came from David's reputation. Becca had helped to build this reputation, but it belonged to David. She had plenty of friends in the industry and was owed many favors, but they weren't really "her" friends. Everyone just saw her as the less glamorous half of the David-Becca partnership. The only people she could confide in about the problems were Ruth and a former client named Fiona, whom she'd become close to. Over coffee one day, Fiona had mentioned that David reminded her of her ex-husband. The conversations that followed led to them to forming a close friendship.

It took a while for Becca to feel ready to escape from the trap. She wanted the agency to be "established," and it was. It had been running well for four years. The staff was terrific, and they had work booked months ahead. Becca was still worried but, at the age of 33, she felt like she was running out of time to have a baby. She was sure she could combine her work with motherhood but was reluctant to talk about it with David. When Ruth phoned to tell her she was pregnant again, she made up her mind. She stopped using contraception and three months later, she had a positive pregnancy test. She had no idea how David would react to the news of approaching fatherhood, but she decided that she didn't care. It came as a shock when he simply saw their baby as a problem that could be solved by Becca having an abortion. He became frustrated and then angry when she refused to "get rid of it." He claimed to have no memory of the many conversations they had had about starting a family once they were established. Being established meant they could now enjoy life, he pleaded. Then he got angry again and told her that he liked life as it was, and if she wanted change, she could leave. That night Becca stayed with Fiona and the next day she made an appointment to see a lawyer. On the phone to Ruth, she was able to tell her the good news about her pregnancy at the same time as the news that she was getting divorced. Ruth privately thought the divorce was good news too.

Becca's daughter, Laura, was born five months after Ruth gave birth to her daughter, Heidi. The divorce lawyers eventually got David to agree to take out loans to buy out Becca's half-share of the business. By the time the money came through, Laura was 18 months old and Becca and Fiona had become a couple. They bought a house and raised Laura and Fiona's son, Adam, as brother and sister. Once Laura was ready for preschool, Becca was ready to start work again. She had come to despise the glitz and deception of most advertising. Instead, she started an agency that concentrated on publicizing the fundraising campaigns of nonprofit organizations.

Once same-sex marriage became legal, Becca and Fiona's relationship was formally recognized in a ceremony that everyone agreed was very romantic. Becca wasn't sure how her parents would receive Fiona as a daughter-in-law, but they were fine about it. Her mother said that "being married to a woman is better than being married to a man who wanted to kill my granddaughter." Their wedding was her father's last family event. He died five months later, and the twin's mother moved into an apartment in a senior community. (*To be continued.*)

Plotting a course on an ocean of choice

By giving us more choices, the cultural transformation we're going through may be amplifying the genetic differences between us. Ruth and Becca are sisters, which means that about half of their genes are identical. Differences in the other half partly explain why non-identical twins aren't identical. But the different experiences they have are also important. If people have a lot of paths to choose from and their genes influence the choices they make, then siblings are likely to grow further and further apart. The different paths they choose will mold their development. For example, if she had been born in an earlier generation, Becca wouldn't have been able to choose to move away and go to university in the city. She would have had to stay nearer home and her life would have been more like Ruth's.

It's also very possible that one, or both, of the twins wouldn't have survived to adulthood if they had been born in an earlier generation. The cultural transformation brought new technology that allowed us to surmount more and more of the limitations of our slow, weak, apish body. We can travel further and faster and communicate with people over long distances. We can survive many of the illnesses and accidents that cut short the lives of our ancestors. And we have more choice over how we tackle the age-old human problem of raising our difficult, large-brained offspring.

Having a baby is now far less risky for mothers. Before the transformation began, about 1 percent of pregnancies were fatal for the woman, either as a result of the pregnancy itself or the process of giving birth. By the middle of the 20th century, maternal mortality had dropped to less than one in 1,000 pregnancies in some Western countries. In wealthier countries today, it's close to one in 10,000. The risks of pregnancy are higher in poorer countries, but even in the poorest families, the odds of a women surviving pregnancy are considerably better than 99 percent.

The improvement in the survival of babies and children has been even more dramatic. Throughout our ancestors' time on Earth, child survival varied a great deal, depending on circumstances. But it was not uncommon for half the babies born in a population to die before reaching adulthood. Today, child survival is better than 90 percent in all but the very poorest families. In most wealthier countries, survival is better than 99.5 percent. New medical technology has given us new ways of controlling when, and if, we become pregnant. Contraception allows us to have sexual intercourse without conceiving a baby, and this potentially increases our choice of how often and with whom we have sex. Immunization, antibiotics, and antiviral medication reduce the risk of damage from the infections likely to spread when people's network of social and sexual interactions gets larger.

Many Westerners of Ruth and Rebecca's generation have not only survived into their seventies, but they're also in quite good health and anticipating enjoying another decade or more of life. Vaccinations protect them from the infections that had once killed a lot of elders, such as flu and pneumonia. Pneumonia was often referred to as "the old man's friend" because it usually brought people a gentle death just as they were becoming too frail to work and beginning to be a burden on their family. In the middle of the 20th century, it was still quite common in the West for elderly relatives to live with their children and be nursed by them, but this is now much rarer. Elder care is another family job that has been taken over by professionals.

People's beliefs and expectations are also changing. As a result, even though some aspects of life are now much easier, humans don't seem to be much happier. Objectively speaking, most of our ancestors lived more brutish, shorter, and less interesting lives than we do. But their culture provided them with beliefs that helped them cope with this. Their expectations were lower, so bad outcomes were less of a shock. What's more, populations shared culturally evolved beliefs, such as faith in a divine purpose and life after death, which helped them to continue to make the most of whatever life they had. The improved survival of babies that occurred during the 20th century caused couples to anticipate a happy outcome to their pregnancy, and this vastly increased the misery for the small number who lost their baby. The change in people's expectations of child survival was so rapid that beliefs and emotions couldn't keep up. Women such as Ruth and Rebecca's Nanny, who had grown up when the death of babies was common, often had trouble understanding the profound grief that younger women started to feel when they had a miscarriage or stillbirth. It took a generation or so for culture to evolve the understanding and to start providing women who had experienced this tragedy with the support and sympathy they needed.

So is it possible for humans to be happier and more content? Or might the transformation be making us feel worse? There are certainly plenty of people who fear that today's "unnatural" lifestyles are unsustainable, not just for the environment but for our emotional well-being as well. Some politicians and pundits argue that these fears are greatly exaggerated, pointing to vast amounts of data which, they claim, clearly show that things are getting better and better. So should we just try to concentrate on the bright side? Is this even humanly possible? Some psychologists have suggested that humans might have a hard time putting together an objective view of the conditions they face because evolution has programmed our brain to ignore good news and focus on problems and how they might get worse.[2]

Understanding our ancestors' lives gives us another explanation of people's continued discontent and, in some cases, utter misery with modern life. Our ancestors lived as part of a family and their security depended on being members of a family team. Feeling pain when they were isolated, ignored, and unvalued by their family improved the survival of both our ancestors and the families they belonged to. Experiencing these unpleasant emotions encourages people to work at being valuable to their family.[3] The role of family has changed as part of the transformation, but the feelings of wanting to belong and be valued are part of our genetic inheritance. They persist and cause us to feel miserable when we believe we don't fit in at school, or work, or with a group of people we admire. It can be hard to escape this misery because the way out isn't always obvious. The people we admire may be very different from us, so communication and trust are difficult. And the members of the groups we want to belong to don't necessarily care about us and may not have our best interests at heart.[4]

In the modern world, working out how to fit in can be a hard and sometimes impossible problem. Modern society seems to provide us with many choices, but we often can't work out which choice will put us in a situation that will lead to us being valued and supported. People are different and will be competent at doing different things. Very few children can grow up to be football stars or actors. And the choices that are best for us might not look good at first. Some people are only attracted to the options that aren't realistically available to them. Some are suffering the consequences of having made bad choices. And many feel alone, yearning for support and appreciation and not knowing how to get it. There are ways of coping. We can keep pets that appreciate us, or try to do things that attract "likes" on social media. Arts and entertainment can distract and inspire us. Therapists, drugs, and alcohol provide a kind of support. But modern cultures need to evolve more ways of preventing and alleviating the loneliness and mental health problems that this new way of life seems to generate.

Ruth and Rebecca had some hard times, but they were successful in the sense that both they and the people around them felt their lives were meaningful and useful. In terms of passing on copies of their genes, however, their lives weren't very successful. Ruth had two children, and Becca had only one biological offspring. For the people who are part of modern culture, life often feels competitive and stressful, but, in the basic biological sense, we aren't competing very hard. Our culture has evolved social tools and stories that encourage us to see a wide range of goals and activities as potentially important. Being a parent is simply one of the choices available to us.

Looked at in terms of the biological struggle to pass on our genes, most of the goals and activities we get stressed about have more in common with childhood games than real competition. Our cultures give us goals and activities that can make our lives meaningful, and they can also make us miserable.[5] Some misery is inevitable because we can't "win" all the time. But, ideally, we can do well enough to keep on enjoying the "game." Problems arise when people feel they can't achieve anything meaningful or when they lack the skills to put on a respectable performance. Modern cultures must strive to evolve ways of making all its members feel they are part of something worthwhile.

Modern cultures must also evolve better ways of dealing with the conflicts that can arise when people do choose to devote effort to producing the next generation. Becca's desire to be a mother conflicted with her husband's desire to continue to enjoy his child-free life, so the marriage had to end. She then had to find a way to get him to provide support for their daughter. Forming a partnership with another woman provided her daughter with a family. Over the history of our species, females have used their ingenuity to find ways of getting help to raise their children. They still do this today.

Part three

The twins turned 70 in the spring of 2020. They had planned a big family get-together to mark their birthday, but the restrictions put in place to slow the spread of COVID-19 made that impossible. Rather than trying to set up some kind of electronic gathering, they decided to postpone the reunion. As they were now officially elderly people, Ruth and Rebecca stayed at home with their spouses. They went out to exercise and do some shopping, but they got a lot of things delivered. They worried most about their mother, who was now in her 90s and still living independently in her small apartment. They feared she would feel lonely staying inside on her own. But she rapidly figured out how to use Skype and Zoom. She regularly "visited" with members of her family and some of her friends.

When the lockdown restrictions were finally lifted, they all met in Paul and Ruth's house. They laid on a lavish feast with the adults sitting around one table while the twins' six grandchildren had a table to themselves. Before they started eating, their mother gave a sweet speech, saying how proud she was of her daughters and all they had accomplished, especially in raising such a wonderful crop of children and grandchildren. She also told them that, as they were only 70, she's expecting to see many more accomplishments. And then she begged them to try to avoid talking about the pandemic for an hour or so. She said that she was now just tired of talking about it and how the world had changed.

After dessert, the children went off to spend time with their phones and tablets. Thomas told the adults about the results of the birthday present he had bought for his mother and aunt. He had collected saliva samples from them and sent them away to one of the genetic testing companies. He'd then spent quite a bit of time reading up on what the findings meant. The most interesting result, he told them, is that they have different versions of the DDR4 gene. He explained that this is thought to affect the balance of neurotransmitters in some parts of the brain. Becca's genome has the version associated with people who take more risky decisions, he said, while Ruth has the version common in people who are more cautious.

"So, is that why our mother left home and moved to the city and ended up marrying a woman while your mother stayed home and has a more conventional life?" Laura asked. "Is it all down to the difference in this one gene?"

"It's certainly true that Ruth was always the 'sensible' one," Becca said. "When we were kids, Ruth would always prevent me doing fun things that she thought were too dangerous."

"Like the time we broke into the Beecham place?" Ruth asked. The twins laughed as the family demanded to know more. Then they described some of their childhood escapades with only slight exaggerations.

"Grandma, weren't you worried about your daughters clambering about in building sites, herding cattle, and trespassing on the estates of well-heeled criminals?" Laura asked.

"No, I don't think I was worried," the old lady replied. "Thinking about it today, it does seem like I was a terrible mother, but it was just the normal thing to do back then. Children were supposed to play outside. You had to let them learn to look after themselves and each other."

"And we loved her for it," said Ruth, "but something changed by the time we were parents. I just couldn't imagine letting Thomas or Heidi do the things we did as children."

"I couldn't either," said Becca. "I may be the risk-taking twin, but it would have been impossible for me to let Laura or Adam take the kind of risks that we did. They now say that grubbing around in the dirt is good for children's immune systems. Perhaps if I hadn't kept you so clean, Laura, you wouldn't have developed asthma.

"A lot of things have changed," said Ruth. "Like . . . until we were teenagers no one ever talked about sex. Nowadays people seem to talk about nothing else."

"Oh, come on," said Thomas, "That can't be true. Adults talked about sex but not to children."

"No, she's right," his grandmother said.

Seeing that everyone was looking at her, she continued, "If your grandfather had even mentioned sex to me, I wouldn't have married him. For years after we got married, we still didn't talk about it. But after we lost our baby son, he said that he wouldn't put me through *that* again. I knew that he was taking care when we made love so that I wouldn't get pregnant. But we didn't talk about it."[6]

"Sex is a normal and natural part of life, Grandma," Laura said. "Why didn't you feel you could talk about it?"

"What married couples do that can bring new life into the world seemed to me to be too important and sacred to talk about. We didn't have the words because these words weren't spoken by 'nice' people. Then, in the 1960s, lots of people started talking and writing about it, so we learned the words. That made sex seem easy . . . and sort of cheap."

"But it isn't easy," said Becca, "For all our talk about sex, people still don't understand it or agree on what's appropriate and what's harmful when it comes to sex."

"Like we don't agree on how we should bring up our children," Ruth added.

All eyes turned to the living room where the children were sitting, each engrossed in what was happening on the screen they were holding.

"I'm so sorry!" said Ruth. "That sounded critical and I actually think that you're all wonderful parents. I must admit though, that I used to get rather annoyed with the way young people are so obsessed with their devices. It's so different from our childhood. But I am now very grateful for our devices because they allowed us to stay connected when fear of that damned virus kept us apart."

Hey, Aunt Ruth, we're not supposed to be talking about the pandemic," said Adam. "And don't worry. We know what you mean, and it's hard to get to grips with the way things are changing, even something so important and basic as how we raise our children."

"It is scary," said Heidi, raising her glass. "So, let's just toast the future and having faith that together we can handle it."

What is driving this cultural transformation?

The modernization process has now begun in almost all human populations. It has revolutionized and continues to revolutionize our ideas about what it is to be human.[7] There's no better illustration of this than the way humans behaved in 2020 during the COVID-19 pandemic. In a few months, the spread of a virus totally changed people's lives. This was an event that is unique in human history, but its uniqueness has nothing to do with the novel coronavirus that appeared in 2019. New viruses are emerging all the time. Only a tiny proportion of them are as infectious and deadly to humans as this one, but the emergence of this virus was hardly unique. New virulent pathogens appeared again and again in human history and prehistory.

What's special about COVID-19 is the population of humans that caught it. If this virus had emerged before the modernization process had got underway—say in 1719 rather than 2019, its impact would have been small. In fact, it might not have spread beyond the local population. For a virus to spread, it must infect people who travel and join new populations, and very few people did this in 1719. Those who had a nomadic lifestyle traveled in family groups and mostly kept apart from strangers. The farmers and artisans who lived settled lives seldom left the region where they were born. The virus might have spread if it happened to infect someone who traveled and interacted with strangers for a living, such as a merchant. But it would have spread at walking pace, taking months or years to spread around the 1720 world. In 2020 it only took a few weeks.

If a traveling merchant who was infected with the virus arrived at a market town and began peddling his merchandise, he might have passed it on to people who visited his stand, but in 1720, people tended to treat strangers with suspicion and didn't get too close. They may not have been able to explain why they were wary, but people who kept business with strangers "at arm's length" tended to do better. Those who were too trusting and friendly were more likely to be swindled, robbed, or become ill with germs that their immune system wasn't familiar with. Few people would copy their outgoing ways. Avoiding close contact with a strange peddler might have protected people from the viruses that infected him, but a couple of his customers in the region might still have caught the virus. Then it would have spread through the region over the next weeks and months. As is the case with the virus today, most infected people would have had mild symptoms or no symptoms at all. A few would have become very ill, and some would have died and been mourned by their friends and families in the usual way.

Plenty of 18th-century death records mention "fever" as the cause of death. But no one was collecting, analyzing, or publishing public health data in the 1700s. No one would have worried about a shortage of hospital beds, because there were no hospitals. People were nursed at home by their family. Doctors and other healers visited people on their sickbed but not everyone could afford their fees, and there was little these medics could do anyway. Their beliefs about the causes of fevers were wrong and their idea of "intensive care" might have involved using leeches to remove some of their patient's blood.

Your ancestors might have noticed that a few more people had fever in the year that the new virus arrived, but 1720 wouldn't have been remembered as the year a pandemic struck. The "Black Death" plague pandemic, which sickened Eurasia and northern Africa during the middle years of 1300s, probably killed over a 100 million people at a time when the entire human population was only about 500 million. Entire towns and villages were wiped out.

The coronavirus caused a pandemic in 2020 because modernization had made people very different from their ancestors who lived only a few generations ago. By our standards, the people of 1720 lived in unsanitary conditions and were poorly nourished. But, as a population, they would have been less vulnerable to the COVID-19. The members of the 21st-century population who are most at risk from the disease are the elderly, obese, or those already suffering from medical conditions like heart disease, cancer, and diabetes. The 18th-century population had far fewer of these people. There was no knowledge of how to prevent or treat any of the infections that regularly spread through their communities, such as cholera, yellow fever, malaria, measles, and smallpox—to name but a few.[8] Infected cuts could be lethal. As a result, most people died before they became elderly. Those who had developed heart disease, cancer, or diabetes were unlikely to be still alive. And, of course, starvation was more of a health problem than obesity.

Modern people are also different because we don't expect people to die from infections. We know so much more than our 1720 ancestors and possess many more technological and social tools. When new and unexpected problems arise, we get a frightening glimpse of the sort of insecurity that was normal for our ancestors who lived only a few generations ago. The people of 1720, whether they were in China, Europe, Africa, or the Americas, lived lives that were similar in some fundamental ways, and were also similar to the lives of their ancestors going back thousands of years. The people living today are different from them, and as the modernization process continues, we're getting more and more different. It's not just the ever-changing technology and the greater prosperity. Our beliefs about how to understand the universe, what humans are like, what life is for, and much more have changed and keep changing.

Why is this? Most of us can think of reasons why our own opinions on some issues have changed. But often we can't be sure. And we certainly can't be sure why other people think like they do. In some ways, humans all over the world have become similar. Except for a few isolated clans in South America, we're all in contact with the modern global culture. But increased contact has also made us more aware of differences between cultures. And often the people closest to us have beliefs and behaviors that are very different from ours. Members of the same family can seem to inhabit entirely different cultural universes.

Few of us would want to exchange our lives today for the sort of life people lived 300 years ago, but the cultural diversity and rapid cultural change are challenging. It sometimes feels like we're trapped in a bus that's hurtling down the road through alien landscapes. Sometimes the scenery is ugly, and everyone wishes they could go faster. But even when things are going smoothly and the journey feels worthwhile, it still seems risky. Will unforeseen bumps in the road, like a global pandemic, cause the bus to crash? Is anyone even driving this bus? The idea that everything is being controlled by a shadowy international conspiracy is almost more comforting than the thought that no one is in control of this cultural transformation. Does anyone have an idea about the route we're taking? Who set the bus on the road and what is fueling our wild ride? Are there any brakes?

Many stories are told about how the modernization transformation started. Most see certain technological changes as being very important, such as the development of printing and improvements in transportation.[9] These allowed ideas and information to spread faster and more widely. Trade, exploration, and migration brought economic benefits. The stories also mention influential thinkers, like John Locke, who wrote about human rights all the way back in the 1600s, and talk about leaders, such as the men who wrote the US Constitution in the 1700s. But stories that concentrate on ideas, inventions, events, and historical figures can only partly explain what's happening to our cultures.

That's because culture belongs to entire populations. It's not just the "important" people and major events that influence the way culture evolves. We're all important, and so are the minor events. The culture bus doesn't seem to have a driver because we're all driving. Long dead people from past populations are also influencing our journey because they helped build the bus and road. The human journey took a new course a few hundred years ago, and this is probably because of the profound change that occurred in the networks that carried the cultural information.

A change in the shape of the information network

Our ancestors lived and raised their difficult, large-brained offspring in families long before they could really be called "human." For tens of thousands of years, people were embedded within a family, and their connection to the cultural information network was through their family. Few Westerners alive today have experienced growing up as part of an old-style strong family. Life was hard except in the wealthiest of families. Children weren't the center of attention—the elders may have enjoyed watching the young ones at play, but they knew not to indulge in too much joy. Children needed to help and they learned as they helped their elders, but their learning was limited. Growing up in a farming family could only prepare them for life on a farm, so peasants mostly remained peasants. A farming childhood couldn't provide experience that would allow someone to pursue another ambition. To become a skilled craftsman, a boy had to be born into a family of craftsmen or be adopted by a craftsman's family. Sometimes arrangements like apprenticeships could be made, but this was often costly for the boy's family.

And, of course, no matter what kind of family a girl was born into, she was expected to serve by producing the next generation. Girls helped raise younger relatives and, once they were grown up, they were expected to be wives, mothers, and eventually grandmothers. The lack of opportunity that people, especially women, faced in the past is often seen as oppression that was imposed by powerful men or religious organizations. But, as we explained in chapter 7, the cultural evolution of such restrictions, and the beliefs and habits that enforced them, was inevitable given the social environment. The population was made up of families that were cooperating in many ways but still competing in the most fundamental biological way. Families had to raise new members in order to survive.

Things could only begin to change when people were able to bypass their families and connect independently to a cultural information network. When it became possible for them, as individuals, to become part of other social groupings, such as workers in a factory, pupils in a school, or worshipers in a congregation, they could connect with more sources of information. Beginning in Europe in the 1700s, increasing innovation and trade created new opportunities that brought benefits to many families. They were able to raise more children than could be supported in their community. The excess youngsters had to find ways of supporting themselves by taking advantage of other new opportunities. They got jobs in mines, factories, and construction. They learned crafts and did handiwork, or they bought, sold, and delivered things in a growing consumer market. Some were employed as servants by wealthier families.

To find and create these new opportunities, youngsters usually had to leave the small community where they were born. Still, they tried to stay near people they knew, traveling with friends or family members and settling in factory towns or urban neighborhoods where relatives already lived. But it wasn't the same as being part of a family. Having been brought up to be wary of strangers, most young adults found the move difficult. Many suffered debilitating symptoms that today might be called "depression" or "fatigue." Doctors of the time called it "homesickness" and sometimes "nostalgia." And they saw these as genuine medical conditions.[10] Often their patients' suffering was so severe that doctors recommended they go back home. This usually cured them, but few could afford to see a doctor, and most didn't have the option of going home. They had to find other ways to relieve the homesickness symptoms. Learning to read and write helped because it allowed people to exchange messages with loved ones back home. And they could work at controlling their anxiety and try to make new friends. Clubs appeared in areas that had a lot of migrant workers, uniting diverse people who nevertheless had interests in common. Joining a club could give a person the feeling of being part of a caring group. Some of the clubs were religious, some political, and some allowed people to enjoy sports and hobbies together. Some clubs were just places where people could eat, drink alcohol, and gamble.

These new groupings were the beginning of a restructuring of the human population. It began in Europe but was soon happening in other populations, and it continues to this day. In some parts of the world, young adults are only now starting to leave the place where their family feels it belongs—the place where their kin have lived for as far back as anyone can remember. It's a story being played out all over the world, most notably today in China, India, the Middle East, and countries throughout, South Asia, Africa and Central and South America. Today's young migrants have the same mixed feelings about leaving home as the European youngsters who made the journey a few centuries ago. Once they leave familiar places, they will be surrounded by new information and their minds will start to absorb it.

Once people are part of a community with a different structure, the internal supports of the ancient family-based culture start to bend. Homesick youngsters gathered in places of employment have a brain that is genetically endowed with the ability to form bonds with other people. The strangers didn't stay strangers. Men who worked together in mines felt happier and safer if they started to see their workmates as "brothers." Women working in textile mills watched out for their younger "sisters." Men and women who attended religious services on their day off began to feel a kinship with the other worshippers. Groups that shared food and drink began to feel bonded. Alcoholic drink may have encouraged this bonding, at least temporarily.

Forming bonds with strangers challenged the bonds people had with their real family. For example, a man who started to attend a church that preached the evils of drink may have felt less strongly bonded with his brother who insisted on getting drunk with his pals every Saturday night. The new groups may have felt a bit like families, but they were different. When people spent time in these new groups, they exchanged different information. Within families, the information communicated tended to have a pro-family spin; it promoted cultural elements that furthered the interests of the family, such as obedience to elders and the raising of the next generation. The information shared in the new groups promoted a different set of interests. Workmates wanted to get the job done and earn money. Sporting groups wanted to improve their fitness and skills and win competitions. Political groups wanted to change the way things were organized and controlled. Religious groups wanted to achieve objectives defined by the beliefs they shared. And so on.

Once people were connected in new groups, the pro-family spin wound down, making it possible to ignore the age-old beliefs about loyalty and duty to family. The cultural elements supporting the family were too tightly tangled with the fabric of people's lives to be unwound quickly, but they could begin to loosen. Over time they became looser and looser. The modernization process is the result. New ideas started to emerge as the beliefs and practices of the family-based system unraveled. For example, people started to talk about "human rights." In the premodern European family-based system, it was *families* that had rights. These rights had been negotiated and sometimes fought for, and then they were passed down the generations. In many parts of Europe, rights to own land or to use land had been tied up for centuries, granted by ruling families. As families weakened, individual family members could be, and needed to be, in a stronger position. It became possible for a man to start his own business and become prosperous. It seemed reasonable that this man should be able to buy property in his own right, no matter what rights had been negotiated long ago by ruling families.

Once the idea of men having rights existed, it could evolve. When European intellectuals in the 1700s started declaring that men have rights, they didn't mean *all* men. Some of these intellectuals owned slaves. But the idea of "rights" started to take on a life of its own. In 1948 the United Nations adopted the "Universal Declaration of Human Rights," which states that "All humans are born and remain free and equal in dignity and rights."[11] To many this declaration seems like a great achievement, but others are skeptical. People in different populations have different arguments about rights. Do men have the right to beat their wives? Do woman have the right to wear the clothes they want without the fear that men will harass them? Do people have the right to healthcare, to own guns, or to marry someone of the same sex?

For most of human history, sex and marriage were controlled by the family. They were weapons in a family's fight for survival, helping to ensure that daughters would get the support they needed to raise their children. It changed once young adults were living and working away from home, because they now had the chance to meet potential marriage partners without their family being involved. It's easy to see how the idea would emerge that if two young people had formed a bond ("fell in love"), their families shouldn't prevent them getting married. Thus, Europeans gradually became more and more committed to the idea of marriage being the union of two people who are in love. Their descendants' beliefs about marriage would be further still from those maintained by strong families.

Modernization began in towns and cities, but soon the family-based culture began to weaken in rural areas too. Many country people had relatives working in the town and increasing numbers of people were learning to read and write. Letters, journals, and literature from the 1700s and 1800s provide plenty of evidence of the way people's beliefs and feelings were changing. New ideas spread in letters between friends and relatives and also through printed books and pamphlets. Philosophical and political essays, sermons, pornography, and scientific reports were all available, but works of fiction were the most popular. A number of best-selling novels were published in Europe in the 1700s.[12] In some cases, the presses couldn't run fast enough to keep up with demand. Books were rented out by the hour. Readers were astonished at how close they felt to the characters in the stories and how much they cared about them. Some had trouble believing that they weren't real people. Friends and family members read the books aloud together, weeping through the sad bits and celebrating happy endings. It was a new experience.

Novels didn't just make people feel bonded to the characters, they also created a community of people who would never get together but had something important in common. They had enjoyed a shared experience. Large numbers of them had walked the same imagined dark corridors, held their breath at the same terrifying events, and felt that they knew the heroine better than they knew their own cousin. Knowing that they had shared all this drama gave readers the feeling of a personal connection with a wider population. It was feelings like this that made it possible for people to start feeling that they belonged to a nation and were citizens of a state.[13] They eventually saw themselves as being part of humanity.

A mysterious change in behavior

The most profound and consequential change that occurs during the modernization process is a reduction in the number of children women have. The demographers who study this change call it "the fertility decline." This phrase isn't meant to suggest that people become less able to have children. They can still have children, but they choose to reduce the number that they have.[14]

The decline happened first in populations of European descent and has now happened (or is happening) in almost all populations (see Figure 8.2). There are large cultural and economic differences between many of the populations, but the course of the decline is remarkably similar in all of them:

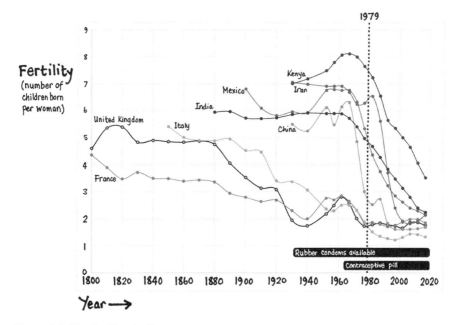

Figure 8.2 The fertility decline

The fertility of almost all populations has changed substantially during the last couple of centuries. Because modern people are part of this change, we tend not to find it mysterious. But reasons often given for the decline aren't consistent with the evidence, as this graph shows.

Couples who lived in European cities were the first to decide to limit the number of children they had. It was in France that the idea of limiting family size first spread to the wider population. The idea spread more slowly in European countries than in countries where the decline began more recently. This may be because communication technology was less advanced in the 19th and early 20th centuries. Some ups and downs in fertility are linked to events. For example, there was a period of low fertility in many European countries between the First and Second World Wars. After World War II, there was a temporary rise in fertility in some countries, which is often referred to as the "baby boom."

In 1979, China began to strongly discourage most of its population from having more than one child. Many Westerners believe this policy caused Chinese fertility to decline, but the evidence clearly shows this isn't true—the decline was already well underway. The year 1979 was also the year of the Iranian revolution. The fertility decline evidence disputes another belief held by many Westerners—that the Islamic government made Iran less modern and forced women back into traditional roles. In fact, it was only after the revolution that most couples began to limit the number of children they had, which allowed women more time to do other work.

Advances in contraceptive technology, such as the rubber condom, may have allowed fertility to fall a little lower than would have otherwise occurred, but contraception didn't cause the fertility decline. It was a new desire to limit pregnancies that caused modern contraceptive technology to be invented.

Before the decline: Couples behaved as if they were trying to raise as many children as possible. Marriage customs differed but, in most cultures, women became pregnant shortly after or just before they married. (In some, a marriage was considered invalid if the wife failed to become pregnant after a certain period.) The spacing between births varied in pre-decline populations, but usually, if a woman's baby survived, she became pregnant again when it was between one and three years old. The spacing tended to increase after the woman reached her late thirties, but healthy women continued to produce babies until they reached their mid-forties. It was common for a woman to give birth to eight or more children during her lifetime.

After the decline: Women usually stop becoming pregnant after their second or third child, even if they're only in their twenties. The transition from high to low fertility isn't gradual. It usually occurs in a single generation. People who grew up with eight or more brothers and sisters only have two children themselves.

Because we (the authors of this book) are part of the modern global culture, we understand perfectly well why people limit the number of children they have. (One of us had two children and the other had only one.) BUT, anyone with an understanding of evolutionary theory must see that this sudden change in human reproductive behavior is, from an evolutionary perspective, very strange. We're descended from ancestors who worked hard to raise as many children as possible. We inherit both our genes and our culture from these ancestors and yet our behavior is very different. How could human behavior change so suddenly?

Few of the stories about human evolution that are told to the public make any mention of the mysterious fertility decline. This is no doubt because few of the people reading these stories see the decline as mysterious. To modern people, low fertility seems normal and natural, and most of us assume that our ancestors felt roughly as we do.[15] Many people imagine that their ancestors wanted to limit the size of their family but didn't know how to do it. Without looking at the evidence, they assume that most of the babies born in the past were a side effect of their ancestors' uncontrollable desire for sex. They believe it was the invention of contraceptive technology that caused fertility to decline.[16]

The timing of the fertility decline in European populations provides indisputable evidence that this explanation is wrong. In many parts of Europe, couples began to limit the number of children they had long before there were advances in contraception.[17] And, in many of the populations that went through the fertility decline later, couples continued to have large families for a while, even though contraceptives were available.[18] Fertility starts to fall not when contraception becomes available, but when people decide they want to limit the number of children they have.

There's no public record of what Europeans were doing to prevent pregnancies, because sexual matters weren't publicly discussed. But the private letters and secret diaries written by Europeans in the 1800s and 1900s clearly show that people were aware of ways of having sexual pleasure that didn't involve releasing semen into a woman's vagina.[19] It seems that most just didn't think married couples should behave like this.[20] And because they didn't leave written records on what they thought on these matters, we simply don't know what they thought. It's possible that they didn't explicitly "think" about it at all.

There are well-known accounts of men who were desperate for a male heir. (King Henry VIII of England is an obvious example.) But the writings of Europeans who lived before the fertility decline don't give the impression that they went into marriage with goal of having a family of a certain size. Perhaps they saw their feelings of sexual desire and love for their children in a sort of spiritual way. If the purpose of marriage was to serve life, married couples might have thought that they didn't have the moral right to decide how many lives they should serve. This doesn't mean they believed women should become pregnant irresponsibly. A mother-to-be needed to be in a situation where she could get the support necessary to raise her child. Only women who were married (or soon to be married) should become pregnant. Once a woman was married, she should be sexually intimate with her husband in what their families saw as the "normal and natural" way. If this resulted in a child being conceived, the couple should do their best to raise it.

For a person who belongs to a modern culture, it seems crazy that a couple who knows how to prevent pregnancy would conceive another baby when their family is struggling to raise the children they already have. What were they thinking of? We can't question our long-dead ancestors, but during the 20th century, Western anthropologists and aid workers talked to many people who lived in poor countries with high birth rates to try to understand why they continued to have children. People everywhere said the same sort of thing. A woman called Sitan, who was interviewed in the West African country of Mali in the early 1990s, is typical. Sitan, already the mother of several children, had heard of the birth control pill but hadn't used it.

Interviewer: Sitan, how many children more would you like to have?
Sitan: Ah! That is for God to decide.
Interviewer: You yourself, how many would you like to have in your whole life?
Sitan: I don't know the number. . . . It's when God stops my births.
Interviewer: How many boys and how many girls would you like to have?
Sitan: It's God that gives me children, since it is God that gives or not. You, you can't make a choice about your children.[21]

In chapter 7, we explained why, in a population of families that is competing for reproductive success, the shared culture will maintain a set of rules that encourages family members to behave in ways that produce surviving children. In such a population, natural selection would also favor families who believe they shouldn't try to limit how many children are born but just obey the rules and accept what happens. They should not only accept the children that are born to them, but also accept the likelihood that some of those children, usually the weaker ones, won't survive.[22]

The endurance of this fatalistic and rather glum view of life makes sense when you remember that evolution is all about copies of traits being passed on to the next generation. Cultural traits are passed on because people learn from the other people in their lives. In populations that are divided up into families, children may learn from other people in their community, but they mostly learn from other family members. A simple thought experiment shows why fertility remained high:

Imagine that most people in the population believe that married couples should accept the children they conceive and not try to prevent pregnancies. Let's call this cultural trait "*accept.*" Now imagine that, every few generations, some couples get the idea of limiting the number of children they have so that their life is less of a struggle. Let's call this cultural trait "*limit.*"

Table 8.1 gives the likely reproductive outcomes of couples with the two cultural traits.

If the *limit* couples had much better lives than the *accept* couples, then a few of the *accept* couples' children might be tempted to adopt the *limit* trait. But the difference is unlikely to have been that great. Life was hard for everyone. So as long as children identified with their own family and its ways, they "inherited" their parents' cultural trait. Natural selection would have worked against the spread of the *limit* trait.[23] Children in the *accept* family are less

Table 8.1 Reproductive outcomes of *accept* versus *limit* traits

	Accept couples	Limit couples
Fertility	Produce 8 children on average	Start to prevent pregnancies after giving birth to 3 children
Mortality	2 children die of fever	1 child gets severe fever, but mortality is halved due to good nursing
	1 child has a fatal accident	Parents pay more attention to their children, so no fatal accidents happen
Surviving children	5	2.5 (on average)

likely to survive, but their parents still have twice as many surviving children each generation. What's more, the *accept* children get plenty of parenting practice by helping to look after younger family members. When they marry, they will be both willing and able to produce babies to be part of the following generation. The *limit* children will have had little experience of caring for a baby or young child, so being a parent will be more of a struggle for them. They may decide not to have children at all. In a few generations, the *limit* cultural trait will likely disappear from the population.

The *accept* cultural trait will still be favored by natural selection after modernization starts but, once young people become connected to nonfamily groups, the *limit* trait has a better chance of being adopted. Members of nonfamily groups share different goals and become accustomed to making choices about how to spend their time. Living away from their family gives them more control of their lives, but most people continue, for a while, to follow the rules and customs they learned at home. They try to teach their children to respect the old values. But, eventually, some will begin to wonder why life has to be so hard. In a population with the new modern structure, where the pro-family spin is diluted, the idea that a woman doesn't need to accept pregnancy after pregnancy can begin to spread.[24]

In chapter 7 we argued that, like genetic evolution, cultural evolution can "runaway," allowing copies of a trait to spread in a population and become more extreme. The classic example of a genetic runaway characteristic is the elaborate tail of the peacock. Magnificent feats of monumental architecture are the cultural equivalent. Mostly, the extremity of characteristics is limited by natural selection. A peacock encumbered by a very elaborate tail is vulnerable to predators. A population that devoted a great deal of effort to the building of temples was vulnerable to attacks from neighboring populations who had put more effort into making weapons and raising boys to fight in an army.

The change in the shape of social networks made modern culture immune to natural selection, at least temporarily. Technology and trade make it easier to have enough to eat, and because people in modern cultures have very few children, they can put effort into all kinds of other activities. The goals that people choose to pursue can drive culture to run away. For example, a popular goal that seems worthwhile, such as "getting a good education," runs away as people compete. The purpose of getting the education can be lost as people spend more and more time trying to achieve more and more certificates, diplomas, and degrees. For the professors we know, the care and feeding of graduate students competes with the care and feeding of children.

Modern populations are currently nearly immune to natural selection on fertility because there is scant variation in family size for it to work on. Even immigrants from higher fertility populations mostly converge on modern fertility in a generation or two. In many parts of the world the family-based culture has only just begun to break down and fertility is only now beginning to decline. Young people can't find work in the community where they grew up and are migrating to centers of employment.

Exceptions to modernity's rule

The modernization process seems relentless, unperturbed by the complaints of people who claim that our lives would be healthier, happier, and more sustainable if we lived in simple old-fashioned communities. But some groups have managed to resist modernization. As modernization gained momentum in European populations in the 1800s, many religious leaders expressed concern that the changes they were seeing were almost certainly contravening God's will. A few people took these warnings to heart.

Protestant Christians who belonged to sects known as Amish, Hutterite, or Mennonite were especially worried that modern culture would damage their relationship with God. Some, who came to be called "Old Order Anabaptists" chose to limit their contact with the wider world and live in small family-based communities. Their communities continue to thrive in many parts of rural North America, but their connections with the wider world are almost entirely economic. Old Order Anabaptists earn money as farmers or skilled manual workers and use it to buy things they judge to be necessary, including modern medical care. But they eschew consumerism and avoid activities that would allow modern ways of thinking to infiltrate their culture. They don't watch television, see movies, or read books that portray modern culture. Some don't own cars or make friends outside their community.

During the 1800s, members of the Jewish faith were also troubled by the way modernization was changing life and behavior. Some groups chose to reject modernity, convinced that modern behavior contradicted Jewish law. They became known as "Haredim," although they're often referred to as "Ultra-Orthodox" Jews (see Figure 8.3). Their family-based communities are now established in North America, Europe, and Israel. Even though many Haredim live in cities and are surrounded by modern culture, they keep separate by maintaining tight communities and limiting their contact with outsiders and media. Children often attend special schools where boys and girls are educated separately.

Religious leaders in the Muslim, Hindu, and Buddhist populations that began to modernize more recently have also criticized modern culture, and

Figure 8.3 Some of the rules members of Haredi communities follow give members a characteristic appearance. For example, married women must not allow their hair to be seen in public, and males aren't allowed to cut the hair on either side of their face.

these populations may also generate religious groups that continue to reject modern ways. The Old Order Anabaptist and Haredi communities show that such groups can live peacefully alongside modern populations, but this co-existence is fragile. They like to think of themselves as independent, but their survival depends on the wealth, tolerance, and protection of the much larger modern population that surrounds them. Their relationships with the larger populations are now at risk because their own populations are growing so fast. By retaining the traditional family-based structure, the Old Order and Haredi communities have also retained the rules and customs that encourage high fertility. It's common for women in these communities to have more than seven children. For decades, their population has grown at between 3 and 4 percent per year. In three and a half generations, a handful of Amish Anabaptist communities, mostly in the American state of Pennsylvania, grew to be a population numbering about a third of a million. Amish now live in at least 31 of the 50 American states. Haredi Jews are even more numerous. It's thought that about half a million live outside Israel and just short of a million live in Israel. Currently, they make up about one in nine Israelis but, in the future, the proportion of Haredim will grow rapidly. Among the Israelis who are under 20, one in five is Haredi.[25] If the Haredi and Old Order Anabaptists can manage to adapt to becoming numerous and still retain the high fertility, they might eventually replace modern populations. ("Blessed are the meek: for they shall inherit the earth," Matthew 5.5.)

The spread of modernity and family limitation

Once members of a population start to see their family as only one of the social groups they belong to, their attitudes to many things start to change, and eventually limiting family size becomes part of the culture of the population. But the amount of time it takes for people to adopt the family limitation idea varies. Differences in the timing of its adoption have had and will continue to have wide-ranging consequences for populations. The changing fortunes of France, where the idea of limiting pregnancies first became widespread, is a good illustration of this.

In 1750 the country ruled by the French king, Louis XV, had the largest population of all the countries in Europe and was at the forefront of the changes that were starting to sweep the continent. The European powers were competing to form connections with the rest of the world. Merchant vessels were sailing back and forth across the oceans. They were trading with people and trading *in* people, transporting humans to places where they were sold and forced to work. France had colonies in the Caribbean, India, and the Americas. In North America, "New France" stretched from the Gulf of Mexico to northern Canada, a far larger territory than the strip of land the English colonies occupied on the eastern coast. French soldiers and their Native American allies were constantly fighting off English colonists who were trying to expand westward. And in 1756 the Seven Years' War began, with the French leading a coalition of European states fighting a coalition led by Great Britain.

New kinds of connections were also being formed within France's borders. People were starting to develop a shared "French" identity and philosophers like the Abbé de Mably were arguing that all the French people, even servants from poor peasant families, were citizens of France who had rights and deserved respect. The poor peasants remained mostly illiterate, but people from higher "stations" in life were eager to educate themselves and their children. Estate owners, craftsmen, and businessmen all over the country wanted to be *au courant* with discussions in the cafés and clubs of Paris. They read the works of intellectuals like Voltaire, Diderot, and Rousseau and discussed them in their own clubs. Letters written by women reveal that they were often just as interested as their husbands.[26] The lives of their parents and grandparents had revolved around family and the local community, but many of the new generation shared a love for and loyalty to the nation they were trying to build. They discussed setting up a better government to replace the royal and aristocratic families. It would take a violent revolution to unseat the ruling families, and this didn't begin until 1789.

Optimism about increasing liberty and fraternity was tempered by fear that disintegration of the old order might lead to moral decay. There was condemnation of fashionable *Parisiennes* who sent their babies away to live with a wet nurse in the country. And "foundling" hospitals in the cities were said to be overflowing with babies that had been abandoned by their unmarried mothers.

In the 1770s a young man called Jean-Baptiste Moheau[27] took it upon himself to find out what was really going on with the French people. He was employed as the personal secretary of a high-ranking public official, Baron Auget de Montyon, and, through his boss, he was able to get access to a great deal of information, including the records of births, deaths, and marriages collected from the regions of France at the request of the royal administration. Scientists like Isaac Newton and Antoine Lavoisier had used mathematics to describe the behavior of the physical world, and Moheau realized that he could use mathematics to describe the behavior of the social world of the people of France. Scratching away with a pen made from a goose feather, he reported the patterns in the data that he had found—associations between conditions in a community and the vital statistics of the community.

Modern people take for granted the idea that governments should organize and serve the citizens of a country, and to do that they must collect and analyze information about their citizens. Moheau was decades ahead of his time in seeing the sorts of things that governments need to know, and he figured out how to obtain that information using mathematics. Inevitably, though, he was also a man of his time, and the way he discusses his finding shows how ideas were evolving about what "a French state" should be like and the effect this was having on people's behavior and ideas.

In 1778 Moheau published his findings in a book called *Recherches et considérations sur la population de la France*. In English, the title would have been something like "Empirical Studies on the Population of France, and Their Interpretation."[28] The book is nearly 300 pages long and has dozens of tables that summarize quantitative differences between the regions of France. The second part of the book looks at factors likely to influence the "progress" and "decay" of the population, and in the chapter called "On morals" he abandons science and his opinions emerge. He begins by pointing out that raising children is an expensive and onerous burden, and so, he says, "logic and a calculating mind would not lead to the propagation of the species." If the species is to continue, he says, morals are necessary:

> [A] sense of principled responsibility must exist to persuade one to assume this domestic burden. The love and obedience of children must also offer a prospect of happiness for parents in their old age. These feelings and dispositions are the result of morality.

Moheau regarded "France" as a giant family, in which people were bound by brotherhood (*fraternité*). Everyone had a duty to the state in the same way that everyone had a duty to their family. It's easy to see why people living at this time would think like this. Up to this point, human populations had been divided into families. When the evolution of their culture forced them to look beyond their weakened families and see themselves as belonging to a nation or state, it was easiest to imagine this new grouping as a scaled-up family. There was also talk of "Mother Russia," the "German fatherland," and America's "founding fathers." The word "patriotism" comes from the Latin word *pater*, which means "father."

If someone sees their state as being like a family, children are important. The family's ultimate purpose is to produce children, and Moheau thought that the same must be true for the state, because a large, strong, healthy new generation will ensure that the state survives and grows. Moheau warns that France may have a problem here because of reports suggesting that "libertine"[29] behavior was becoming increasingly common:

> [R]ich women, for whom pleasure is the greatest interest and the sole occupation, are not the only ones who regard the propagation of the species as a dupery of olden times; already the disastrous secrets unknown to any animal but man[30] have penetrated in the countryside: nature gets cheated even in the villages.
>
> If these licentious practices, if these homicidal tastes spread more, they will be no less disastrous for the State than the plagues that devastated it in the past. It's time to stop this secret and terrible cause of depopulation that stealthily undermines the nation; it might soon be too late to control it.

To a modern mind, these paragraphs seem to be the tirade of a dangerous lunatic, but they reflect what many French people were thinking in 1778, especially those who were fans of Jean-Jacques Rousseau. Rousseau feared that modern society was corrupting the innate sense of morality he believed that humans possess when they are living in a "state of nature."

But set aside thoughts of morality and cultural differences and think practically. If the state is like a family that is competing with other families, then Moheau was justified in his concern. As it turned out, the "secret and terrible cause of depopulation" couldn't be stopped, and this did undermine France's competitive position in the world.

In 1800, women in France and the United Kingdom were producing babies at about the same rate, but then fertility dropped in France and rose in the UK (see Figure 8.2). Figure 8.4 shows the result of this. The birth rate remained high in the UK and other parts of Europe for most of the 1800s. Infant and child mortality remained high, especially in the cities, but the birth rate was high enough that the population grew. People in many parts of Europe were becoming more prosperous. More people could afford to get married and couples were managing to raise more children. In France, however, the idea

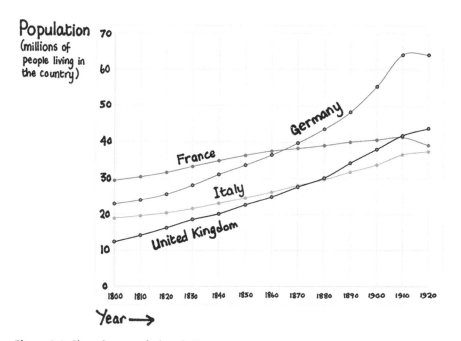

Figure 8.4 Changing populations in Europe
At the beginning of the 1800s, France was by far the most populous European country. But couples in France began to limit how many children they had more than three generations earlier than people in other parts of Europe. The population grew rapidly in other European countries even though millions of people were leaving these countries to settle in the Americas, Australia, and southern Africa. France's low fertility meant that the population fell after 1914 because of mortality during the war and the deadly flu epidemic that followed it.

that it was good to limit family size spread to more and more of the country, and fertility declined further.

At first, the idea didn't spread beyond the French border. English women continued to believe that, once married, they should accept and do their best to care for the children that came along. Even wealthy English women like Emma Wedgewood, who had visited France in her youth, saw this as her duty. Emma married Charles Darwin in 1839, when she was 30. They had 10 children, the last born when she was 47. It was the same for the English Queen Victoria and her German husband. They had nine children and Victoria might have become pregnant again if her husband hadn't died when she was only 42. It wasn't just wealthy couples who were managing to raise many children. People from all "stations" in life were better off and able to get married at a younger age and raise more children. In England, and then in Germany and other northern European countries, employment opportunities grew and grew as the Industrial Revolution transformed people's lives. These countries became manufacturing powerhouses and the size of their trading empires exploded. France was industrializing too, but it happened more slowly in France, where it was harder to find people to work in the factories. After two generations of low fertility, the French population didn't have so many young adults looking for employment.

The transportation technology that was making international trade easier was also making it easier for Europeans to migrate to other countries. As the population of European countries increased, so did the number of people leaving Europe to see if they could start a new and better life in another part of the world. South and Central America were the destination of many people from southern Europe—about two million from Spain, 3.7 million from Italy, and a million and a half from Portugal. The Portuguese emigrants mostly settled in Brazil. Nearly two million people from the Austro-Hungarian Empire also went to South America. Northern Europeans were more likely to migrate to North America—about 5 million from Germany, 3.6 million from Poland, 2.7 from Scandinavia, 3.2 from the Austro-Hungarian Empire, 5 million from Italy, and 2.2 million from Russia (see Figure 8.5). Another 10 million Russians went east, colonizing Siberia and Central Asia. The largest stream of migrants came from the two large islands off the coast of Europe where the people spoke English. Thirteen and a half million British and Irish went to North America and another three million went to southern Africa and Australia. When they got there, they raised a lot of children. The fertility of the English-speaking people at a key stage of modernization goes a long way to explaining why English is currently the main language of diplomacy, science, and global business.[31]

Figure 8.5 Emigrants boarding a US-bound steamship in the German port of Hamburg. This sketch was published in the American political magazine *Harper's Weekly* in 1874.

Because the people of France were early adopters of the idea that it's good to limit births, they contributed only a trickle to the river of European emigration. The seven or so million people in eastern Canada who speak French today are descended from the few thousand colonists who settled there in the 17th and early 18th centuries. In 1759, France and Britain fought a battle over the northern part of "New France." France lost, the territory became part of the British colony of Canada, and the French-speaking people living there became subjects of the British Empire. The southern part of New France was sold to the United States in 1803 for $15 million in a deal known as the Louisiana Purchase. The money was used to help fund an army that the French ruler Napoleon would use to invade other European countries. Napoleon's armies were eventually defeated. Throughout the 1800s, France's population declined relative to the populations of other European countries. So did France's power and influence.

It wasn't until the latter part of the 1800s that the idea of limiting births began to catch on in European countries outside France. In some parts of the Netherlands, Belgium, Scotland, and northern Germany, declining birth rates were detected before 1880. Birth rates were dropping in most parts of England by 1900. In these countries, news that the birth rate was declining triggered a fair amount of public concern. Some public figures argued, like Moheau had done, that their country needed a large and growing population to remain strong and vigorous. But over time, this view was replaced by a new worry.[32] It was noted that wealthier and more publicly influential people were having fewer children while the birth rate remained high among poorer families and immigrants. By the early part of the 20th century, many better-off people were expressing concern that the population of their country was deteriorating because children born to people like them were being outnumbered by children they regarded as inferior. Some feared that these children would genetically inherit weakness and inadequacy from their parents. Others worried that growing up in poverty and squalor would make them weak and sickly adults. Early birth control campaigners like Margaret Sanger, in the United States, and Marie Stopes, in the United Kingdom, argued that their country would be weakened if poor families continued to produce so many children. Some Europeans also had a problem with "foreigners" who had immigrated to their country, mostly from other parts of Europe. It seemed to many that, if their nation was a scaled-up family, then recent immigrants or people who look or behave differently shouldn't be part of it.

Within a generation or two, almost all Europeans, rich and poor, were controlling the number of children they had, and by the time the Second World War ended in 1945, fertility was low all over Europe, except in a few isolated regions.[33] Fertility was also low in the populations descended from the European migrants who had colonized other parts of the world. In some of these populations, a small and temporary rise in fertility occurred between 1945 and 1970. It became known as "the postwar baby boom." After that, the fertility of all these populations fell to very low levels, and it has remained low ever since. The development of contraception technology and the availability of abortion have made it easier for women to avoid having a baby.

The fertility of a population always declines after their culture starts to modernize. The first non-European country to experience fertility decline was Japan. Japan had been trading with European countries since the 1600s, and during the 1800s industrialization and social mobility increased in Japan as it had in European countries. Japanese fertility remained quite high until the 1940s, and then it dropped rapidly. By 1970 it had begun to decline in most other Far Eastern countries. By 1985 the decline was well underway in Iran, Bangladesh, and large parts of India. As the 21st century began, fertility was still high in Afghanistan and some parts of Africa, but since then it has begun to fall in most of these countries too.

All populations go through similar changes and face similar challenges as they begin to modernize. People become exposed to new ideas and start to question the ways of their elders. At first populations grow rapidly because new connections bring greater prosperity. Life often gets chaotic and dangerous as families no longer exert as much control over their members. For life to become more peaceful and organized, the newly formed collections of strangers have to become communities and evolve ways of organizing services that families had always performed. Food and clothing must be available. The sick and injured need to be cared for. Institutions are needed to keep the peace and to punish and prevent crime. In the old strong family system, ruling families were expected to keep order by controlling their own members and forcing other families to do the same. But the new populations that gather in places of employment have to do what their bosses tell them. And they must live and work among people they don't know. It isn't always easy to get on with strange people who have odd customs.

In the populations that were first to modernize, the historical record tends to emphasize the positive—describing the development of order rather than the chaos, misrule, and corruption that lasted for many years (see Figure 8.6). The evolution of new customs is a process. People must experiment with ways of developing and enforcing rules that will keep the peace and organize trade and public services. Before the rules can work well, people's minds must change. They have to come around to respecting the authority of strangers and the rules they make. They also must be willing to give financial support to new institutions.

The most visible sign of people's changing minds is the decline in fertility. This provides an economic boost that can have a long-term effect on the quality of people's lives.[34] Instead of struggling to raise many children, men and women can work to increase security, comfort, and pleasure in their own lives and the lives of the small number of children they have. Eventually the effect of this boost diminishes because lower fertility means that, after a couple of decades, there are fewer young adults ready to enter the labor pool. The average age of the population increases, and a larger and larger proportion of the population is made up of elders who need to be supported.

Compared to living in a village with a family-based culture, life in a town or city with a modernizing culture is vibrant and exciting. People are connected to a vast information network. These connections can, potentially, make people kinder and more tolerant.[35] Although modern people compete in all sorts of ways, their low fertility means that they're no longer competing in the fundamental biological way. We may be destroying habitats and gluttonously consuming resources, but, in evolutionary terms, we have become the most unselfish living organism to have ever inhabited the planet. If current trends continue, within a few generations, the population of humans on Earth will start declining. And, if current trends continue, this decline will not happen because we poisoned our environment, ran out of space, or started killing one

Figure 8.6 The British painter William Hogarth's famous depiction of drunken chaos on the streets of 18th-century London was produced to support the passage of a law that would make gin and other strong alcoholic drinks more expensive. The Sale of Spirits Act (commonly known as the Gin Act) was passed by the British Parliament in 1751. It was one of the many new social tools that improved life in London.

another in massive numbers. It will happen because of the cultural changes that make us want to have fewer children.

Many of the people living in cultures in the earliest stage of modernization are suffering brutal and miserable lives, but modernization is a work in progress. It's still evolving, even in populations of European descent. For many generations, the cultures of these populations have been experimenting with ways of forming a government, organizing economic activity, and reducing ignorance, poverty, crime, and conflict. Their efforts do seem to be getting somewhere. Most populations of European descent are relatively stable and prosperous. But their cultures have absorbed lessons from a turbulent past, when feelings of national superiority led them to oppress others and fight among themselves.[36] They're gradually coming to terms with the fact that Europeans make up a small and shrinking part of the population of humans on Earth. During the first half of the 20th century, while their fertility was falling, other populations were growing rapidly.

Despite having had very different histories, religious beliefs, and customs, non-European populations go through quite similar changes once they become connected to the wider world. They proceed through the modernization process along roughly the same path as Europeans. Populations that are at an early stage of modernization, such as those in parts of Africa and South and Central America, are still growing rapidly, but the young people in these populations are starting to adopt the idea that their lives can be better if they limit how many children they have. They develop a hope for a better life as they witness the disintegration of the structures and beliefs that their parents took for granted. A confusion of forces that work together and against one another try to impose order or take advantage of the disorder. There are powerful families, gangs, tribal nationalism, various religions, armies, militias, business interests, and endless ranks of outsiders. To members of European-descent cultures, the chaos and corruption found in some countries seem insoluble and even inhuman. But it was only a few generations ago that Europeans were enduring the early stages of modernization. There were periods of similar violence and discord. Even so, people survived and even thrived. It simply takes time to evolve new ways of keeping the peace and providing services. Aid, advice, and interference from outsiders add complexity to this evolutionary process, and the effects are difficult to predict.

Meanwhile, many people are migrating (or trying to migrate), hoping that living in another place and being part of a different population will allow them to have a better life. Migration and the mixing up of cultures can cause problems. It takes time for immigrants to learn the language and customs of their new population. Most immigrants to more modern populations readily adopt modern ways, and almost all their children work hard to fit it. But there are some well-publicized exceptions, and this is influencing the way some people in modern cultures feel about immigration. The modernization transformation widens people's view of the world, and this tends to make them more tolerant of cultural differences. But it takes time for people's minds to change.

To make our more connected-up world work better, populations must evolve new social tools. This also takes time. Large cultural differences exist and there are many challenges ahead. But modernization is making humans more similar in important ways. Almost everyone in the world considers themselves to be a citizen of a country and sees the leadership of their country as affecting their lives, for good or for ill. We all use money to buy things that are traded internationally. And even though people continue to see themselves as belonging to a family, tribe, nation, or ethnic group, most see "humanity" as a single species.

Epilogue

Stories are supposed to have an ending—ideally it will be a "good" ending that leaves readers feeling satisfied. This isn't possible with our story. It's finishing at a moment of high drama with many unanswered questions:

- Are humans now destined to become more and more modern? If so, what will "more modern" look like? Will the different modernizing populations become more similar culturally? Or will they grow apart and perhaps form warring factions?
- What will happen to the huge and unwieldy global network of trade and information-sharing that humans have created? Will an uncontrollable event such as the coronavirus pandemic cause it to fragment? Or will it become stronger as humans evolve social tools to make it work better?
- Evidence suggests that human activity has caused and is continuing to cause the average global temperature to rise. A mass extinction of species threatens. This will change environments in many ways. But what exactly will this change be like? And how will the different cultures respond? We have invented weapons of mass destruction in a world where nations and their leaders often behave belligerently, incompetently and insanely. Is a catastrophic accident just a matter of time? We can't answer these questions but, on the basis of evidence available today, we can talk about some of the things likely to influence human evolution in the next few decades.

If things continue pretty much as they are, the composition of populations in the near future can be calculated with reasonable accuracy. In 20 years, today's five-year-old children will be 25-year-old adults. Most of today's 80-year-olds will be dead. But things won't continue as they are. There will be unpredictable events. A pathogen might emerge that is far more infectious and lethal than the coronavirus that appeared in 2019. Treatments to counter aging might be developed. An explosive volcanic eruption could fill the upper atmosphere with ash and gases that reduce the amount of sunlight and warmth reaching the Earth's surface. This would temporarily disrupt weather patterns, reduce crop yields, and potentially cause widespread famine. Acts of violent aggression by some humans might also bring disaster. A disaster that strikes a single major city will have global implications.

During the early stages of modernization, human populations grow rapidly. Larger, more densely packed populations are more vulnerable to shocks and maintaining the modern way of life requires complex infrastructures to transport food and water and dispose of waste. These infrastructures are vulnerable to sudden and unexpected changes. We now have "societies" instead of families and countries have evolved social tools of eye-watering complexity to keep these societies working. We call them "systems"—political systems, justice systems, healthcare systems, economic systems, and so on. Everyone agrees that these systems are far from perfect but reaching agreement on how to improve them may be beyond our grasp. If our systems collapse, as those of the Soviet Union did in the last decade of the 20th century, the suffering will be enormous.

There is reason to hope that, by connecting us together, modernization may have made us more resilient. Like our ancestors who survived the last ice age, we might be able to find ways of seeing ourselves as one people. This might encourage us to work more closely together rather than seeing different people as competitors. Today, billions of us are connected. The fundamental question is: How well *will* we work together? The answer depends on how local cultures and our shared global culture evolve.

If future disasters don't have a large death toll, the human population will continue to grow for a few more decades. And as long as no cure for old age is found, it will probably peak at around 11 billion. The rate of population growth has been decreasing since the 1960s, but our numbers will continue to rise for a while longer. This is partly because human life expectancy has increased—on average, we're living longer. But the increasing human population size is mostly the result of the earlier increases. About a third of the people alive today are young and haven't begun to have their children yet. If current trends continue, most of them will have two or fewer children, and by the time these children are elderly, the human population will be declining. Over the forthcoming decades, more and more cultures will have to adapt to the rapid aging of their population. There will be fewer children to educate and more old people to care for. Finding young men willing to fight for a cause might become so difficult that old-fashioned human-fighting-human warfare will become impossible. And instead of trying to limit immigration, wealthier countries might be competing to lure young adults to join their population.

We (the authors of this book) believe that the new story of human evolution we're telling allows us to be more optimistic about the survival of our descendants than those old stories about apemen competing in the savanna. If competition for women and resources had saddled us with a "human nature," then our options would be much more limited. Notions of human nature have a long history in Western cultures. You can find these ideas everywhere, from lofty discussions ("original sin") to idle chitchat ("boys will be boys"). "Human nature" is mostly used to explain "bad" behavior, so stories that talk about it tend not to end well. These stories also gloss over human variability. Some humans do behave as if they have an innate desire for power and status. Some act like they're incapable of moderating their greed or lust and no doubt genes have some influence on this behavior. But that doesn't mean that humans are *genetically programmed* to behave like this. The evidence of our ancestors' lives gives us no reason to believe that such programming is an inevitable human inheritance. Most variation in human behavior is culturally not genetically transmitted and a lot is individually developed. No doubt a few of our forebears were cheating renegades or violent bullies (for whatever reason), but for over a million years, most of our ancestors were valued members of hard-working families. We inherited the genes of people who, as young children, learned how to make older family members care for them. As they reached adulthood, they learned to behave in ways that were appreciated by their peers. And as they became parents and grandparents, they learned ways of making younger family members respect them.

Experiences in today's modern world are different from those of our ancestors, but they still inform our beliefs about how to fit in. We might want to dress up as a princess, make fun of our overweight classmates, or buy a new Lexus that we can't afford—or we may decide not to. Humans are influenced by their culture, just as we're influenced by our genes, but we aren't dupes who simply swallow the information we get fed. Consciously or unconsciously, we decide what to do, even if we end up just doing pretty much what our friends do. The decision-making process of modern people is particularly complicated and unpredictable because we're exposed to such a vast amount of cultural information.

We (the authors of this book) believe that it's time to abandon the idea of "human nature." If we do, there'll be a downside. We'll no longer be able to use it to excuse behavior that is damaging to ourselves or others. The fact is, we're not stuck with endlessly acting out behavior that was shaped long ago by the impersonal forces of natural selection. We're active participants in the evolutionary process that shapes our behavior. The decisions that we, as individuals, make may have a modest effect on how our culture evolves. But, collectively, these decisions have a huge impact. For example, for thousands of generations, family-based culture tightly constrained sexual behavior and, as the influence of families diminished, this constraint loosened allowing ideas about sexual behavior to evolve rapidly and become very diverse. Not surprisingly, this has generated confusion and conflict. Some individuals came to believe that that they could fulfil their sexual desires by imposing their will on people they had power over. If we think this is wrong, and we're in a position to complain and gain support, we can reduce this behavior by changing our culture.

Abandoning the idea of "human nature" will deprive us of the comfortable notion that we possess instincts that will guide us through basic biological life events—such as finding a mate, mating, and the all-important task of raising offspring. It would be nice to believe that our offspring are born equipped with a brain that will, with the minimum of input, develop into an organ that generates "normal" human behavior. But it isn't so. The evidence suggests that the input youngsters receive as they grow up is vitally important. Human genes provide us with a brain that develops the ability to connect to a network. This network delivers our cultural inheritance, and we develop behavior based on the information we receive. Most of our ancestors grew up surrounded by a large family that had culturally evolved to provide youngsters with a network, experiences, and responsibilities appropriate to their age and stage of development. Modern children grow up sharing a home with a relatively tiny family, but they can connect with a very wide world. They go out to attend educational institutions that employ teachers who expose them to information and experiences believed to be age appropriate. Some children who go through this process thrive and develop into adults who find a place in the modern world and make valuable contributions. But not all children thrive, and it's important to gain an understanding of why this is.

When our ancestors organized their social lives around reproduction, their evolutionary path could be explained in Darwinian terms. Humans were unusual because, instead of competing for reproductive success as individuals, it was families that competed, and family members worked as a team to raise the next generation. The restructuring of our cultural networks has freed us from the strong cultural drive to raise the next generation. As cultures modernize, they redirect their members' lives and redeploy the emotions that encouraged our ancestors to produce and care for children. We now use sexual behavior for a wide range of social purposes, but rarely to conceive offspring. And many of us dedicate a great deal of parenting behavior to the care of dogs, cats, and other appreciative species. It seems likely that the cultural norms that supported strong families will continue to erode, but it's not possible to predict what new cultural norms will emerge to replace them. It depends on which ideas run away with people's imagination.

If you're stuck on the idea of "species," then all evolutionary stories end in tragedy. Species go extinct. Only a handful evolve into new species that carry the project of life forward. But every living organism today, whatever its species, is descended from a lineage that survived. Evolutionary stories are like adventure yarns or computer games. Lineages evolve in complex environments where they're confronted with a never-ending series of challenges. A few million years ago, the world began to change, challenging living things with colder, drier, and increasingly variable environments. Many species went extinct, but the survivors possessed adaptations that allowed them to cope with the new variability. Our own ancestors leaned heavily on evolving a large brain and the social tools that enabled them to create and share cultural solutions. Most human lineages did suffer extinction, and our own lineage seems to have come close to it some 70,000 years ago. After that our numbers grew steadily until the last few centuries, when we rather quickly became the Earth's dominant species.

Perhaps it's healthiest to think of the future as an adventure that we are perpetually embarking upon. Unlike other animals, we tell stories about our past adventures and make elaborate plans for our new ones—even if unforeseen events never allow things to go according to our plan. The prime adventure of most of our ancestors was raising the next generation. Modernity has allowed many of us to devote our lives to lots of other challenges, ranging from literal adventures like space exploration to abstract quests in politics, business, sports, the arts, and science. Looking into the future, we can see looming obstacles of our own collective making, including climate change, loss of biodiversity, and political instability in this nuclear age. The exact dimensions of these threats are obscure, and the best way of coping with them is controversial. But then, adventures are always beset by uncertainties and threats.

Our ancestors tackled their challenges as part of a collective or team. Our interconnected world and the problems we share encourage us to see all humans as part of the same team. Perhaps our biggest challenge is developing social tools that will make this a reality, with everyone on the human team sharing the opportunities, responsibilities, and rewards of an adventurous life.

Notes

Chapter 1

1. In her book, *Creatures of Cain: The Hunt for Human Nature in Cold War America*, Erika Milam (2019), a historian, discusses the early development of the human evolutionary story and how it was influenced by (and influenced) what people believed their fellow human beings to be like.

2. New evidence and ideas on human evolution have been described and discussed in a number of books that have been written for a wider readership. We may not agree with everything written in these books, but we nevertheless recommend the following:

   - *The Righteous Mind: Why Good People Are Divided by Politics and Religion by Jonathan Haidt.*
   - *Mothers and Others: The Evolutionary Origins of Mutual Understanding by Sarah Hrdy.*
   - *The Story of the Human Body: Evolution, Health, Disease by Daniel Lieberman.*
   - *Behave: The Biology of Humans at Our Best and Worst by Robert Sapolsky.*
   - *The Archaeology of the Mind: Neuroevolutionary Origins of Human Emotions by Jaak Panksepp and Lucy Biven.*
   - *The Goodness Paradox: The Strange Relationship Between Virtue and Violence in Human Evolution*
   - *Richard Wrangham*
   - *How We Do It: The Evolution and Future of Human Reproduction by Robert Martin*
   - *This View of Life: Completing the Darwinian Revolution by David Sloane Wilson.*
   - *From Bacteria to Bach and Back: The Evolution of Minds by Daniel Dennett.*
   - *Sense and Nonsense: Evolutionary Perspectives on Human Behaviour by Kevin Laland and Gillian Brown.*
   - *The WEIRDest People in the World: How the West Became Psychologically Peculiar and Particularly Prosperous by Joe Henrich*
   - *Survival of the Friendliest: Understanding Our Origins and Rediscovering Our Common Humanity by Brian Hare and Vanessa Woods*

   For a textbook on human evolution, we recommend *How Humans Evolved*, now in its 8th edition, written by our friends and colleagues Rob Boyd and Joan Silk.

3. Ideas about the importance of stories for humans are examined in *The Storytelling Animal: How Stories Make Us Human* (2012) by Jonathan Gottschall.

4. In this book we try to tell stories about episodes in human evolution that are consistent with the evidence. The problem is that the evidence is inevitably fragmentary. As Charles Perrault shows in *The Quality of the Archaeological Record* (2019), fossils and durable artifacts are sketchy traces of lived lives. Archaeologists, paleoanthropologists, historians, comparative biologists, and comparative ethnographers make as much sense out of these traces as they can, but the actual events and processes that gave rise to the complex

beings we call "us" were incredibly complex. Some traces lead to pretty firm conclusions we might dignify with the term "facts," but more often they require interpreting using theory. Theories are unstable, and liable to be upset by new evidence and new theoretical arguments. Scientific facts and arguments are a constantly moving target. Some science fiction writers tell speculative stories about the future, endeavoring not to violate any current science as of the time of writing. The past is nearly as opaque as the future, so if we are to tell a connected story of our evolution, it has to look a bit like this kind of science fiction.

In interpreting the sketchier traces of our evolution, we lean on what is called the *theory of gene-culture coevolution*. The basic tenet of this theory is that human culture, a product of the fact that humans learn more from each other than other species, is an extraordinarily important part of our evolutionary story, and that it became increasingly important over the history of our lineage. Summaries and examples of this approach to understanding our story can be found in: Shennan (2002), Richerson and Boyd (2005), Boyd and Richerson (1985), Mesoudi (2011), Bettinger, Garvey, and Tushingham (2015), Henrich (2016), and Baum (2017). Other authors want to tell a more gene centered story. For review, see: Richerson (2018).

5. American psychiatrists Teicher and Sampson (2016) reviewed evidence that experiences resulting from childhood maltreatment alter brain development and saw neglect as being a form of maltreatment. More evidence of the link between childhood trauma and neglect, brain development, and mental health is being published all the time. See, for example, Opel et al. (2019).

6. Sarah Hrdy has devoted her career to countering male bias in research in human evolution. Her book *The Woman Who Never Evolved*, first published in 1981 (with a later edition in 1999) is a classic. But women began trying to warn their male colleagues about this bias much earlier. See, for example, Sally Linton (1971).

7. Today, many women are doing anthropological research from an evolutionary perspective. The women who began working in the field in the 20th century, like Laura Betzig, Elizabeth Cashdan, Kristen Hawkes, Sarah Hrdy, Debra Judge, Jane Lancaster, Donna Leonetti, Bobbi Low, Ruth Mace, Monique Borgerhoff Mulder, Rebecca Sear, Mary Shenk, and Polly Wiessner have blazed a very wide trail.

8. See Volk and Atkinson (2013) and Volk and Atkinson (2008).

9. See Lewens (2012).

10. Darwin's ideas about humans are much misunderstood both by his contemporaries and by recent scholars. Only a few Darwin specialists, such as Robert Richards, have carefully read the *Descent of Man* (Richards, 1987). It seems to us that Darwin's treatment of the role of culture in human evolution was rather sophisticated, and that the failure of the social sciences to incorporate his ideas was a blunder of historic proportions. It left some of us who stumbled upon the same ideas decades later with an abundance of career-making low-hanging fruit to harvest. But it left the social sciences lacking critical tools for three-quarters of a century! This is argued in Richerson and Boyd (2010).

11. Wikipedia has an excellent and well referenced entry on Wallace, at https://en.wikipedia.org/wiki/Alfred_Russel_Wallace.

12. From Wallace's essay on the origin of the races of man (Wallace, 1864).

13. Charles Darwin, *The Descent of Man and Selection in Relation to Sex* (1871).

14. Herbert Spencer (1820–1903) was another Englishman with an impressive beard. He considered himself to be a philosopher, anthropologist, and sociologist, as well as a biologist. He was inspired by reading Darwin's *On the Origin of Species* and coined the phrase "survival of the fittest," which many people now believe Darwin invented. But Spencer's ideas of evolution were quite unlike those of Darwin. Darwin's theory of natural selection proposed that living things changed in response to environmental change. Spencer promoted the idea that evolution is progress from the simple to the complex and from the less good to the better. This "progressive" idea of evolution was old even then, and many people still believe it today. See Delisle (2019). Darwin's theory of evolution proposed that there is no drive to greater complexity or diversity except when it is favored by environmental circumstances. For example, the shelly shallow-water marine invertebrates that fossilize well are more diverse today than at any time in the Earth's history. According to a classic paper by Valentine and Moores (1970), this is merely because plate tectonics has divided the earth's present land mass into several relatively small continents with long coastlines spanning many degrees of latitude. Each long coastline has its own distinctive suite of species and this has driven diversity to record levels. Spencer saw progress everywhere—in the physical world, the biological world, human societies, and the human mind. The public loved his flamboyant way with words, and his optimistic outlook suited the middle part of the 19th century. But even though technology was making machines and human life more complex and (for many) more comfortable, there is no evidence that this is part of some cosmological march of "Progress." As far as physics is concerned, Spencer's big idea was akin to the second law of thermodynamics, but backward! We talk more about this in Chapter 7.

15. Jemmy Button continued to live with his people at the tip of South America, but he also never forgot the English people that he had met and the experiences he had had in his early teens. He was called upon several times to help English-speaking missionaries trying to "help" his people and teach them about Christianity. At least two books have been written about his life: Hazlewood (2001) and a children's book called *Jemmy Button* (Vidali, Uman, and Barzelay (2014)).

16. There are, of course, some similarities between our behavior and the behavior of other animals, especially apes. Darwin looked for and found evidence of similarities in his 1972 book *The Expression of the Emotions in Man and Animals*. These similarities are the result of humans sharing ancient emotional circuits with other mammals. For a very readable description of this, see Panksepp and Biven (2012). Darwin did recognize that there is a great gap between the minds of even the most untutored humans and the cleverest nonhuman animals and describes it in chapters 3 and 4 of his 1871 book *The Descent of Man*.

17. Primatologist Frans de Waal is a strong advocate for the mental capabilities of primates and other animals. See, for example,. De Waal (2016).

18. B. Hare, J. Call, and M. Tomasello (2001).

19. The most famous (so far) example of animal generosity is provided by the mouse-sized common vampire bat (*Desmodus rotundus*), which lives and hunts in the livestock-grazing areas of South and Central America. They fly out from their roost every night to find an animal (usually a cow) that they can land on, bite a small hole in, and cling onto long enough to lap up the blood that flows out. In populations that have been closely monitored by biologists, about one in five of the bats returns to the roost without having managed to

get a meal. When this happens, it's common for a bat in the roost who has been luckier to regurgitate some of the blood in its stomach and allow the hungry bat to lap it up. Often the bats that help each other are relatives, such as two sisters or a mother and her pup, who is weaned but still inexperienced. But sometimes these bats provide food to hungry nonrelatives they have been roosting with for some time and who have perhaps helped them in the past. See Wilkinson et al. (2016).

20. For a good review of mothering behavior, including the mothering of other people's children, see Hrdy (2009).

21. This is discussed in Frost and Richerson (2014).

Chapter 2

1. This story of baby ape life was partly inspired by the many books we have read and the wildlife documentaries, lectures, and talks that we have seen. It's also inspired by conversations with primatologists like Bill Mason, Meg Crofoot, Joan Silk, John Mitani, Roman Wittig, Isabel Behncke, and Sarah Hrdy. A number of books have been published recently that summarize results of research into the bodies and behavior of humans and talk a lot about how humans are similar to or different from apes. Here are some of the ones that we have read and recommend to anyone who wants to know more: Hrdy (2009), Lieberman (2013), Martin (2013), Sapolsky (2017).

2. This is something a human female can't do when she gives birth. A human baby has to turn as it travels out of the womb to squeeze its head through the opening in its mother's pelvis. As a result, the baby faces backward as it emerges from the vagina and cannot be pulled out by a mother bending forward.

3. Oxytocin is a chemical that can be synthesized and used as a drug. Delivered intravenously, it can be used to "induce" labor by triggering contractions of the uterus, but oxytocin delivered this way doesn't enter the brain. Some scientists suggest that if a way was found to deliver it to the brain, it might be possible to use oxytocin as a drug to influence behavior. Some studies have suggested that people may behave in a more trusting and trustworthy way if oxytocin is squirted up their nose, but most studies have shown no effect. Oxytocin made naturally within the brain can have a powerful effect on the emotions of many mammals by attaching to certain receptors. Synthetic oxytocin doesn't appear to reach these receptors if it's given by injection or administered nasally. See Nave, Camerer, and McCullough (2015).

4. Mother chimps and, to a lesser extent, bonobo mothers have good reason to fiercely protect their young. See notes 8 and 12.

5. Diets in wild apes are hard to study quantitatively. See Thompson and Wrangham (2008). The avidity with which chimpanzees feed on ants and termites, and manufacture tools to catch them, suggests that they might be an important source of fat, protein, and micronutrients to supplement the ripe fruit that is their dietary mainstay. Similarly, chimpanzee males are avid, if opportunistic, hunters of smaller vertebrates. However, in the few quantitative estimates we have been able to find, the role of animal resources appears to be quite minor compared to ripe fruit. The primates for whom insects are a major source of nutrition are small-bodied forms. Chimpanzee reproductive rates are regulated by energy availability and possibly protein availability. At the least, the fact that chimpanzees invest considerable effort in exploiting animal foods suggests that they are resource-limited. See

Deblauwe and Janssens (2008), Oelze et al. (2011). It seems likely that the last common ancestor was similarly keen on eating animals when possible (Raubenheimer and Rothman (2013).

6. Hinde and Milligan (2011).

7. Pontzer et al. (2016).

8. Pusey et al. (2008).

9. Primatologists observing them in the wild have recorded many instances of aggression, killing, and cannibalism. Observations are summarized and analyzed in Wilson et al. (2014).

10. Feldblum et al. (2014).

11. Arcadi and Wrangham (1999).

12. Chimpanzee mothers bring their infants into a hostile social world. See Thompson (2013), Pusey and Schroepfer-Walker (2013). Infanticide by males rising in the dominance hierarchy is widespread in social mammals and was first brought to the attention of the primatological community by Sarah Hrdy (1974). Females may engage in infanticide as a particularly rough form of competition. Victims of infanticide are commonly eaten, so the perpetrators may gain a direct subsistence benefit. See Walker et al. (2018). In such a world, chimpanzee mothers with newborn infants have been observed to take "maternity leave," absenting themselves from social contact, especially contact with males that pose the greatest risk. See Hitonaru and Nakamura (2018), Lowe, Hobaiter, and Newton-Fisher (2019). Young, inexperienced mothers are at greatest risk of losing infants to infanticide and other hazards. Consequently, males prefer to mate with older, experienced mothers. Bonobo rates of infanticide and other forms of violent conflict are much lower, so we are free to imagine the last common ancestors as "rough customers" like chimpanzees, a bit more gentle like bonobos, or somewhere in between. Of course, our shared ancestor was different from all of us, but we can't know in what ways.

13. Muller, Thompson, and Wrangham (2006).

14. Young chimpanzees are almost exclusively reared by their mothers. Fathers and female relatives play at most very marginal roles. See Hayashi and Matsuzawa (2017). Compared to humans, the social network of young chimpanzees is very restricted. When acquiring their culture, children see that people are not all the same and often behave very differently. At a very young age, they begin to be selective about whom they learn from. (More about this is Chapter 5.) Chimp infants are pretty much constrained to learn from their mother. If a mother chimp doesn't know some useful trick, her offspring are unlikely to learn it. See Lonsdorf (2006).

15. The genes of humans, chimpanzees, and bonobos are quite similar, despite major differences in anatomy and behavior. See Prüfer et al. (2012). We would really like to know the details of how genes underpin the differences in the anatomy and behavior of the three species. See Laland, Odling-Smee, and Myles (2010) and Ross and Richerson (2014). Understanding this is a much more difficult problem than estimating the similarities in gene sequence, because the functional significance of most of the genetic variants is not well understood. Most genes are part of complex functional and regulatory circuits. It's proving complicated to work out how genetic variation found in sequences of DNA translates into the anatomical and behavioral differences seen in whole organisms. But, in a few cases, it has proved relatively simple. One example of this simplicity is provided by the genetic changes that cause some humans to continue to produce the enzyme which breaks down the milk sugar "lactose." We mention this in chapter 7.

16. The *Ardipithecus ramidus* find was reported in a series of papers in *Science* magazine in 2009. See White et al. (2009). It is worth noting that finding a fossil trove sufficient to infer most of the anatomy of a creature like this is very rare. This at least partly reflects preservation problems. As noted in the text, acid soils characteristic of tropical forests are very unkind to bones. Studies of how the bones of dead animals do or do not get buried to potentially form fossils suggest that most skeletons are scavenged, scattered, or trampled before being fossilized, as with *A. ramidus*. Skulls and especially teeth are the most durable parts of animals and are over-represented in fossils. We also need to worry about how common the animals of interest were among the fauna in their living community. Of the total vertebrate fauna at the Afar, Ethiopia, *A. ramidus* sites, only 1.5 percent of the vertebrate fossils recovered were from this species. Of course, this site was intensively dug only because the hominin fossils were discovered. We might entertain the hypothesis that our ancestors were generally somewhat rare until rather recent times. Estimates based on human genes suggest that australopithecines were tolerably common, but that members of our own genus *Homo* were rather rare. See Li and Durbin (2011) and Scerri et al. (2018). When populations are large, considerable genetic diversity is preserved, but less is preserved when they are smaller. In one famous site, Swartkrans in the Cradle of Humankind cave complex in South Africa, excavated laboriously by C. K. "Bob" Brain (2004) and many collaborators, they discovered 4 specimens of our genus *Homo*, 39 of australopithecines, and 39 of various species of monkeys. The total number of birds and mammals recovered was 704. Stories tend to suggest that ancient humans were successful in proportion to the degree to which they were like us. This is a mistake.

17. We've been arguing that our last common ancestor with chimpanzees and bonobos must have generally resembled the apes living today, but we have to be somewhat vague about how their social lives were organized. There is no standard form of ape social organization. As the sociologist Alexandra Maryanski has pointed out, each of the living great apes has its own distinctive form of social life. Orangutans are mostly solitary except for mothers and their young, but even if orangs live separately, they are aware of the others of their kind living in or visiting their neighborhood. Gorillas live in "harem" groups with a single male and two, maybe three females with their young. Chimpanzees live in larger groups of males, females, and young. In both gorillas and chimps, males are larger and "dominate" females. The males in chimpanzee groups are often related and form a dominance hierarchy. Bonobos also live in multi-male, multi-female groups like chimpanzees, with males in a dominance hierarchy. But bonobo females cooperate to limit male dominance and sometimes dominate the males. In both chimpanzees and bonobos, males inherit their fathers' territories and females usually move to a different group when they reach sexual maturity. Modern humans are also technically apes, and we live in a great variety of societies, some formally organized by kinship, some not, some with formalized dominance hierarchies, some more egalitarian.

We can maybe infer from this variation that ape social organization evolved quite readily compared to that of other primates. Consider the many species of monkey that live in Asia and Africa. All of them live in multi-male, multi-female troops with male dominance hierarchies. It's the females that inherit their mother's territory and the young males that disperse at maturity. Closely related female clans collectively compete for dominance over other clans. See. Maryanski (1992).

18. Methodological note: Our reconstructions of human evolution depend on several sources of data. Classically, such reconstructions depend upon fossils and, in the human case, durable artifacts like stone tools. Comparative biology often gives important clues. Sometimes traits are pretty conservative in a lineage, or they vary predictably as a function of environmental variables. In the previous footnote we argued that ape social organization is neither conservative nor does it vary in any obvious way with ecology. The genomes of living populations contain a huge amount of information about history, and we can even recover ancient DNA from bones tens of thousands of years old (so far). The advent of inexpensive gene sequencing has begun to unlock this treasure-trove of information. The complexity of the mapping between genes and whole organisms still makes interpreting evolutionary events challenging. Natural selection, for example, falls in the first instance on the capabilities of whole organisms, or major parts of them, and only secondarily on the genes that influence bodies. Paleoecology is another important source of information about the past. It is past environments that generated the selection pressures that shaped bodies and physiology, and, in the human case, our cultures. In the best-case scenario, the evidence from all these domains point in the same direction. If only! Still, we have come an immense distance since Darwin wrote *The Descent of Man* 150 years ago.

The evolution of the human hand is an example of an adaptive change that is reasonably well understood, courtesy of a combination of comparative functional anatomy and the fossil record. See Young (2003), Horns, Jung, and Carrier (2015), Kivell et al. (2016), Williams-Hatala et al. (2018). The hands of the other apes have longer curved fingers and relatively shorter thumbs than modern human hands. Human hands also have wider articulatory surfaces. Chimpanzees use their hands for precision gripping tasks, make tools from wood, and use hammer stones to open nuts, but their forelimbs are also essential for locomotion, especially for climbing in trees. Hence their long, curved fingers. They cannot, however, knap stone tools, in part at least because the precision power grip that humans use to accomplish this task requires the modifications seen in our hands. They also have trouble generating the power grip used to grasp tool handles and spear shafts. Although complete fossil hands are rare, the few that exist suggest that the later bipedal australopiths and early members of our genus already had hands at least partly evolved in the modern human direction. Not coincidentally, the earliest knapped stone tools seem to have been made by australopithecines sometime before the first *Homo*. Selection for increased toolmaking abilities began before our brains began to expand, as early as 3.3 million years ago. Stone tools of some sophistication appear by 2.6 million years ago.

The paleoclimatic and paleoecological background for the evolution of humans is gradually becoming better known. See Sánchez Goñi et al. (2017) and Sánchez Goñi et al. (2018). Over the last 70 million years, the Earth has on average become cooler, drier, and more variable. Deserts and grasslands replaced evergreen forests over much of the Earth's surface, leading to the evolution of grazing mammals like modern horses and ungulates. See Zachos et al. (2001). The Earth has experienced considerable climate change during the last seven million years, especially during the Pleistocene (2.6 million years ago to 11.7 thousand years ago). Climates became still cooler and drier and exceedingly variable. Adapting to Pleistocene variability has influenced the evolution of many animals. For example, observations of fossil skulls suggest that brain size has increased in many mammalian lineages in this period, plausibly in response to increasing environmental variability beginning about 2.6 million years ago. See Jerison (1973). Although apes have long relied on learning, those that lived seven million years ago may well have had a smaller brain with less cognitive capacity and poorer social learning skills than the today great apes. Many human evolution stories told to the public have portrayed it as a series of breakthroughs in which growing brains drove the evolution of culture. It makes just as much (if not more) sense to imagine that climate variation drove the evolution of culture as a mode of adaptation to variability, and that the more sophisticated culture required bigger brains. See Sherwood, Subiaul, and Zawidzki (2008).

19. Some patterns of variation seem to be transmitted from generation to generation by epigenetic inheritance. Several systems of inheritance besides gene sequence differences exist. We concentrate on the cultural system based on imitating others or being taught by others in our social network in this book because it is so important in our lineage. Biological systems of epigenetic inheritance involve such things as the modification of chromosomes to cut or enhance the expression of genes. See Jablonka (2013) and Bošković and Rando (2018).

20. Research on animal friendships is reviewed in Seyfarth and Cheney (2012).

21. Isabel Behncke Izquierdo (personal communication).

22. For example, mice don't need to have had a scary experience with a cat in order to be afraid of them. It's been found that the smell of cat triggers the emotion of fear in mice, even mice that have never been near a cat. See Panksepp and Biven (2012). It could be that apes experience similar triggers, and perhaps also humans. But ape (and certainly human) responses are also shaped by what we learn from experience. Even reactions that feel and look "instinctive" may be shaped partly or even entirely by learning.

23. We should take care not to fetishize brains. Brains are important organs for acquiring and coordinating behavior, but the whole body participates in the acquisition and execution of behavior. The fact that our legs are long compared to our arms, and our hips allow us to balance over them, virtually guarantees that we will learn to walk upright. Some humans have balance disorders so extreme that they learn (awkward) quadrupedal walking. See Ozcelik et al. (2008). Similarly, the architecture of our hands ensures that babies will learn to grasp and manipulate objects. Other apes can grasp with their feet and, despite our foot being extensively modified for walking, individuals without functional hands can become rather "handy" with their feet. Similarly, across the body. Our digestive system influences what we find satisfying in food. Our shoulders allow us to learn to throw accurately. These effects are called "embodied cognition." See Wilson (2002).

24. When a female chimp mates with several males, no one knows who fathered the baby, and if a male is less likely to attack an infant when he may be its father, then having sex with many males may decrease the chances of her baby being killed by a male. It is, however, impossible to know what a chimpanzee "knows" or doesn't "know." It's hard enough to observe what they do. See Murray et al. (2016).

25. It is impossible to be certain that the last common ancestor with chimpanzees and bonobos was male philopatric or not, as we have remarked earlier (see notes 1 and 11). It seems like a good guess but a phylogenetic analysis found that strict philopatry plausibly evolved after the chimpanzee-bonobo line diverged from the lineage that led to humans. See Duda and Zrzavý (2013).

26. Bădescu et al. (2016) and Lonsdorf, Stanton, and Murray (2018).

27. For a review of work on great apes and culture, see Whiten (2017).

28. The way that brains and bodies interact in the course of development is an area of active controversy. The dominant theory for many years has been one in which the mind (or brain) is elaborately structured by genes so that it can interpret sensory information and produce outputs to the action systems of the body. Books by E. O. Wilson, Steven Pinker, and others have popularized this view. See Wilson (1978) and Pinker (2002). But see also Edelman (1987). Edelman attacked this innatist picture of cognition on the basis that the neocortex of humans and other large-brained species was far too complex to be wired in any detail by genes. More recently, a number of authors have enlarged on Edelman's idea, including the following: Panksepp and Biven (2012), Anderson (2014), Krubitzer and Stolzenberg (2014), Ellis and Solms (2017), d'Errico and Colagè (2018), Baum (2017) and Heyes (2018). These authors argue that the brain's "neocortex" is an organ of phenotypic flexibility. Functional cognitive modules are organized, contingent on experience, during development. Cortical resources are organized into functional circuits so that functional cortical circuits typically make use of a number of cortical regions, and any given cortical bit typically participates in a number of functional modules. Thus, for much of the cortex, the phrenological picture of the brain as a collection of regions each devoted to a particular cognitive function seems to be wrong. This makes sense energetically. Brain tissue metabolic overhead is high and it should be used as efficiently as possible. Any given small bit of cortex should be busy as often as possible, so neurons are extensively recycled from one task to the next as one goes about daily life. The cortex has a number of powerful mechanisms for self-organization, including social learning, and reinforcement-based learning. Reinforcement-based learning is particularly important in connecting cognitive development adaptively to environmental contingencies. Reinforcement is generated by the emotional and appetitive circuits in the brain stem and adjacent structures. Eating when you are hungry is reinforcing. Pleasant social interactions, likewise. Fear is unpleasant and tends to extinguish behavior that causes it. The emotion circuits are an ancient and highly conserved part of vertebrate brains, but they project extensively to the neocortex and receive projections from it. This mutual modulation means that the emotional circuitry can have a big impact on what is learned, but also that what is learned can modulate the emotions and appetites. Thus, rather general reinforcement mechanisms can shape learning and cultural evolution, but culture and individual learning can reshape reinforcement. For example, in many cultures, capsicum, the pain-causing compound in red peppers, comes to be experienced as rewarding by most people. This picture of the neocortex as a highly flexible knowledge acquisition system guided by reinforcement matches neatly the adaptive analysis of culture as an adaptation to environmental variability that Robert Boyd and Peter Richerson made in 1985.

On the evidence developed by the authors mentioned above, cognition is substantially built on the basis of principles discovered long ago by behaviorists. Genes can certainly act on the ancient, conserved emotional and appetitive circuits of the brain stem and related areas. For example, humans are not as fearful of other people as chimpanzees are of strange chimpanzees. Our docility in turn makes our affiliation with distant kin and even strangers comparatively easily reinforced, as Herbert Simon (1990) noted our more complex culture and our societies are built on the foundation of a down-regulated fear of strangers. With regard to diet, our appetites are tuned to find (and make) foods rich in fat, protein, and calorie-dense carbohydrates satisfying. Chimpanzees, and our last common ancestor, with their longer guts, were quite satisfied with less nutrient-dense plant foods—foods that we still might try in desperation in a famine.

In our view, the "cognitive revolution," started by Noam Chomsky's famous critique of B. F. Skinner's picture of language acquisition, made an epic mistake. It suggested that there could *either* be innate cognitive explanations *or* behaviorist ones based on reinforcement. This error was compounded by some authors suggesting that behaviorists and other social scientists imagined cognition to be created on an implausible gene-free "blank slate" idea.

29. Alan Rogers (1989) introduced a simple model of the evolution of culture in a temporally variable environment. Individuals in the model were of two types: individual learners and social learners. Sure enough, in his model social learners increased in the population when rare because social learners did not have to pay the cost of individual learning. However, as they became more numerous, social learners more and more often learned from other social learners rather than from individual learners. In a variable environment, abundant social learners would often end up with outdated behaviors. At equilibrium, the fitness of social learners and individual learners would be equal. Culture evolved, but social learners were semi-parasitical information scroungers who didn't increase the fitness of the population. Could it be that much if not most nonhuman culture is of this nonadaptive type? In the case of humans, culture does seem to be the sort of thing that often does increase mean fitness. Humans are selective social learners and teachers. For example, if something we learn from others doesn't seem to work well, we'll seek out someone with a better technique or try figuring out the problem for ourselves. On this, see Boyd and Richerson (1985 and 1995). We had incorporated such considerations in our models of human culture and found that this made an adaptive system. But it took a bit of time to figure out exactly what was "wrong" with Rogers's model.

That's because this model isn't exactly wrong. For example, patent law arose because inventive people's ideas could easily be copied by others. If invention is costly, inventive people will be discouraged from investing time and effort in inventing unless they can capture some of the public benefits of their inventions. Patent law attempts to maximize the total benefits to society of invention by giving inventors a fair cut of the benefits. In simpler human societies, the prestige that accrues to skilled craftspeople may act as an informal sort of patent. Community members let them know that they are especially valued members and this may cash out as extra care when they are sick or make them a higher draw in the marriage market. See Henrich and Gil-White (2001).

Humans aside, it was possible to imagine that most nonhuman social learning was just information scrounging that was not fundamental to their adaptation. But these studies were at the beginning of a golden age in the study of nonhuman social learning. By now, the evidence suggests that social learning is an essential adaptation in many species. Laland (2007) and colleagues have worked on a diverse set of species, including stickle-backed fishes, for whom such things as foraging and prey avoidance are socially transmitted. Hal Whitehead and colleagues have reviewed the many feeding adaptations in whales and dolphins that are a product of cultural evolution (Whitehead and Rendell (2015)). Andy Whiten (2017), meanwhile, has reviewed the evidence that great ape culture is consequential for their lives. Human culture stands out for its complexity, degree of cumulative complexity of cultural traits, and the degree to which our adaptations are dependent on it. But it is clear by now that the differences between humans and other social learners are quantitative, not qualitative.

30. On the importance of mothers in chimpanzee socialization, see Lonsdorf (2006). Nevertheless, it's likely that chimps sometimes gain information by watching other individuals. This would account for the spatial patterns of cultural variation they exhibit. See Whiten (2017).

31. Whiten et al. (1999).

32. Cibot, Sabiiti, and MacLennan (2017).

33. See note 27.

Chapter 3

1. In 1976 paleontologists looking for fossils in Tanzania, in western Africa, found footprints made by three individuals that had been preserved in the ash of a volcanic eruption that occurred about 3.7 million years ago. They look no different from footprints that would have been made by three small humans, but they have been eroded and don't show much detail. More detail might show that australopithecines, with their longer arms and shorter legs, walked slightly differently from us. See Leakey and Hay (1979).

2. In 2006 the South African paleontologist Lee Berger (2006) reexamined the very first australopithecus bone found—the skull of a juvenile known as the "Taung child," found in 1924 by Raymond Dart. Dart had worked out that it was likely the victim of a predator, and Berger concluded that the child had been killed by a large bird of prey. The skull has marks on it similar to those found on the skull of a monkey that has been killed by a crown hawk eagle.

3. A kind of monkey known as the "vervet monkey" lives today in Africa in the same kinds of habitats as those occupied by our australopithecine ancestors. Vervet monkeys have a set of predator alarm calls with different calls for eagles, snakes, and leopards that require quite different escape strategies. See Cheney and Seyfarth (1990).

4. This was part of a slow process of climate change that had begun millions of years earlier. Fifteen million years ago, much of the Earth's land area was covered in tropical and subtropical evergreen forests, and in those days, apes lived as far north as where Moscow is today. At the same time, more rapid change processes were occurring in the eastern part of Africa. The gradual movement of continents was slowly tearing Africa apart, creating a "rift" that runs north-south in eastern Africa. Even though the movement of continents is slow, the volcanic eruptions and earth movements that it triggers can bring sudden changes to, for example, the course of rivers or the size of lakes. See Maslin,. Shultz, and Trauth (2015).Evidence of the change in vegetation of Africa is given in Uno et al. (2016).

5. The physical anthropologist Daniel Lieberman (2013) gives good descriptions of the bones and teeth of australopithecines and early humans, and what they tell us about their lives.

6. There are some groups of chimpanzees with territories on the edge of wooded areas that do sometimes go out into the drier grassland and savanna habitats. Primatologists, notably Bill McGrew and his colleagues, have closely observed these chimps to try to understand what limits their excursions into the habitats where our ancestors thrived. They found that chimps do go out into these areas but don't forage efficiently, and have concluded that they lack cultural knowledge of these areas, what foods they contain, and how to access these foods. This is what would be expected if mothers and young didn't spend much time out in these areas. See McGrew, Baldwin, and Tutin (1981, 1988).

7. Sarah Hrdy (2009) spent many years studying the way that animals care for their young and the help that mothers receive from others (known as "allomothers"). She has compiled a great deal of evidence that ape or early human mothers starting to raise their young together is an important part of the human evolutionary story.

8. Primatologists observing chimpanzees report that a mother's older offspring may help a bit by playing with their younger sibling for a while. See Bădescu et al. (2016).

9. Pair-bonding and male parental care are uncommon but not absent in mammals (Woodroffe and Vincent (1994)). Pair-bonding occurs in gibbons, which are quite closely related to great apes (and are sometimes called "lesser apes"), but the father provides little or no care for the baby apart from working with the mother to defend the territory in which they forage. Titi monkeys, which live in the South American tropical forests, pair-bond, and the fathers do provide care. In titi monkeys, it is the males rather than the females that carry the baby, and the pairs are inseparable. The female forages and the male follows around after her carrying their baby. When the baby is hungry, he gives it to its mother to suckle for a while.

10. See Gray and Anderson (2012). The suggestion that males brought back food for the family were very popular in the middle part of the 20th century; see Lovejoy (1981). Many people still assume that if mothers receive help raising their offspring it must have come from the father of the offspring, because, in evolutionary terms, fathers have as much of an interest in seeing their offspring survive as mothers do. In fact, care from fathers is very rare in mammals and is only found in species that are pair-bonded. There is no evidence that australopithecines were pair-bonded, and the size difference suggests that they were not. When males are considerable larger than females, as they are in other apes, this is usually because males expend most of their spare effort squabbling among themselves for access to females. The size dimorphism in australopithecine skeletal material suggests that they were following the standard ape pattern. See McHenry (2005) and Shennan and Steele (2005). Lovejoy and colleagues have cast doubt upon some of the data that are usually interpreted as evidence of sex dimorphism, and all agree that the skeletal sample is too small to be completely confident of the dimorphism hypothesis. See Reno et al. (2003).

11. Bell, Hinde, and Newson (2013) make the argument in mathematical terms.

12. If you are interested in learning more about the social lives of sperm whales, we recommend a book written by Hal Whitehead (2003). The application of the sperm whale life history to australopithecines may seem far-fetched, but we think that, given ape flexibility in regard to social organization, it is advisable to think outside the box regarding the life histories of our ancestors. For another possible model, consider Mattison et al. (2019). Some ethnographically known human societies follow an extreme family formation system in which men remain living with their mothers and sisters. They mate with women in visits to other female-dominated households. Men contribute a little to their natal household economy and only provide a few gifts to their consorts and the children they have probably sired in other households. Males spend most of their effort competing for consorts rather that working to support either household. Mattison and colleagues call this the *expendable male hypothesis*, and it seems to occur in economies where women simply don't need much help from men, for one ecological reason or another. We know from bonobos that females can cooperate to dominate males even when they are not related to each other. Australopithecine cooperative breeders might have tolerated sons and brothers hanging around as long as they didn't cause too much trouble and provided a little help. Big males

would have been useful to deter predators, for example. For homework try to think up another plausible model of australopithecine family life!

13. Australopithecines had moved into environments where the availability of food was likely to be different at different seasons of the year, and so they may have become fatter during the wet season and skinny in the dry. Spending less of their time climbing trees meant that they could carry around more fat. More about this is covered in Heldstab, van Schaif, and Isler (2016).

14. For more information, see W. Trevathan (2010).

15. Nina Jablonski (2008) provides a wonderful description of human skin and how it evolved.

16. The emotions that make a mother mammal feel bonded to her infant are the result of communication between different parts of the brain and the hormones that circulate in her body. But the same systems are also behind other feelings of bonding—the attachment that pair-bonded mates experience, the attachment between friends, and even the attachment that we feel for a pet dog and the dog feels for us. Scientists are understanding these systems in increasing detail and also learning more about the genetic variation linked to variation in the strength of bonding and what triggers it. See Feldman (2017).

17. See note 3.

18. How much the brain is specifically reorganized to support human-specific cognitive achievements is controversial. Some authors argue that the anatomical reorganization of the cortex in human evolution was substantial. for this view, see Tobias (1987) and Holloway, Broadfield, and Yuan (2003). Other authors argue that mammalian brains mostly follow rather tight patterns of correlation between parts, and that gene-based reorganization was modest. On this view, see Krubitzer (1995), Finlay, Darlington, and Nicastro (2001), and Semendeferi, Lu, Schenker, and H. Damasio (2002). We are inclined to think that the modest reorganization hypothesis is likely to prove correct. If you are interested in going into more detail about how the brain of humans and other animals develops the ability to think in complex ways and organized complex behaviors, we recommend several books that discuss this, approaching the problem in different ways. See note 28 in chapter 2.

19. For a review of this, see Hublin, Neubauer and Gunz (2015).

20. See note 18.

21. Cathy Hayes and her husband Keith were determined that Viki would spend her whole life in a human home and not go to a zoo or research facility, but unfortunately her life was not very long. Viki died of viral meningitis when she was seven. The Hayes' experiences raising a chimpanzee are lovingly described in *The Ape in Our House* (Hayes 1951).

22. Andy Whiten (2017) and his many colleagues and collaborators have investigated social learning and how it influences the evolution of culture. They have carried out some very clever experiments comparing social learning in humans with that of other primates. The experiments have shown that chimpanzees are often slow or reluctant to abandon ways of foraging that have worked for them in the past and adopt more effective methods. Humans seem to find it easy to compare two methods and choose the more effective one, but it may be harder for other animals, even for apes, who are very good learners. Young chimps living in the wild spend the first years seeing their mother forage, and potentially learn a great deal from this observation, but they seldom if ever get the chance to closely watch other experienced foragers. It's therefore not surprising that chimps don't always make comparisons and replace a familiar foraging technique with a better one. In their social

environment, natural selection would not favor a chimp that had such mental abilities. Some experiments, however, have revealed apes to be more flexible. It may be that apes can learn how to be more efficient at comparisons. See also Harrison and Whiten (2018).

Research reveals humans to be adept at making choices about what to learn and whom to learn from. The ability can be observed in very young humans. See Rakoczy, Warneken, and Tomasello (2009), Chudek et al. (2013), and Poulin-Dubois and Brosseau-Liard (2016).

23. Chimpanzees living in captivity have very different experiences than chimps living in the wild. They don't have the problem of foraging for their food, but they have other problems to solve, especially if they live in an ape research facility. Researchers give them a wide variety of puzzles to solve to learn about their mental abilities. Chimps that are born in research facilities are much better at these puzzles than chimps that were born and grew up in the wild. But, as you would expect, those living in the wild are much better at foraging. Behavior of captive-bred chimps is very different from that of those in the wild, and they have been observed to behave in ways that seem quite human. Sometimes one will intervene in a conflict to between two other chimps—discouraging aggression, as we imagine that Australopithecine mothers would have done. This shows that apes are capable of behaving like this. But in wild groups of chimpanzees, this kind of third-party intervention isn't often seen. See Miller et al. (2017).

24. Wallace, (1864).

25. What Darwin (1874, p. 179) said was: "A tribe including many members who, from possessing in a high degree the spirit of patriotism, fidelity, obedience, courage, and sympathy, were always ready to aid one another, and to sacrifice themselves for the common good, would be victorious over most other tribes; and this would be natural selection."

26. Many evolutionists are now convinced that one of the most important components of human evolution is selection for friendliness (aka "prosociality"), although there are many different ideas about when and in what form this selection took place. See Boyd and Richerson (1982), Simon (1990), Richerson and Boyd (1998), Boehm (2012), Wilson (2012), Richerson et al. (2016), Hare (2017).

27. Wrangham (2019) explains the domestication syndrome and suggests that an important element in human evolution is humans domesticating themselves. We recommend this book even though the story he tells is rather different from ours. We suggest that the actions of carers discouraged aggression in youngsters to the extent that very aggressive youngsters often never made it to adulthood. Wrangham suggests that when an aggressive male did reach adulthood and tried to dominate other males, the lower-ranked males ganged up on him and killed him. We have discussed stories with Richard and agree that there is merit in both stories. Probably a bit of both went on. For a genetic explanation of the domestication syndrome, see Wilkins, Wrangham, and Fitch (2014). The fox domestication study is summarized in Belyaev (1979).

Chapter 4

1. The scientists who found and described the remains of long extinct human-like animals have tried to categorize them into species that seem more or less like us. The remains have been given names like *Homo habilis*, *Homo ergaster*, *Homo rudolfensis*, and *Homo erectus*. It's impossible to know, based on the evidence found so far, how the individuals who left

these remains were related, either to each other or to us. We have decided in this book not to concentrate on these details but to look instead at what the remains reveal about what early humans were doing. Klein's (2009) textbook is a good source for details about the stones and bones. *Scientific American* and *National Geographic* magazines have produced some well-illustrated materials.

2. The brain of humans today is bigger still, adding another 50 percent or so. We're deliberately being vague about exact sizes of brains because a small increase in brain size doesn't necessarily mean much. Body size is a big factor because larger bodies can support larger brains. The brains of some australopithecines may have been bigger than the average size of a chimp's brain, but in some cases their bodies were also bigger. Also, the amount of blood flow to our ancestors' brains increased as they became more human and less ape. It's been calculated that a brain size increase of about 3.5 times could have increased processing power by a factor of 6. See Seymore, Bosiocic, and Snelling (2016). Evidence from the remains of ape-like animals living between two and one-and-a-half million years ago shows that a significant change in brain size was taking place during this time. See Park et al. (2007).

3. There is great debate about when humans began to talk, as we discuss later in this chapter. It's reasonable to speculate that humans living 1.5 million years ago had some sort of language, but no one supposes that it was like the languages humans use today. Some scholars think that if early humans had language at all it would have been extremely rudimentary. See Coqueugniot et al. (2004). Merlin Donald (1991) speculated that people living at this time could have been talented mimics, but unable to speak. Moderately sophisticated stone tools were being made by 1.5 million years ago, implying substantial cognitive abilities that might well have also supported a fairly sophisticated spoken language, as we imagine. See Gibson (1991) and Toth and Schick (1993).

4. One and a half million years ago, hyenas were larger than they are today, and they hunted in packs. See P. Palmqvist et al. (2011).

5. The "Turkana Boy" skeleton is categorized by some as "*Homo erectus*," but there is quite a bit of controversy about how the remains of early human-like apes should be categorized, so we have decided to simply refer to them as "early humans."

6. Hrdy (2009).

7. Mattison, Quinlan, and Hare (2019).

8. For more information about this, see Tomasello (2019).

9. Zupancich et al. (2019).

10. See note 3 above.

11. Richerson and Boyd (2010).

12. The skull with one tooth was one of five early human skulls found in layers of sediment beneath what is now a small village call Dmanisi, which is on a pass going through the Caucasus Mountains on the Georgian side of the border between Georgia and Turkey; see Rightmire, Lordkipanidze, and Vekua (2006). One of the skulls found there was not just lacking teeth, but the sockets that had once held the teeth had been "resorbed"—they had filled in and the jaws healed over. This is how archaeologists know that this early human had lived for some time after they had lost their teeth. The Dmanisi remains have been dated to about 1.8 million years ago. Stone tools about 2.1 million years old have been found in China; see Zhu et al. (2019). Early humans seem to have spread to all over Eurasia shortly after they evolved in Africa. The five skulls and the other remains found at Dmanisi

were quite diverse, and the size of the brain held by the skulls varied from 600 to 730 cc. Their diversity has made scientists think again about how they should categorize early humans, wondering if what the individuals they had once categorized as belonging to different species might belong to a single very diverse early human species. The have also begun to think that early humans with different-sized brains might have lived and worked together.

13. Richard Dawkins (1976) in *The Selfish Gene* explains how a "gene-centered" approach to understanding evolutionary theory explains the uncompetitive behavior seen in some animals, especially social insects. This is one of the classic books explaining developments of evolutionary theory in the 20th century as biologists learned more about the way genes carried information from parents to offspring.

14. Richerson and Boyd (1999).

15. Many entomologists would say that humans are not the only animal that has evolved language, because forager honeybees use a sort of language (the "waggle dance") to communicate the local source of food to other foragers in their hive. They also have a complex chemical communication system, and they share food and the care of their hive's offspring. See M. Trhlin and Rajchard (2011).

16. Mollica and Piantadosi (2019). Interestingly, the largest share of this information relates to vocabulary and only a relatively small part to grammar. A long-standing argument by many linguists is that grammar is unlearnable because kids don't get enough information in the speech that they hear to figure it out; see Steven Pinker's (1994) *The Language Instinct*. The vocabulary has to be learned because it differs so dramatically from language to language. Mollica and Piantadosi's finding supports the skepticism of many other linguists regarding the "poverty of the stimulus" argument regarding language learning. See also Moerk (1983) and Evans and Levinson (2009).

17. A potato, for example, is a part of the potato plant (called a tuber) where energy is stored as starch, which the plant uses to regrow at the start of next growing season. An onion is a bulb (made up of a stem protected by a covering of leaves), where the onion plant stores starch. A carrot is a root, where the carrot plant stores starch. A nut is a seed, which contains energy in the form of starch and oil, which, if the seed germinates, will be used by the baby plant as it grows its first roots and leaves.

18. More details of this can be found in Richard Wrangham's (2009) book *Catching Fire: How Cooking Made Us Human*.

19. Karkanas et al. (2007), W. Roebroeks and Villa (2011) and Shahack-Gross et al. (2014).

20. Bozek et al. (2014).

21. We now have a reasonably good understanding of why these changes occur. Energy from the Earth's core slowly tears the planet's surface apart and puts in back together. Over millions of years it shifts the position of continents, creating mountain ranges, earthquakes, and volcanos as the land masses split, move, and collide. The study of this is known as "plate tectonics."

The position of continents and mountains affects the flow of currents of water in the oceans and air in the atmosphere. Carbon is stored in the form of limestone and fossil organic matter and recycled by erosion and volcanoes. This, in turn, affects the Earth's climate and, through climate, the evolution of living organisms. The sun's energy shines more strongly on the Earth's equator than on the poles, but this energy can by distributed all over the planet by the movement of water in the oceans and air in the atmosphere. But about 20 million years ago, the position of the continents began to restrict the flow of warm water to the poles and ice began to build up there. Ice also began to build up on mountain ranges where warm air could not flow. With more and more water trapped as ice, there was less water circulating between the oceans and the atmosphere. This caused a reduction in rainfall and a lowering of sea levels. See Zachos et al. (2001). All this is explained in an entertaining, if somewhat out-of-date book, by the late great Wallace Broecker (1995), which you can find on the internet to download for free.

    The climate started to become less stable about two and a half million years ago when the flow of warm water and air was impeded to the extent that the climate became far more sensitive to small fluctuations in the way the sun's energy shines on different parts of the planet, which are caused by unevenness in the Earth's orbit and tilt. The ice sheets started to grow and shrink in response to the fluctuation, and the sea levels started falling and rising. These changes were happening over thousands of years, which seems slow to us but was much faster than the rate of climate change that organisms on the planet were used to. Those most able to adjust were most likely to survive. Recent evidence has shown that during the times when the ice sheets are large and sea levels low, the climate is often particularly unstable on millennial and submillennial time scales. See Martrat et al. (2007) and Loulergue et al. (2008). Climate change in turn drives other ecological changes, such as the composition of the vegetation, which in turn affects the animal communities the vegetation can support. Theory suggests that human cumulative culture is an adaptation to the high amplitude millennial and submillennial scale variation characteristic of the Pleistocene. See Sánchez Goñi et al. (2018) and Richerson and Boyd (2013).

22. Muscles are made up of many fibers containing molecules called "myosin," which use energy to contract. The way a muscle can perform (its strength and the speed of its contractions) depends on the kinds of myosin it contains. Chimpanzee limb muscles are higher performing because they have a higher proportion of the myosin molecules that contract rapidly. The downside of this is that they use more energy, so chimpanzees can only use their speed and strength in short bursts. Human limb muscles have more of the slower-contracting myosin, which makes them slower and weaker but more efficient, which means that, with training, we can run (slowly) for hours. See O'Neill et al. (2017).

23. The nose also works to warm and moisturize cold dry air, which can be just as hard on the delicate membranes. See Wolf et al. (2004).

24. Genetic evidence also shows that many of us have some ancestors who weren't in Africa at this time—who were part of populations living in Eurasia (including those called "Neanderthals"). More about this in later chapters. Apologies about how vague this sounds. Calculations made by some geneticists have encouraged them to make more precise or definite statements about the size of the population and when it lived, but there continues to be discussion about the accuracy of some of the assumptions made in these calculations. It seems better to be cautious and vague. See Harpending et al. (1998), Atkinson, Gray, and Drummond (2008) and Schiffels and Durbin (2014).

Chapter 5

1. It's likely that all humans who lived 100,000 years ago had skin that was quite a bit darker than people of European descent. Light skin may be a very recent adaptation, only appearing within the last 10,000 years. It might have been an adaptation to farming in

northern regions. If farming families had a mostly vegetarian diet, it wouldn't have contained much vitamin D. Humans don't need a diet containing this vitamin because we can make it within our bodies. The problem is that this is only possible if our skin is exposed to enough sunlight. The chemical reactions in the body that make vitamin D are powered by energy from sunlight. People who farmed in warm sunny places would have been able to make enough vitamin D even if their skin was dark. But it was more difficult for the people who began to farm further north, where winter days are short, skies are often cloudy, and skin is often covered with clothing. Lighter-skinned people had a better chance of surviving in these conditions because their skin could absorb more of the energy from a small amount of sunlight. See Jablonski (2008).

2. It's also important to remember that Africa is a huge and very diverse continent. Evidence of human habitation has been found in many parts of Africa. Detailed discussions of the evidence and ideas about our ancestors at this time can be found in *Modern Human Origins and Dispersal* edited by Sahle, Reyes-Centeno, and Bentz (2019) and containing chapters authored by some of the leading researchers in the field.

3. These rare and remarkable artifacts were found in a German peat bog in an excellent state of preservation. See Schoch et al. (2015).

4. It was only in the 1990s that evolutionists started to think seriously about the tremendous costs associated with growing and maintaining a large brain. In the years since, they have looked in more detail at the nature of the costs and how different animals handle the trade-off between the benefits and costs of increased brain size. See Aiello and Wheeler (1995), Isler and van Schaik (2009), and Heldstab, van Schaik, and Isler (2016): 25–34.

5. More recent remains of small-brained humans were found on the Indonesian island of Flores in 2003. The ancestors of these people may have traveled there during an ice age about 700,000 years ago and become trapped when the ice melted, causing sea levels to rise. Their descendants managed to survive until about 50,000 years ago. They were trapped on the island, but also isolated and protected. Environments where food and competition are limited tend to favor animals of smaller size, which may be why these humans evolved a very small stature. Their remains show them to be just over a meter tall and they have been assigned to their own species, *Homo floresiensis*. For some years, they were referred to by the nickname "the Hobbit." See Gómez-Robles (2016).

6. We do not mean to imply by this that predictions of temperature change over the next century and its effects are not important.

7. Richard Potts (1996) of the Smithsonian National Museum of Natural History in Washington, DC, has provided many convincing and readable accounts of the role climate variation played in human evolution. Note 21 in chapter 4 provides more sources of information on climate variation during the Pleistocene. Concern about future climate change due to increased levels of carbon dioxide in the atmosphere has increased scientific interest in past climate changes. Concern has long been expressed that that climates of the future will not just be warmer, they are also likely to be more unstable. See, for example, the United States' National Research Council (2002) document. Today's complex civilizations won't be able to adapt to climate variation as readily as our nomadic hunting and gathering ancestors. See also Whitehead and Richerson (2009).

8. In traditional societies, older siblings play an important role in the care of children. Kramer and Veile's (2018) survey suggests that about 43 percent of direct child care is provided by allomothers and that this care is a significant time savings for mothers. They go so far as

to say that children are specialized for caring for younger children. A study of teaching in two forager groups has revealed that children do most of the teaching. See Lew-Levy et al. (2020).

 For more information about children in small-scale societies, see Whiting and Whiting (1975), Kramer (2010) and Hewlett (2017).

9. Infants turn out to be quite active participants in their own enculturation. They are rather skilled social learners by 12 months of age. See Carey (2009). As young as three months, before their own motor skills allow them to be active agents, babies are sensitive to the causal intent and efficacy of adults. See Liu, Brooks, and Spelke (2019). By one year, babies are quite sensitive to moral issues, disapproving of and punishing misbehaving puppets. See Bloom (2013).

10. Toddlers are quite helpful and are ready learners of social norms. See Warneken and Tomasello (2009), Tomasello (2019), and Chudek and Henrich (2011). Felix Warneken (2018) reviews data suggesting that both toddlers and apes are able to cooperate to produce benefits but they perform badly on tasks that involve sharing out the rewards of cooperation. Older children are better and by three to five years of age they begin to divide cooperatively acquired resources using rules of fairness, which are likely to be cultural norms that they learned socially. Apes do not seem to be motivated to divide the rewards of cooperation. For example, individuals who are in the top positions in the hierarchy don't treat partners fairly. As a result, individual lower down in the hierarchy won't cooperate with them to produce joint benefits.

11. Child development studies have concentrated on studying Western children. Indeed, most psychological studies have been carried out on Westerners. This has resulted in researchers generating many questionable findings about human psychology. It's still too soon to know how severe the problem is, but it's quite severe. Henrich, Heine, and Norenzayan (2010) colleagues found that an over-reliance by psychologists on research participants from "WEIRD" (Western, Educated, Industrialized, Rich, and Democratic) countries gives a very distorted view of humans, because childrearing in other cultures is very different. Even in Western societies, modernization over the last few generations has radically changed childhood. The small but lively and determined group of cross-cultural psychologists has long noted these differences. See Vygotsky (1978), Werner (1979), Rogoff (1990), Greenfield et al. (2003), Keller (2013), and Kline, Shamsudheen, and Broesch (2018).

12. See Tomasello (1996). Whiten et al. (2009) conducted a series of experiments comparing the social learning of chimpanzees and human children. Chimpanzees are good at learning from others, but children are better. In particular, children "over-imitate," copying functionally irrelevant aspects of a demonstrators' behavior, whereas chimpanzees concentrate on the functional aspects. Over-imitation may help children master causally opaque aspects of a performance that a chimpanzee will not notice. The hypothesis is that the human willingness to accurately observe and copy performances allows cultural complexity to accumulate in a way that is impossible with the less accurate social learning of other animals. In comparative experiments, children can master complex social learning tasks substantially better than chimpanzees and other primates. This is partly because successful children use demonstrations and verbal instructions to help others master the task. See also Dean et al. (2012).

13. See Kinzler et al. (2009) and Kinzler, Corriveau, and Harris (2011). Many populations are culturally diverse and it's necessary to tolerate behaviors and looks that seem "foreign." But

we should not expect this to be easy. It makes sense that humans should have evolved the inclination to fear or avoid people who look and behave in ways that are unfamiliar. Even if the foreigners aren't bad people, interacting with them is problematic because they will likely not know or understand the rules that govern the behavior of people in your particular group. The children in your group should definitely not be influenced by them. It doesn't follow from this that people should have an "innate" tendency to be racist. Being racist implies a sensitivity to physical differences such as skin color or facial features associated with what people today see as defining "race." There is no reason to expect that humans would have evolved to identify "foreigners" based on these characteristics. Our ancestors saw plenty of physical differences among their own family members based on age and gender, but few would have seen people of different "races" because, for most of human evolutionary history, people could only travel on foot. Even though they could (and did) make long migrations, most of them spent their lives within a few hundred miles of where they were born. Humans are much more variable culturally than genetically, and even today most neighboring societies are very similar genetically. See Bell, Richerson, McElreath (2009). The Danes and the English, for example, are similar to look at but speak mutually unintelligible languages and operate their societies in rather different ways. The foreigners our ancient ancestors met were people from the next valley who spoke a different language or dialect, and it was these differences that they needed to be sensitive to. Neighboring societies often have different ecological circumstances and incompatible social arrangements. To the extent that this is so, one should be wary of imitating people with different cultures. Not surprisingly, cultures have evolved many ways to categorize themselves and others. See Boyd and Richerson (1987), F. J. Gil-White (2001), and McElreath, Boyd, and Richerson, (2003). Symbolic differences like dialects may indeed emerge to protect us from the inappropriate ideas of neighbors. See Richerson and Boyd (2010). At the same time, neighbors may often have ideas that are useful, and the diffusion of innovations from one society to the next is a well-known phenomenon. See Rogers (1995). Even children need to be, and are, rather sophisticated shoppers in the marketplace of ideas!

14. Research into the workings of brains is really exciting at the moment, and many books have been written that attempt to explain this research to nonspecialist readers. Inevitably, interpretations differ, as they do in any fast-moving field. Invasive experiments on "lower" organisms like rats, mice, and even invertebrates have resulted in great advances in our understanding of neurobiology in general, and in the function of vertebrate brains specifically. Noninvasive brain imaging techniques like functional magnetic resonance imaging (fMRI) allow us to better understand what goes on in human brains.

Two rather different arguments have been developed by evolutionary social scientists to explain how the human mind/brain is organized to make the adaptations we see in humans possible. Edward O. Wilson (1978) and Lumsden and Wilson (2005) argue for a "human nature" theory in which selection on genetically coded "epigenetic rules" constrained learning and culture to conform to fitness enhancing ends. He argued that cultural evolution could not play any fundamental role in human evolution because culture only became important in the last few thousand years and because genes wired the human infant's brain in "exquisite detail" before the child acquired any culture. Some evolutionary psychologists have proposed a similar theory describing the brain being composed of hundreds or thousands of specialized, encapsulated modular elements that evolved in the Pleistocene to support our hunting and gathering lifestyle. See also Tooby and Cosmides

(1992) and Frankenhuis and Ploeger (2007). These authors proposed that social scientists have greatly exaggerated the importance of transmitted culture relative to the innate information endowed to us by selection on genes in the Pleistocene. They claim that these genes "evoke" certain behaviors and call this "evoked culture." Stephen Pinker (2002) has been one of the most ardent proponents of this idea.

The second hypothesis is the one we are describing in this book. It stems from an evolutionary functional analysis of human culture. This view is described in Boyd and Richerson (1985) and Richerson and Boyd (2005). In this view, culture is an adaptation to spatially and temporally variable environments, especially environmental variation that occurs on relatively short time scales ranging from a generation out to perhaps a few hundred generations. Rapid variation within a generation mainly relies on individual-level phenotypic flexibility, although rapid cultural diffusion among peers may spread useful innovations quite rapidly. Changes that are quite slow on the generational time scale are tracked well enough by selection acting on genes. The human brain is a metabolically expensive organ, prone to damage and a cause of obstetric difficulties for mothers and infants. The complexity of culture across primates correlates well with brain size, suggesting that our very large brain is necessary to support our extraordinarily complex cultures. Cultural evolution, in this view, simulates ordinary organic evolution by rather accurately transmitting a large volume of information by teaching and imitation from large social networks. Individuals select the information that they acquire and teach, and this selection is based to some extent on the usefulness of the information. Selective teaching and imitation in turn causes cultural evolution to be faster than genetic evolution and hence capable of tracking rapidly changing environments in time and fine-scale changes in space (Reader, Hager, and Laland Kevin (2011) and Perreault (2012)). It is perhaps no accident that human brains and the culture they support evolved in the hypervariable climates and ecologies of the Pleistocene (Richerson and Boyd (2013)).

In our view, a theory of what the brain does has to explain how human behavior is so diverse and yet, at the same time, so exquisitely adapted. For example, the rather small digestive system of humans requires that we eat a diet with a much higher concentrations of energy and protein than our ape relatives, and like our ape relatives, we need a diet that contains vitamin C. Humans living in a vast range of habitats manage to get a diet that meets these criteria. Our cuisines are adapted to our environment and also to our body. For example, the people who lived in the High Arctic had to live mostly on meat and had very limited access to plant sources of vitamin C. The heat of cooking rapidly degrades the small amount of vitamin C in meat. High Arctic people were able to survive because they consumed much of their meat raw.

A plausible account of the flexibility of brains begins with the observation that the neocortices of large-brained mammals have so many cells and connections between cells. There are far more than there are genes in on our chromosomes (Edelman (1987)). It is hard to see how Wilson's hypothesis that infant brains are wired in detail by genes would actually work. If the organization isn't laid down by genes, neocortices must somehow organize themselves during development. Much evidence suggests that our neocortex is a highly flexible organ that dynamically allocates and reallocates resources in adults. This flexibility is not unique to humans but is a general feature of mammalian neocortices (Anderson (2014) and Krubitzer and Stolzenberg (2014)). In the view of cognitive neuroscientist Cecelia Heyes (2018), general purpose reinforcement-based learning, aided by culture, is largely responsible for constructing the functional cortical circuits in the brain.

A plausible account of why culture is generally adaptive was proposed by Jaak Panksepp (2012), a neuroscientist interested in the emotions. Panksepp's work on the ancient, evolutionarily conserved, subcortical structures that generate our emotions and appetites suggest that these relatively simple parts of the brain generate the reinforcement that tends to keep behavior aligned with fitness. For example, our appetites for foods express the imperatives of our digestive physiology. Our short guts demand foods that are nutrient dense, such as meat, fat, and concentrated starch and sugar. Short guts can make limited use of complex carbohydrates like cellulose. Starchy staples like bread, potatoes, and rice are the basis of many agricultural cuisines, but cattle and their relatives find grass a perfectly good source of calories, using their complex stomachs as fermentation vats for cellulose-rich foodstuffs. Human cuisines are wondrously inventive in concentrating nutrients and making otherwise indigestible foods edible using techniques like cooking, grinding, and fermenting. Reinforcement shapes our cultural cuisines in ways that adapt what we eat to our short guts. The role of reinforcement in human evolution is a topic of much current interest. See Baum (2017) and Panchanathan and Baum (2019).

15. Laureano Castro and Miguel Toro (2004) argue that teachers' ability to encourage or discourage particular cultural variants is a key feature of human cultural transmission. William Baum (2017) notes that human are unusually sensitive to social feedback compared to other animals. Maciej Chudek and Joe Henrich (2011) review evidence that children readily learn and act on social norms that others model for them.

16. Playing and watching team sports have powerful effects on social emotions and attachments. See Lee and Kim (2013) and Campo, Mackie, and Sanchez (2019).

17. The human ability to work together has been studied from the perspective of development, and our behavior compared to that of other apes by Mike Tomasello (2019) and his collaborators. Social psychologists in the social identity theory tradition have investigated how we come to incorporate group memberships into our individual identities and how this identification influences group behavior. See, for example, Haslam (2004) and Hogg (2016).

18. People have long recognized that humans partly owe their success to our ability to learn right from wrong and to feel bad when we've done wrong. It's the idea behind the story in the Bible of Eve being tempted by a serpent to eat an apple and getting cast out of the Garden of Eden. The story was probably told and handed down in some form for many centuries before it was written down and found its way into the Jewish writings that form the basis of the Old Testament. In the story (as told in chapter 3 of Genesis in the King James version of the Bible), Satan (the serpent) urged Eve to disobey God and eat the fruit by telling her, "For God doth know that in the day ye eat thereof, then your eyes shall be opened, and ye shall be as gods, knowing good and evil."

The story seems to be saying that God gave the first humans the opportunity to be different from other animals and disobey him. Without mentioning the downside of being disobedient, Satan (the serpent) told Eve about the long-term benefits—that they would gain a new perception that would make them able to be like gods. As Satan predicted, when she and Adam ate the fruit, they did immediately begin to see things differently: "And the eyes of them both were opened, and they knew that they were naked; and they sewed fig leaves together, and made themselves aprons." When God saw that they were acting like they were ashamed, he knew that they had been disobedient. He told them that life would now be a lot harder for humans, saying things like "In the sweat of thy face shalt thou eat bread, till thou return unto the ground; for out of it wast thou taken: for dust thou art, and unto dust shalt thou return."

God was right, knowing right from wrong did make life harder, because it forces people to work together. They feel bad if they let others down by not doing what is expected of them. But Satan was also right in saying that this would make their descendants like gods. Chapter 11 of Genesis tells a story of what happened many generations later. Humans, who at that time all spoke the same language, said, "let us build us a city and a tower, whose top may reach unto heaven." God saw this and said, "the people is one, and they have all one language; and this they begin to do: and now nothing will be restrained from them, which they have imagined to do." But he had a plan to stop them reaching heaven and said, "let us go down, and there confound their language, that they may not understand one another's speech." It was a cunning move because, by breaking down their ability to communicate, God put an end to the possibility of humans working together. He "scattered them abroad from thence upon the face of all the earth."

19. A few percent of humans are diagnosed by clinical psychologists as psychopaths. Psychopathy is a complex disorder that can be dissected into two or more sub-elements. Behavior geneticists attribute a good bit of the variation in psychopathic behavior to genes using comparisons of identical to fraternal twins to estimate the influence of genes. See, for example, Babiak and Hare (2006) and Dhanani et al. (2017).

20. Dan Everett (2016), an anthropological linguist, argues that much of the content of culture is not easily accessible to our consciousness. It is a common experience of field anthropologists that the people they study just don't think like university-educated Westerners. The people Everett studied in the Amazon region of Brazil do not like to talk about things for which they have no direct evidence. Hypothetical goings-on or what might happen in the future don't strike them as worth talking about. The philosopher Michael Polanyi (1966) popularized the dark matter idea under the term "tacit knowledge." Anthropologist Joe Henrich and his cross-cultural psychology colleagues Steve Heine and Ara Norenzayan (2010) chastise psychologists for focusing their studies on the weirdest people in the world: Western university undergraduates. See also Richard Nisbett's (2003) great book on the topic. On "tight" versus "loose" cultures, see Gelfand et al. (2011). See also Hofstede's (1984) classic book on cross-cultural differences in workplace values. One of the themes of the book is that the modern values that we and most of our readers hold make it difficult for us to understand even our own recent ancestors, much less living non-modern people or ancient ancestors. Tacit knowledge provides us with tools for living our lives, but it also places shackles on our ability to understanding the lives of other people.

21. People prefer to live and work among people who are like them, and this makes practical sense. It's much easier to work with people who have internalized the same norms. Chances are, you'll understand each other, have the same values, and follow the same social rules. Considerations like this can lead to the evolution of symbolic markers of groups that make it easier for people to group themselves on the basis of shared norms. See McElreath, Boyd, and Richerson (2003). Francisco Gil-White (2001) argues that humans use the same essentialist categorization strategy for ethnic groups as we use for species. Both evolutionary biologists and most social scientists, however, point out that essentialist categorization systems do considerable violence to actual patterns of behavioral variation in both ethnicities and species. Nevertheless, as a practical matter, differences between norms of ethnic groups are often sufficiently profound to make intimate interactions awkward. Gil-White's 2002 (unpublished) ethnographic analysis of customary differences between neighboring

Kazakh and Mongol groups in western Mongolia is an interesting case study. There is a huge literature in the social sciences on ethnicity and ethnocentrism. Classically, authors argue that there is an ethnocentrism complex that suggests that the basic pattern is one of in-group cooperation and out-group distrust or conflict. See also LeVine and Campbell (1972), Choi and Bowles (2007), and Brewer and Campbell (1976). Marilynn Brewer (2007), one of the most prolific and acute writers on ethnocentrism, paints a much more nuanced picture. Ethnic conflicts loom large in the news, but in fact most ethnic interactions are peaceful most of the time. If this were not so, the world's great multiethnic cities—London, Singapore, Los Angeles, New York—could hardly exist, much less be well-functioning and prosperous.

22. The anthropologist Bill Irons (personal communication) studied Turkmen communities in northern Iran in the 1960s when some Turkmen still lived in small, family-based nomadic groups. He recalls that when he asked them how many children they wanted, they stared back at him with the same kind of puzzlement as he would get if he asked a Westerner how many years he wanted to live.

23. In many traditional societies, when family members don't think their prospects for raising a new baby are good, the baby is often killed (Schiefenhövel (1984)). Sarah Hrdy (1981) describes in considerable detail how mothers and their helpers are far from instinctually nurturing baby-raising automatons. They make highly strategic decisions based on their own survival and health as well as their infant's. Wealthy modern societies indulge in romantic notions of motherhood that ignore such imperatives, not infrequently driving desperate mothers in tough situations into acts deemed criminal today but which were considered tragic necessities in the past.

24. Humans use strategies of starting, stopping, and spacing births, as well as abortion and infanticide, to actively adjust births to economic and ecological circumstances. See Knodel (1987) and Landers (1995). In early modern Western Europe before the Demographic Transition, fertility was regulated by marriage. Couples could not marry until their economic situation was secure. By education or inheritance, prospective grooms had to become "eligible bachelors." They had to acquire a farm or trade that paid sufficiently well to support a family in the sort of comfort they were familiar with. This typically meant relatively late marriage even in good times, and even later marriage in tough times. A fair proportion of the population never married, especially in hard times. This tradition contrasts with a more common pattern of universal early marriage in traditional agrarian societies, which meant that population was regulated by relatively high mortality, which rose and fell as economic conditions changed. On the decline of fertility in Europe since the 18th century as a chapter in demographic history. See. Coale and Watkins (1986).

25. A technique known as comparative phylogenetic analysis suggests that the practice of arranging marriages goes back at least as far as the first modern human migrations out of Africa. See Walker et al. (2011).

26. Sear and Mace's (2008) review of 45 studies of child survival in small-scale societies, the authors found that the presence of certain relatives was important to help in child survival, but who provided help varied. They found that in some cultures, the absence of a child's father had no effect on its survival, and that, in general, the contribution of fathers to child survival was very small. Males' efforts tend to be substitutable. In hunting and gathering societies, men tend to provide game for the whole camp rather than directly for their own families In kin-based societies, if the father is dead, uncles and grandfathers can easily step

in to fulfill the father role. Complete mother substitutes are much harder to find, especially for nursing kids.

27. Robert Martin's (2013) book is an excellent review of human reproductive biology in primatological perspective.

28. A number of psychologists have speculated about the evolution of human "mating strategies," and some have conducted research to test hypotheses about how men and women choose who to mate with. It has been suggested, for example, that humans are somehow genetically primed to see some members of the opposite sex as attractive, and that people look for different characteristics depending on whether they want a "short-term mate" or a "long-term mate." This idea has been published in scientific papers and summarized in many articles, books, and other media aimed at a wider audience. See, for example, Gangestad and Simpson (2000) or Miller (2011). We feel that these ideas, which were all developed and almost entirely researched by Westerners, do not take sufficient notice of the vast individual and cultural diversity in human mating and parenting behavior. The mating systems of contemporary university campuses in the age of easy contraception hardly resemble at all the mating systems of our ancestors! Among other things, in most traditional societies, including hunter-gatherers, young people are not free to make autonomous mating decisions. Parents are typically involved, since marriages are often alliances between families. In this regard, see Apostolou (2007) and Walker et al. (2011). Some non-modern societies do have high rates of nonmarital sex, but men who father children in such liaisons usually still have parental responsibilities. The concept of humans cooperatively raising their young entails that men are recruited to their alloparental duties by one institutional means or another, except in those few societies where men are surplus to economic needs. See Walker, R. S. (2012), Scelza et al. (2020), and Mattison, Quinlan, and Hare (2019). As we have stressed in this book, the ultimate evolutionary payoff is children successfully raised. Humans differ from most mammals in that males (in most societies) act as allomothers—they help women to raise their children. Institutions like marriage that regulate lust in the service of allomaternal effort are a major part of the *human* story of mating.

29. Pair-bonding has evolved in a number of different kinds of mammals, including gibbons, which are closely related to great apes and are sometimes referred to as "lesser apes." The basic mechanisms in the brain that cause the bonding emotions are similar to those that generate the bonding emotions mother mammals feel for their babies. But it the case of pair-bonding, it is usually copulation that triggers emotions. The bonded pair stay close to each other and their offspring once they're born, but are hostile to others. The family often share a territory that they defend. The pair-bonding rodent known as the prairie vole, which lives in grasslands in the central United States, has been the subject of much of the investigation of the neurochemistry of pair-bonding. Its small size and rapid breeding make it an ideal laboratory animal. A closely related vole species known as the "montane vole" doesn't pair-bond. Investigating genetic differences between the two species has led to a better understanding of how social behavior may be influenced by genes that affect the distribution of receptors for certain neurochemicals in cells in certain regions of the brain. See Feldman (2017), Tabbaa et al. (2017), and Stetzik et al. (2018).

30. The practice of uniting males and females in socially recognized reproductive units may have a long history in human cultures, but the form of marriage we experience today in modern cultures, in which a couple marries after having fallen in love, has a very short history. See Coontz (2006).

31. Evolutionary approaches are promising to help bring a better understanding of obesity. See Wells (2006) and Nettle, Andrews, and Bateson (2017).

32. Hardy (1960) and Morgan (1982) are the strongest proponents of the aquatic ape hypothesis. Wikipedia provides an extensive review of the arguments and history of at https://en.wikipedia.org/wiki/Aquatic_ape_hypothesis.

33. The link between brain size and fatness has been explained by Heldstab, van Schaik, and Isler (2016) and Pontzer et al. (2016).

34. The link between fatness and female fertility is also well described (Frisch (2004) and Hakimi and Cameron (2017)). Variation in average fatness in human populations is often attributed to adaptive challenges. For example, Polynesians are taller and fatter than most human populations. This has been attributed to their long-distance voyaging, which might favor having large body reserves to survive long hazardous voyages where starvation was a risk. See Kirch (2017) and Bindon and Baker (1997). But some scholars are sceptical of attempts to find adaptive explanations for observed population differences. For example, Gosling et al. (2015).

35. Summaries of human behavior ecology literature include Winterhalder and Smith (2000) and Mace (2014). Regarding life history theory applied to humans, see Kaplan et al. (2000), Stearns, Allal, and Mace (2008), and Del Giudice, Gangestad, and Kaplan (2016). Our take, presented in Richerson and Boyd (2020), is that the human life history evolved to take advantage of the properties of cultural adaptations to temporally and spatially variable environments.

36. This was measured by comparing brain scans of children and adults. Children's brains require about 43 percent of their total daily energy intake, supporting the hypothesis that children's growth is slow because the brain competes for energy with the growth of the rest of the body (Kuzawa et al. (2014)).

37. For more information the uniquely human pattern of growth, see Bogin (1999) and (2015).

38. For more information on human lifespan, menopause and growing old, see Sievert (2015).

39. A cooperating family doesn't just provide care for children. The sick and injured also tend to be treated generously (Gurven et al. (2000)).

40. The adaptive role of menopause is much disputed. Hawkes and colleagues (1998), based on their observation of foragers argued that resource production by grandmothers was an important form of alloparental care. Kaplan and colleagues (2000) countered with evidence from a different group of foragers. Men don't experience a relatively abrupt and complete cessation of fertility as women do, but they do also have age-related declines in fertility. The ability of a few older men to attract mates and produce children could be a significant form of sexual selection, according to Bribiescas (2018). For example, if men with high cultural accomplishments among our ancestors acquired prestige and could attract mates at advanced ages, this form of female choice might have favored the growth of innate capacities for social learning and teaching.

41. The observation that life had evolved over time was proposed long before Darwin wrote *Origin of Species*, but many people saw evolution as progress toward greater complexity. A book written by the journalist Robert Chambers, published in 1844 (15 years before *Origin of Species*), described both the evolution of stars and the evolution of living things and suggested that the evolution was driven by a universal law. He proposed that a natural principle causing things to become more complex was created by God when he created the universe. The book, called *Vestiges of the Natural History of Creation*, was extremely

popular. Abraham Lincoln is said to have read the book and been impressed by the idea of a universal law. Darwin was grateful for the book and believed that it prepared the public for his own ideas on evolution. But he disagreed with the idea of a universal law that caused everything to become more complex. He saw evolutionary change to be driven by competition to survive in a local, changing natural environment. If you want to know more about this, we recommend Secord (2003).

42. For an extended discussion of Neanderthals, see Churchill (2014).

43. Eurasians, North Africans, and people descended from Eurasians are mostly of *Homo sapiens* descent but have a small amount of Neanderthal DNA. Those from parts of Southeast Asia and Oceania also have up to 5 percent Denisovan DNA. Sub-Saharan Africans are descended from several fairly distinctive lineages, some archaic, that can be detected today in most people of African descent. See Reich (2018).

Chapter 6

1. In more northerly regions, where the temperatures have remained low, the DNA in animal remains degrades more slowly. The remains of people who lived and died in northern North America and Eurasia are, therefore, more likely to yield samples of DNA with intact sequences that can be compared to sequences in present-day humans. DNA has now been recovered from the remains of several individuals that lived between 40,000 and 25,000 years ago. Comparisons suggest that people who made their living hunting mammoth steppe animals are part of the ancestry of many people living today. For more information, see Reich (2018) and Yang and Fu (2018).

2. A spear-thrower (or *atlatl*) is a tool that acts as a lever. By placing a light spear or dart into a thrower, hunters can increase the velocity of their throw. See https://en.wikipedia.org/wiki/Spear-thrower.

3. The climate and ecology of the last ice age can be reconstructed in a fair amount of detail using pollen and other physical and biological proxies found in ice, lake, and ocean cores. See Bradley (2013). Guthrie (2001) gives a reconstruction of the mammoth steppe climate of the last ice age that plays a major role in this chapter.

4. Paleoclimatologists learned about the dramatic variability of climates during the last ice age compared to climates during the last 11,000 years from evidence gained from ice cores raised in the early 1990s from the Greenland Ice Cap. See Dansgaard et al. (1993), Ditlevsen et al. (1996), and Alley (2000). The data in our figure are from a more recent NGRIP core reported by Andersen et al. (2004). Data downloaded from https://doi.pangaea.de/10.1594/PANGAEA.586886.

   The most recent ice age seems to have been more variable than the seven ice ages that preceded it. That is as far back as high resolution data have been published. See Martrat et al. (2007) and Loulergue et al. (2008). Richerson and Boyd (2013) note that increases in human brain size and cultural complexity track the increasing variability of the ice ages on this time scale, as predicted by cultural evolutionary theory. The smaller brains and more limited cultural sophistication seen in some earlier humans suggests that climate change on the millennial scale was more limited in the earlier part of our history. Ongoing unpublished work on even longer cores suggests that this might be true (Maria Sanchez-Goñi, personal communication).

Climate scientists fear that the changes in the Earth's atmosphere caused by our use of fossil fuels will alter the much more stable pattern of winds and ocean that exists on Earth today and bring more extreme climate fluctuations. Much smaller changes in global temperature will have a large effect on humans living today because we no longer live in small bands of nomadic hunters and gatherers who can move when conditions change. Climate change will devastate our farms and the habitats we have built. We suspect that that ice age hunter-gathers were much better adapted to rapid climate change than contemporary farming and industrial societies will prove to be. See Richerson, Boyd, and Bettinger (2001). The U.S. National Research Council was already warning about "Abrupt Climate Change" and "Inevitable Surprises" in 2002.

5. Guthrie, R. Dale, "Origin and Causes of the Mammoth Steppe: A Story of Cloud Cover, Wolly Mammoth Tooth Pits, Buckles, and inside-out Beringia," *Quaternary Science Reviews* 20 (2001): 549–74.

6. Much of this investigation has been done by Russian scientists studying animals found in the Siberian permafrost. A fascinating book by R. Dale Guthrie (1990) reviews the Russian finds and tells the story of how he investigated the body of a giant bison that was unearthed by an Alaskan gold mining operation.

7. Evidence of the humans who spent the most recent ice age in other parts of the world is much sparser, but archaeologists are gradually finding clues to how they lived. For example, early cave art, roughly comparable to that discovered in Western Europe, has been found in Sulawesi (an island in Southeast Asia). See Roebroeks (2014) and Aubert et al. (2019). In North China, there is now a good record of early Upper Paleolithic artifacts contemporaneous with those in Western Europe. See Li et al. (2019).

8. Dale Guthrie's (2005) lovely book on Upper Paleolithic cave art includes a number of iconoclastic ideas. One is that many of the less spectacular images were the graffiti of adolescent boys. They seem to include pornographic drawings (like modern male adolescent drawings) but are lacking in images of war (unlike modern examples and the cave art known ethnographically). Perhaps people living around 30,000 years ago were still in a phase of building larger networks to cope with the relative rarity of humans and did not experience the strong competition from other humans that motivates warfare. Guthrie also points out that most of the art is highly naturalistic, with few images that can be interpreted as related to supernatural themes.

9. These publications give some idea of the geographical range and technical sophistication of the mammoth steppe hunters around 30,000 years ago: H. Pringle (1997), Soffer, Adovasio, and Hyland (2000), Anikovich et al. (2007), and Nikolskiy and Pitulko (2013). Klein's (2009) chapter on the Upper Paleolithic provides an excellent account of the stones and bones evidence. It is now a bit out of date but we don't know of any more recent comparably broad review.

10. R. Dale Guthrie (2005) notes the complete lack of images of warfare in paleolithic art.

11. Even though ice age humans seem to have roamed deep into the mammoth steppe, there is no evidence that they made it as far as North America. The mountains of eastern Siberia may have blocked their eastward migration into Beringia—the land in eastern Siberia and Alaska that is now divided by the Bering Strait. The eastern Siberian mountains would have been exceedingly cold, and they may have been so barren that there wasn't enough vegetation to provide the kindling necessary to start burning greasy bones. Had the mammoth steppe hunters made it over the mountains, they would have been able to walk right

over into North America, as their descendants eventually did after the climate began to warm. See Hoffecker and Elias (2007).

12. For more information about human population changes during the last ice age, see Rogers (1995), Prufer et al. (2014), Sankararaman et al. (2016), and Rogers, Harris, and Achenbach, (2020). This is a fast-moving field, so readers interested in the latest details should look for current papers.

13. It was Noam Chomsky who started a "cognitive revolution" among psychologists in the 1950s, but paleoanthropologist Richard Klein is the most vocal supporter of the idea that a cognitive revolution took place among our ancestors in the paleolithic (Klein and Edgar (2002)). Klein (2009) provides evidence of a dramatic change in human behavior occurring between 40,000 and 50,000 years ago, and suggested it was due to a genetic mutation. Several other scholars have been involved in the development of the idea, it is argued in several books and is part of the human evolutionary story told in the popular "Sapiens" book by historian Yuval Harari (2014).

14. Archaeologists are finding far more evidence of what humans were doing in Africa before the migration of Africans into Eurasia around 60,000 years ago. The evidence doesn't support the abrupt modernization proposed by Klein (1995). But nor does it suggest a simple trajectory of gradual accumulation of cultural sophistication as was hypothesized by McBrearty and Brooks (2000). Will, Conard, and Tyron's (2019) review suggests that many artifacts that later become conspicuous parts of the Upper Paleolithic of Eurasia evolved in Africa before the migration but there were regional variations and complex temporal patterns within regions. Some techniques which archaeologists once believed to be strictly modern are now known to go as far back as the earlier ice age over 200,000 years ago.

   Some of the most interesting work comes from sites in Southern Africa. See Henshilwood et al. (2001, 2002), Jacobs et al. (2008), and Marean (2010, 2015).

15. The importance of network size in human cultural evolution has been addressed in several studies, including Henrich, (2004), Powell, Shennan, and Thomas (2009), Richerson, Boyd, and Bettinger (2009), Kline and Boyd (2010), Derex et al. (2013), Derex, Perreault, and Boyd (2018). The growth of human populations began about 50,000 years ago, more or less coincident with the spread of *Homo sapiens* out of Africa into Eurasia (Atkinson, Gray, and Drummond 2008). It is also more or less coincident with the beginning of the hypervariable part of the last ice age. It seems possible that before the climate became extremely unstable, human hunters did not have a strong advantage over other top carnivores. But when the climate chaos really set in, the ability to work together and develop cultural adaptations allowed our ancestors to thrive. This led to increased population size and the potential for even larger social networks (Churchill 2014).

16. Evidence from DNA extracted from ancient human bones can provide clues to the social life of the ancestors of these long dead people. The DNA carries a record of how much mixing took place between family groups. If it was typical for people to mix with only a few tens of people throughout their whole lives, their DNA is likely to have a much less diverse set of genes than if meetings between hundreds or thousands of people took place. As we have seen, network size is also important for cultural complexity. Analysis of the DNA of two Neanderthal women suggests that they came from groups with very little mixing (Prufer et al. 2014, 2017). Both women lived more than 50,000 years ago and died thousands of miles apart, one in what is now Croatia and the other in what is now Siberia. The Siberian woman's group appears to have been so isolated that its members were highly

inbred. The offspring of matings between close relatives are less likely to be healthy, and most cultures avoid them if possible. But the DNA sequences on this woman's chromosome pairs were so similar that her parents might have had the same mother, and the lack of diversity suggests that mating of close relatives had occurred in earlier generations also. Modern humans a little before 30,000 years ago, by contrast, were both genetically and culturally more diverse, suggesting larger social networks. See Bergström and Tyler-Smith (2017) and M. Sikora et al. (2017).

17. Marshack (1971) became so interested in the meaning of what he thought of as Paleolithic notation that he changed his career and devoted it to trying to gain a better understanding of what the etched marks meant.

18. Many books and papers have been written describing the findings from the investigations on the Sunghir site. The works of Erik Trinkaus and colleagues are good places to start (Trinkhaus et al. (2014) and Trikhausis and Buzhilova (2018)). The Sunghir burials and other features of cultures around 30,000 years ago suggest to some that mammoth steppe hunting populations were socially stratified (Hayden 2001). We depict them in our story as being egalitarian, like most ethnographically known hunter-gatherers. Social stratification and economic inequality has only been observed in a few cultures that make their living by hunting and gathering. The peoples living in the rich environment of the Northwest Coast of North America are one well-described example (Bettinger 2015).

19. Attempting to connect with people on the Internet is often unsatisfying, and sometimes dangerous, because we lack the social tools for meeting and judging people in this new way. The use of cultural evolutionary theory to understand Internet networks is developing. See Acerbi (2020).

20. We have suggested that ritual can be a useful social tool. There is growing interest among scientists of the psychological and social effects of ritual. See Frecska and Kulcsar (1989), Konvalinka et al. (2011), and Dunbar (2020).

21. Durkheim (1915) is often cited as the first person to take a modern academic interest in how rituals influence human behavior. The historian William McNeill (1995) argued that participating together in rhythmic activities stimulates bonding in humans. Anthropologists and psychologists have also contributed much to the literature. See, for example, Wiessner and Tumu (1998), Sosis and Alcorta (2003), Atran and Henrich (2010), Atkinson and Whitehouse (2011), Whitehouse and Lanman (2014), and Curry, Mullins, and Whitehouse (2019).

22. David Reich's (2018) book summarizes the genetic links between the Eurasian hunters and people living today. Descendants of this population eventually became part of populations that gave rise to populations currently living all over Eurasia and North America. More generally, his book is an introduction to the revolution that the recovery of ancient DNA has made possible in our understanding of the population movements and resultant interbreeding of our ancestors. Svante Pääbo's (2014) book provides a fascinating personal story of his and his collaborators long and arduous the hunt for techniques that made it possible to recover ancient DNA sequences.

Chapter 7

1. Our visions of hunter-gatherer childhoods are inspired by the work of Barry Hewlett and Karen Kramer. See note 8 of chapter 5 for citations to their work.

2. David Reich's (2008) book explains how genetic studies have allowed us to make statements like this. But these genetic studies have also introduced some mysteries. For example, some groups of Native Americans from South America have been found to have genetic sequences not present in Native Americans from further north. The sequences are similar to those found in the native populations of Australia, Melanesia, and island Southeast Asia. The findings suggest that two separate populations colonized the Americas, one from northeastern Asia and one from much further south. See Skoglund et al. (2015).

3. Alley (2000).

4. The observation that cooperative behavior is more likely to evolve in harsh environments was first made by the Russian scientist Peter Kropotkin (1902). In an essay entitled "Mutual Aid: A Factor in Evolution," he argued against the idea that inequality and injustice were inevitable results of Darwin's theories, and that, in many circumstances, theory predicted that individuals who cooperated were more likely to survive than those who competed. He supported his theoretical arguments with many examples of cooperation between non-human animals, and also with historical examples of human cooperation. Mathematical modeling suggests that there is something to this idea, and much data suggests that he might be right. See Smaldino, Schank, and McElreath (2013) and Richerson et al. (2016).

5. A fairly recent symposium on the origins of agriculture provides a broad overview. See Price and Bar-Yosef (2011). For a review of archeobotanical evidence of the origins of farming, see Fuller et al. (2014). For a review the origins of animal domestication, see Larson and Fuller (2014). The role of climate amelioration in the timing of the origins of agriculture is discussed in Richerson, Boyd, and Bettinger, (2001).

6. Warfare is often supposed to be an ancient and universal impulse, perhaps even going back to our nonhuman ancestors. Books written in the 1960s about the biological origins of human aggression, such as *On Aggression* by Konrad Lorenz and *The Territorial Imperative: A Personal Inquiry Into the Animal Origins of Property and Nations* by Robert Ardrey, gave a rather nuanced picture of humanity's tendency for violence, but simplified summaries of their books were used to justify ideas about the aggressive nature of our species. See Milam (2019). More recent authors have also commented on the apparent ubiquity of warfare. For example, Keeley (1996) and Gat (2008). Some have suggested that men evolved the capacity to cooperate so that they fight together against a common foe. See Bowles (2009), Boyd and Richerson (1985), and Turchin (Chaplin, CT: Beresta Books, 2015). But in the Chapter 6 we noted that Dale Guthrie found little evidence in Upper Paleolithic art for warlike behavior, in contrast to similar art of Holocene people.

7. It's true that conflict has been observed between groups of chimpanzees, with males from one group attacking and sometimes killing males from the group that live in a neighboring territory. See Wrangham and Peterson (1996). But it doesn't follow from this that hating other group members is a special part of ape (or human) "nature." Living things compete and in many species the competition can become violent.

8. A recent review of the ethnographic literature by anthropologist Richard Lee (2018) concludes that it was cooperation and conflict management, rather than aggression and competition that were the primary drivers of human behavioral evolution. Violent episodes in history may be linked to climate change, but a recent paper (Shaffer 2017) suggests that we shouldn't take for granted that the climate change we will experience in the future will lead to increased violence.

9. Kelly (2005) and Lahr et al. (2016).

10. As children, we learn how to make the sounds of the language of our people. Once we're adults it becomes hard to learn to make new sounds. For this reason, most people find it hard to learn to speak a new language like a native. Speaking a language with a certain accent and jargon or slang words is a very effective way of marking out who is in a group and who is an outsider. Language differences may function to keep us separate from each other. This can be adaptive if other people have ideas that might be harmful to if you were tempted to copy them. See Richerson and Boyd (2010) and Labov (2001).

11. Rush (2005).

12. When people began to herd grazing animals, they probably saw them as no more than a source of meat and materials. But selection for less frightened and aggressive animals would have eventually resulted in flocks with females who were tame enough to be milked. At first, the milk was probably just given to babies to supplement the milk they got from their mothers. Milk would have been less nourishing for older children and adults because, like all mammals, humans normally lose the ability to digest the milk sugar lactose when they get older (see also note 18 in this chapter). About 40 percent of milk calories are in lactose. At some point, our ancestors discovered techniques for processing the milk that both removed the lactose (by allowing it to be digested by bacteria) and allowed milk to be stored (by being made into cheese, yogurt, or some other processed product). In time, it was found that members of some herding populations were able to continue to get nourishment from milk all their lives. We now know that they had a genetic mutation that caused them to continue to make the enzyme (lactase) that breaks down lactose. People with this mutation began to concentrate on dairying and their descendants continue to be able to digest milk. For more on the ways people have adjusted to milk, see material from a talk by Bret Beheim and Masanori Takezawa, "Japanese Milk Consumption: Asymptomatic Lactose Intolerance Following a Recent Cultural Diffusion," available at https://figshare.com/articles/hbes2018-lactose-japan_pdf/6837734.

13. "Father" is in quotation marks because the social father doesn't have to be the actual genetic father, but in many patrilineal cultures a child inherits the rank of the man who is named as his or her father, either through adoption or biological kinship.

14. Low (2015) gives an overview of relations between the sexes in many cultures from a human behavioral ecology perspective. See also Hill and Hurtado (2009) and Starkweather and Hames (2012).

15. Walker, Flinn, and Hill (2010).

16. This is, of course, what Westerners today often refer to as "female genital mutilation." It is a widespread practice that is often criticized by Westerners. See Vogt et al. (2016).

17. The field known as "human behavioral ecology" looks at how cultural differences between populations may be related to differences in how populations make their living and the ecological conditions they face. Ruth Mace and colleagues and Siobhan Mattison and colleagues have looked particularly at links between marriage customs and ecology. See Holden and Mace (2016), Du and Mace (2019), and Mattison, and Hare (2019).

18. The milk sugar lactose has a molecular structure similar to sucrose (table sugar) but a special digestive enzyme called "lactase" is needed to break it down so it can be absorbed from the intestines. Mammals need lactase when they are infants in order to digest their mother's milk but usually stop making it when they reach weaning age. In human populations that herded animals and learned how to extract their milk it became beneficial to be able to continue to digest lactose. Natural selection favored those with genetic changes that

allowed them to continue to make lactase. The story of the evolution of adult lactose persistence in dairying populations is one of the best described examples of culture-led gene-culture coevolution. See Gerbault et al. (2011). A number of others are known, however, and there are probably many waiting to be discovered. See Ross and Richerson (2014).

19. In their influential book *The Major Transitions in Evolution*, Maynard-Smith and Szathmáry (1995) looked at the mechanisms that allowed life on Earth to become more complex. They showed human societies to be just the latest event in a trend toward greater complexity. This can be seen as evolutionary "progress" but it may just a byproduct of the earth's geochemistry gradually becoming more favorable to complex life. See also Ward and Kirschvink (2015).

20. A vivid example of families bound together by culture rather than genetic relatedness is described by the Dutch anthropologist Jan van Baal in his 1,000-page book, published in 1966, reporting on the years he spent studying the Marind, a warlike tribe of headhunters who lived on the western side of the island of Papua, which is now part of Indonesia. The Marind had a ritual called *otiv bombari*, which they believed improved a woman's fertility. It was the custom for woman who had recently given birth to have sexual intercourse with all the men in her husband's sub-clan, often up to 10 or more, in a single night. This was consistent with the Marind belief that the mixing of semen and vaginal fluid brings fertility. The *otiv bombari* ritual was not only ineffective, it might have been the cause of the problem. Marind women had low fertility, and this was probably at least partly the result of damage to their fallopian tubes caused by pelvic inflammation resulting from taking part in the ritual when their genital area had not yet healed from giving birth. The *otiv bombari* custom would likely have disappeared, as their low fertility caused the tribes numbers to dwindle, but for another custom of the Marind. The clans regularly adopted children from other tribes, often kidnapping them during raiding expeditions. The Marind-amin women raised the adopted children as their own, and the children often had no knowledge that they had been adopted. The families and their customs therefore survived even though they maintained their numbers by adding new members who were genetically unrelated. See Van Baal and Verschueren (1966) and, for more examples of the ways cultural and genetic evolution can sometimes diverge, see Robert Paul (2015), whose book describes many examples of conflict between cultural and biological reproductive success.

21. For many Europeans in the 19th century, this "rise of civilizations" seemed like progress and such ideas persist in Western culture. We (the authors of this book) hope that a cultural evolutionary approach will help us to be more objective. Darwin himself was scathing in his denunciation of "civilized" societies that tolerated slavery and genocide. Anthropologists, who typically study simpler, frequently more egalitarian, societies sometimes remark "civilization was a big mistake."

22. Darwin's (1874) main work on human evolution was between the same two covers as his seminal work on sexual selection. This was partly because he thought that sexual selection played a big role in the evolution of many conspicuous human traits, like skin color, due to fads and fashions in what was physically attractive. However, Darwin also thought that cultural evolution, like sexual selection, was an agent-based process, shaped by the choices humans make about what cultural variants to adopt and abandon. This is today the locus of lively debates among evolutionary biologists. See also Laland et al. (2014) and the final chapter of Boyd and Richerson (1985), where we showed that prestige-biased cultural evolution can give rise to similar runaway dynamics as sexual selection.

23. Helena Cronin (1991) does a very good job of explaining runaway sexual selection as well as the evolution of cooperation of social instincts.

24. The first scholars to puzzle over the huge pyramids of ancient Egypt imagined that they must have been built by slaves who were forced to drag massive stones by whip-wielding servants of a crazed Pharaoh. But see Shaw (2003), https://harvardmagazine.com/2003/07/who-built-the-pyramids-html.

25. If you would like to know more about historical linguistics, we recommend Deutscher (2005).

26. This is also explained in Reich's (2018) book, which we have recommended many times.

27. Some people see similarities between stories of a Bronze Age mass migration into Europe from an area near the Caucasian mountain chain and European folklore that was quoted by the Nazis. See Hackenbeck (2019). However, human genetics has generally provided no support to racist beliefs and ideas. It has show, for example, that the white "race," is really rather recently descended from Africans, and all contemporary populations are mongrel mixtures of multiple ancestral populations.

28. The exception is people from the Italian island of Sardinia, also explained in Reich's (2018) information-packed book.

29. One archaeologist who did pay attention to the work of the historical linguists was David Anthony (2010), who began investigating the idea of a great migration into Europe during the Bronze Age years before the genetic evidence was published. His conclusions seem to be borne out by the recent ancient DNA evidence.

30. The Basque people, who claim as their homeland an area around the border of Spain and France, speak a language that belongs to no known language family, and some linguists suggest that it might be descended from the language spoken by cereal growers who had migrated to Europe before the arrival of the Indo-European speakers. See: https://en.wikipedia.org/wiki/History_of_the_Basque_language

31. In March 2019, the magazine *New Scientist* published an article titled "Story of Most Murderous People of All Time Revealed in Ancient DNA," explaining that "starting 5000 years ago, the Yamnaya embarked on a violent conquest of Europe." It is available at https://www.newscientist.com/article/mg24132230-200-story-of-most-murderous-people-of-all-time-revealed-in-ancient-dna/#ixzz5xsfsGGVv.

32. Stephen Shennan's (2018) excellent book on the first farmers of Europe tells this story.

33. Minute analysis of ancient remains, including the DNA they sometimes contain, is leading to the development of new ideas about historical events and population differences in susceptibility to disease. See, for example, Rasmussen et al. (2015) and Prohaska et al. (2019).

34. The graph in figure 7.6 is based on estimates from the History Database of the Global Environment (https://themasites.pbl.nl/tridion/en/themasites/hyde/index.html) and K. Klein Goldewijk, et al. (2001).

Chapter 8

1. According to the *Oxford English Dictionary*, the word *modern* was first used hundreds of years ago and simply means "of the present time." But it took on a new meaning as 19th-century scholars such as Max Weber started describing the changes that were taking place in European society. In the 20th century, as the changes were occurring in more and more populations, sociologists developed "modernization theory." See the *Wikipedia* entry on this, at https://en.wikipedia.org/wiki/Modernization_theory.

2. Some writers attempt to counter this bias by celebrating modernity. These ideas are discussed in Steven Pinker's (2014, 2018) books. See also Ridley (2010) and Shermer (2015). There certainly are things to celebrate about modern societies, but it is also the case that modern societies are presently not sustainable, either demographically or environmentally. It's perhaps not surprising that people are more inclined to draw attention to the problems we face than things which seem to be going well. However, for what it is worth, the United Nations global Happiness Report's high rankings are dominated by the most modern countries (https://en.wikipedia.org/wiki/World_Happiness_Report). Pessimism about modernity can be exaggerated.

3. Social psychologists have studied social exclusion in the modern cultures, how it affects us, and how people counter those effects. See, for example, Turner et al. (1987), Haslam 2004, and Baumeister (2011).

4. Many modern cultures have the stated goal of increasing diversity, tolerance of diversity, and the range of choices available to their members. Attempts to meet these goals has brought controversy. Lukianoff and Haidt (2019) discuss problems in the United States in their book *The Coddling of the American Mind*. Such discussions might help to further the process of evolving social tools to allow us to negotiate the constantly changing modern culture.

5. Judging by the scientific literature on happiness as measured in social surveys, people who live in small prosperous countries like Finland and Switzerland are a bit happier than those who live large prosperous countries like the United States and Germany. People in poor countries are the most unhappy. See https://en.wikipedia.org/wiki/World_Happiness_Report.

6. In the 1990s, Simon Szreter and Kate Fisher (2010) interviewed British men and women who married and had their children during the first half of the 20th century. Their findings reveal how rapidly and dramatically sexual attitudes changed. See also Fisher (2000).

7. Scattered populations in history, like the ancient Athenians, strike us as having been a bit modern, but Athenian democracy and Socratic discussions of an ideal society were superficial and short-lived. In Athens, only "free" men could vote. No one in Athens imagined that slavery might be wrong, not even the slaves. And none of Socrates's pupils were female.

8. Knauft (1986).

9. There are, of course, lots of books about the modernization process, and they emphasize different aspects of what was going on during the transformation. For a summary of the practical changes that took place as global connectedness developed, see Dickinson (2018).

10. It's possible that many of today's psychological symptoms would have been attributed to homesickness by doctors practicing in the 18th or 19th centuries. See Matt (2011).

11. Hunt (2007).

12. *Robinson Crusoe* by Daniel Defoe was published in April 1719 and had gone through four editions by the end of the year. Another staggeringly successful English work of fiction was *Pamela* by Samuel Richardson (1740), about an innocent 15-year-old serving girl who has to fight off the attentions of her rich boss. He finally asks her to marry him. Jean-Jacques Rousseau, the highly influential Swiss philosopher, composer, and essayist was most famous in his own time for his 800-page novel called *Julie, or the New Heloise,* published in 1761. The book, written in the form of letters between a pair of lovers, touched the hearts of readers in ways they had never been touched before, and many readers insisted on believing that the lovers were real people.

13. This idea was developed by Anderson (1991).

14. By "fertility" we mean "total fertility rate," or "TFR." The birth rate of the population is affected by the age structure of the women in the population and their life expectancy. To make comparison easier, demographers calculate a statistic to take these into account. The TFR is the average number of children that would be born to a woman over her lifetime if she (1) experiences the exact current age-specific fertility rates through her lifetime, and (2) survives from birth to the end of her reproductive life.

15. A great deal of historical fiction, even prize-winning fiction acclaimed for its authenticity, encourages readers to believe that people who lived in the past behaved in modern ways and shared our beliefs about parenting and sexuality. For example, *A Gentleman in Moscow* by Amor Towles has a scene set in the 1920s in which the main character, Count Rostov, "crossed the empty sitting room and entered the bedchamber, where a willowy figure stood in silhouette before one of the great windows. At the sound of his approach, she turned and let her dress slip to the floor with a delicate whoosh." A sexual relationship followed. It's not the sexual relationship that is inauthentic. People did have nonmarital sex in the 1920s. But it does seem very unlikely that Count Rostov would have that relationship without giving thought to the possibility that the "aging starlet" who had seduced him might be carrying syphilis, gonorrhea, or some other sexually transmitted disease. These infections were incurable and devastating in the 1920s.

16. We don't know of any evidence collected on current Westerners' beliefs about reproductive behavior in the past, so we have based this on many years of talking to Westerners. Most of the people we've talked to say that people in the past had high fertility because they lacked modern contraceptive technology. Some also say that people in the past may have wanted more children because infant and child mortality was higher, and they feared their children would die. And some suggest they needed large families to have children to work on the farm and look after them in old age. These three explanations also appear in articles aimed at the public and some academic articles. But they aren't supported by evidence and do not make sense in terms of evolutionary theory.

17. The decline is fertility is part of a wide range of changes that occur as populations begin to modernize, and it isn't easy to determine whether an association between two changes means that one is the cause of the other, or if they are both caused by some other change. Causation might also be reciprocal. Education drives up the rate of technological progress, but technological progress increases the demand for education. The decline in fertility is associated with a number of other changes (such as increased wealth, increased education, and increased survival of babies), and each of them can look like a possible cause of the fertility decline. Most analyses of the phenomenon assume *ecological* causality in the sense that current behavior is "explained" statistically by factors such as current GDP per capita, current average years of women's education, and the like. But the purely statistical approach to causality in cases like this is hopelessly muddled because of what is known technically as "multicollinearity." With all the changes going on at once, how is it possible to know for sure what changes are causes and what changes are effects? Without a proper causal model, the statistics say that everything explains everything else!

Causally, we see the changes of modernization as a cultural evolutionary process. Norms and attitudes toward reproduction are influenced by cultural transmission that takes into account changes that happened in the past. Culture is dynamic and influenced by what went on in the past (and is inherited) and also by what is going on now. See Newson and Richerson (2009). Traditionally, historians and social scientists have not made much use of the kinds of dynamic systems models that natural scientists routinely use to hone their intuitions and explain their data in cases where evolutionary causation is active. There is a growing body of literature aimed at rectifying this gap in the historical and social-scientific toolkit. See Romer (1986), Richmond and Peterson (2001), Turchin (2003) and Korotayev, Malkov, and Khalttourina (2006). See also Peter Turchin, "Cliodynamics: History as Science," http://peterturchin.com/cliodynamics/; and the Center for the Dynamics of Social Complexity, http://www.dysoc.org/. For sophisticated methods of empirical inference in the face of complex causal processes, see Efferson and Richerson (2007) and McElreath (2015).

18. When anthropologist Caroline Bledsoe and colleagues studied contraceptive use by women in rural Gambia, they discovered that a woman often chose to use contraception after she had had a stillbirth or miscarriage. They were using contraception not because they wanted to limit the size of their family, but because they wanted to increase their chances of having a baby that survived. Food is limited for these women and their work is arduous, so it takes time for them to recover from each pregnancy. They see a miscarriage or stillbirth as a sign that their body needs more time to regain the strength it needs to produce a healthy baby. By using contraception, they make an adequate space between pregnancies. See Bledsoe, Banja, and Hill (1998).

19. See, for example, Hitchcock (1997) and Kirkpatrick (2000).

20. In his book *An Essay on the Principle of Population*, first published in 1798, Thomas Malthus showed mathematically that giving food to the poor could only relieve suffering temporarily because population would always grow faster than the supply of food. He said that population rise could only be slowed by moral restraint (refraining from sexual activity) and what he called "vice" (engaging in sexual activity that did not risk conception). It's possible that poor communities might have been more tolerant of "vice," especially during times of famine. But when times were good, people seemed to believe that neither moral restraint nor "vice" were necessary. Malthus' dismal predictions about the fruitlessness of helping the poor was true up to the time he wrote, but not later as the Industrial Revolution gathered steam. See Lindert (1985).

21. This interview is reported in Pollak and Watkins (1993). See also Coale (1973). Coale, who set up the European Fertility Project to study the decline of fertility in Europe, said that before people would start to limit the size of their family, they had to realize that the control of fertility was "thinkable," or "within the calculus of conscious choice." This is also discussed in van de Walle (1992).

22. The death of babies and young children was very common until the 20th century. Before the 19th century, most people seemed to accept it with a stoicism that looks like indifference. In some families, parents didn't bother to attend their child's funeral, and young children, even those from wealthy families, had no special commemoration. Their bodies might be buried in the back garden or placed in a mass grave. See Zelizer (1994).

23. Limiting family size may reduce reproductive success (or "fitness") in the short term, but evolutionary scholars known as "human behavioral ecologists" wondered if it might increase fitness in the long term. Their research on this topic has revealed no evidence that this is the case. See Kaplan et al. (1995) and Goodman, Koupil, and Lawson (2012).

24. We are suggesting that a widening of people's social networks and a change to less family networks results in many cultural changes, including a change in the believe that couples should try, if possible, to raise the children that are born to them. Evidence linking

social network change to lowered fertility is reviewed in Newson et al. (2005) and Colleran (2016).

25. There is a fairly extensive literature on Old Order Anabaptists. See Kraybill and Bowman (2001), Hurst and McConnell (2010), Janzen and Stanton (2010), Kraybill, Nolt, and Weaver-Zercher (2010), Nolt (2015), Donnermeyer, Anderson, and Cooksey (2019), and Kraybill (2019). The Haredi have also been studied. See Heilman (2000), Stadler (2009) and Biale et al. (2017).

26. See, for example, Parker (2013).

27. Moheau was born in Paris in 1745 and died there, executed by guillotine in 1794. He had risen to the position of Commissioner of War. His execution was justified on the grounds of corruption, although no evidence was provided for his prosecution. It took place during the time historians refer to as the "Reign of Terror," a time of great revolutionary fervor when many people were publicly executed. On the bicentenary of Moheau's death, France's National Institute for Demographic Studies republished an annotated version of his book, Moheau and Behar (1994).

28. An English translation of part of the book and some background comments is published in Moheau (2000).

29. Recent research supports the suggestion that moving to cities changed the sexual behavior of Europeans. By comparing genetic and genealogical data of men living in Belgium and the Netherlands, it has been possible to calculate how often a man's genetic father (as determined by his Y-chromosome) was different from the man who was married to his mother. It was found, in this population, that the man named as the father of a child was also its genetic father in most cases. Overall, the rate of mismatch between social and genetic father was about 1 percent. The rate of mismatch increased with modernization, however. The highest rate, as much as 6 percent, occurred among poor families that lived in cities. See Larmuseau et al. (2019).

30. There has been quite a lot of discussion among historians about what was meant by the phrase "disastrous secrets unknown to any animal but man" (in French, "*funestes secrets, inconnus à tout animal autre que l'homme*"). It could mean secrets of how to have sexual pleasure in ways that don't lead to pregnancy, or it could just mean illicit sex in general. Before publication, the text had to be approved by the Royal Censor and so couldn't talk explicitly about something as profane as sexual behavior.

31. These numbers are, of course, estimates. Wikipedia has a well-sourced summary of European emigration, at https://en.wikipedia.org/wiki/European_emigration.

32. For example, in Great Britain, a "National Birth Rate Commission" was set up by the National Council of Public Morals, and they published their report, *The Declining Birth Rate, Its Causes and Effects*, in 1916. The report established the existence of a number of practices:

> Conscious limitation of fertility is widely practiced among the middle and upper classes and there is good reason to think that, in addition to other means of limitation, the illegal induction of abortion frequently occurs among the industrial population. (37)

There was disagreement over whether such limitation was moral, but the committee of enquiry did agree on one thing:

> There is no reason to believe that the higher education of women (whatever its indirect results on the birth rate may be) has any important effect in diminishing their physiological aptitude for bearing children. (37)

The committee considered whether or not a declining birth rate could be considered desirable, since a growing population tends to press upon the means of subsistence. They noted that the birth rate was declining as resources were increasing and concluded that the decline was not desirable:

> For a country open to commercial commerce with the world, like Great Britain, the law of Malthus does not necessarily hold at all. It cannot, in fact, be said that there exists any over-population in this country, in the sense that population has actually been growing faster than the available means of subsistence. The consumption of various foods and other materials per head of population has increased.... There is no reason to think that a further reduction in the English birth rate would at the present time give a larger yield of wealth per head. It would attract foreign labour into this country, if it were allowed to enter, and would check the migration of labour to our dominions. It would slightly lower the death-rate, if the reduction occurred in the most prolific and improvident class, where infant mortality is high, but not otherwise.

Twenty-four members of the 27-member committee signed an "Addition to the report" in which they expressed concern because they found the decline in birth rate to be "dysgenic" rather than "eugenic." They pointed out that:

> Restriction (of fertility) prevails most in the classes in which the conditions of family life are most favourable, and the largest families are found under those conditions, hereditary, environmental or both, which are most adverse to the improvement or even maintenance of the quality of the population.

They advised that practical steps, such as improving the welfare of families, be taken to arrest the decline, and asked, "If we value our national type, should we not desire its diffusion?"

33. The European Fertility Project, set up in the 1960s under the directorship of Ainsley Coale at Princeton University published a report detailing the course of the fertility decline in Europe. See Coale and Watkins (1986).

34. This is known as the "demographic dividend," and the boost it gives an economy has been observed in many countries.

35. The books by Ridley, Pinker, and Shermer (see note 2 of this chapter) provide plenty of evidence that modernization is associated with people being nice to one another, although the history of the first half of the 20th century suggests that this association is not law-like.

36. In many societies, people are good at solving problems of cooperation in the laboratory. Herrmann, Thöni, and Gächter (2008) used the experimental technique of asking people to play a "public goods game with punishment" in 16 societies around the world to compare how well subjects could organize cooperation in order to get money for themselves.

Players were formed into mini-societies of four people. In the public goods game, each player has an endowment provided by the experimenter that they can choose to invest each round in a public fund. The experimenters multiply pooled contributions to provide a potential profit for investing in the public good. The fruits of investing in the public good are divided equally among the four players, regardless of how much they contribute. The wrinkle is that each player receives 0.4 payoff units for each token invested. If contributions from other players are high, the public good pays off handsomely to everyone, but each individual will be best off keeping their own tokens regardless of what others do. Earlier work had shown that contributions to the public good started rather high but declined toward zero over rounds unless players had some sort of social tools to work with. See also Ostrom, Gardner, and Walker (1994). Fehr and Gächter (2002) allowed their subjects to use costly punishment as a tool. If enough players were willing to pay from their own endowment to punish noncontributors by reducing their payoff, there was no benefit to not contributing, so almost everyone contributed after ten rounds. This result is replicated widely in Western countries, where most such experimentation is done.

In an extension of the use of experimental games to investigate prosocial behavior cross-culturally, pioneered by Joe Henrich and colleagues, Herrmann, Thöni, and Gächter tested the ability of people in 16 mostly non-Western societies to develop cooperation in 10 rounds of the public goods game with punishment. A number of societies could, but a number of others couldn't, including one (Greece) that counts as Western. The reason was that in some societies there is not just prosocial punishment of low contributors, but also lots of antisocial punishment of high contributors. Players from antisocial punishment societies could not organize cooperation. These are also societies with low trust of neighbors and generally poor economic performance. Contrariwise, societies high in prosocial punishment tended to be high-trust societies with high or rapidly improving economic performance. Once again, cultural diversity seems to be a very real and consequential phenomenon. See also Henrich et al. (2001, 2006, 2010). For econometric studies in a similar vein, see Desmet, Ortuño-Ortín, and Warcziarg (2017) and Giuliano and Nunn (2020).

Bibliography

Acerbi, A. *Culturual Evolution in the Digital Age.* Oxford: Oxford University Press, 2020.

Aiello, L. C. and P. Wheeler. "The Expensive-Tissue Hypothesis: The Brain and the Digestive System in Human and Primate Evolution." *Current Anthropology* 36, no. 2 (1995): 199–221.

Alley, R. B. *The Two-Mile Time Machine: Ice Cores, Abrupt Climate Change, and Our Future.* Princeton, NJ: Princeton University Press, 2000.

Alley, R. B. "The Younger Dryas Cold Interval as Viewed from Central Greenland." *Quaternary Science Reviews* 19, no. 1 (2000): 213–226.

Andersen, K. K., N. Azuma, J. M. Barnola, M. Bigler, P. Biscaye, N. Caillon, J. Chappellaz, et al. "High-Resolution Record of Northern Hemisphere Climate Extending into the Last Interglacial Period." *Nature* 431, no. 7005 (2004): 147–151.

Anderson, B. R. O. G. *Imagined Communities: Reflections on the Origin and Spread of Nationalism.* London: Verso, 1991.

Anderson, M. L. *After Phrenology: Neural Reuse and the Interactive Brain.* Cambridge, MA: MIT Press, 2014.

Anikovich, M. V., A. A. Sinitsyn, J. F. Hoffecker, V. T. Holliday, V. V. Popov, S. N. Lisitsyn, S. L. Forman, et al. "Early Upper Paleolithic in Eastern Europe and Implications for the Dispersal of Modern Humans." *Science* 315, no. 5809 (2007): 223–226.

Anthony, D. W. *The Horse, the Wheel, and Language: How Bronze-Age Riders from the Eurasian Steppes Shaped the Modern World.* Princeton, NJ: Princeton University Press, 2010.

Apostolou, M. "Sexual Selection under Parental Choice: The Role of Parents in the Evolution of Human Mating." *Evolution and Human Behavior* 28, no. 6 (2007): 403–409.

Arcadi, A. C. and R. W. Wrangham. "Infanticide in Chimpanzees: Review of Cases and a New Within-Group Observation from the Kanyawara Study Group in Kibale National Park." *Primates* 40, no. 2 (1999): 337–351.

Atkinson, Q. D., R. D. Gray, and A. J. Drummond. "mtDNA Variation Predicts Population Size in Humans and Reveals a Major Southern Asian Chapter in Human Prehistory." *Molecular Biology and Evolution* 25, no. 2 (2008): 468–474.

Atkinson, Q. D. and H. Whitehouse. "The Cultural Morphospace of Ritual Form: Examining Modes of Religiosity Cross-Culturally." *Evolution and Human Behavior* 32, no. 1 (2011): 50–62.

Atran, S. and J. Henrich. "The Evolution of Religion: How Cognitive By-Products, Adaptive Learning Hueristics, Ritual Displays, and Group Competition Generate Deep Commitments to Prosocial Religions." *Biological Theory* 5, no. 1 (2010): 18–30.

Aubert, M., R. Lebe, A. A. Oktaviana, M. Tang, B. Burhan, A. Jusdi, B. Hakim, J. Zhao, I. M. Geria, and P. H. Sulistyarto. "Earliest Hunting Scene in Prehistoric Art." *Nature* 576 (2019): 442–445.

Babiak, P. and R. D. Hare. *Snakes in Suits: When Psychopaths Go to Work.* New York: HarperCollins, 2006.

Bacci, M. L. *A Concise History of World Population.* Malden, MA: John Wiley & Sons, 2017.

Bădescu, I., D. P. Watts, M. A. Katzenberg, and D. W. Sellen. "Alloparenting Is Associated with Reduced Maternal Lactation Effort and Faster Weaning in Wild Chimpanzees." *Royal Society Open Science* 3, no. 11 (2016): 160577.

Baum, W. M. *Understanding Behaviorism: Behavior, Culture, and Evolution.* Malden, MA: John Wiley, 2017.

Baumeister, R. F. "Need-to-Belong Theory." In *Handbook of Theories of Social Psychology.* Vol. 2, 121–140. Los Angeles: SAGE, 2011.

Bell, A. V., K. Hinde, and L. Newson. "Who Was Helping? The Scope for Female Cooperative Breeding in Early Homo." *PloS ONE* 8, no. 12 (2013): e83667–e83667.

Bell, A. V., P. J. Richerson, and R. McElreath. "Culture Rather Than Genes Provides Greater Scope for the Evolution of Large-Scale Human Prosociality." *Proceeding of the National Academy of Sciences USA* 106, no. 42 (2009): 17671–17674.

Belyaev, D. K. "Destabilizing Selection as a Factor in Domestication." *Journal of Heredity* 70, no. 5 (1979): 301–308.

Berger, L. R. "Brief Communication: Predatory Bird Damage to the Taung Type-Skull of *Australopithecus africanus* Dart 1925." *American Journal of Physical Anthropology* 131, no. 2 (2006): 166–168.

Bergström, A. and C. Tyler-Smith. "Paleolithic Networking." *Science* 358, no. 6363 (2017): 586–587.

Beheim, B. and M. Takezawa "Japanese Milk Consumption: Asymptomatic Lactose Intolerance Following a Recent Cultural Diffusion," https://figshare.com/articles/hbes2018-lactose-japan_pdf/6837734.

Bettinger, R. L. *Orderly Anarchy: Sociopolitical Evolution in Aboriginal California.* Berkeley: University of California Press, 2015.

Bettinger, R. L., R. Garvey, and S. Tushingham. *Hunter-Gatherers: Archaeological and Evolutionary Theory.* Boston: Springer, 2015.

Biale, D., D. Assaf, B. Brown, U. Gellman, S. Heilman, M. Rosman, G. Sagiv, and M. Wodziński. *Hasidism: A New History.* Princeton, NJ: Princeton University Press, 2017.

Bindon, J. R. and P. T. Baker. "Bergmann's Rule and the Thrifty Genotype." *American Journal of Physical Anthropology* 104, no. 2 (1997): 201–210.

Bledsoe, C., F. Banja, and A. G. Hill. "Reproductive Mishaps and Western Contraception: An African Challenge to Fertility Theory." *Population and Development Review* 24, no. 1 (1998): 15–57.

Bloom, P. *Just Babies: The Origins of Good and Evil*. New York: Crown, 2013.

Boehm, C. *Moral Origins: The Evolution of Virtue, Altruism, and Shame*. New York: Basic Books, 2012.

Bogin, B. *Patterns of Human Growth*. Cambridge: Cambridge University Press, 1999.

Bogin, B. "Human Growth and Development." In *Basics in Human Evolution*, edited by M. P. Muehlenbein, 285–293. Boston: Academic Press, 2015.

Bošković, A. and O. J. Rando. "Transgenerational Epigenetic Inheritance." *Annual Review of Genetics* 52, no. 1 (2018): 21–41.

Bowles, S. "Did Warfare among Ancestral Hunter-Gatherers Affect the Evolution of Human Social Behaviors?" *Science* 324, no. 5932 (2009): 1293–1298.

Boyd, R. and P. J. Richerson. "Cultural Transmission and the Evolution of Cooperative Behavior." *Human Ecology* 10 (1982): 325–351.

Boyd, R. and P. J. Richerson. *Culture and the Evolutionary Process*. Chicago: University of Chicago Press, 1985.

Boyd, R. and P. J. Richerson. "The Evolution of Ethnic Markers." *Cultural Anthropology* 2, no. 1 (1987): 65–79.

Boyd, R. and P. J. Richerson. "Why Does Culture Increase Human Adaptability." *Ethology and Sociobiology* 16, no. 2 (1995): 125–143.

Bozek, K., Y. Wei, Z. Yan, X. Liu, J. Xiong, M. Sugimoto, M. Tomita, et al. "Exceptional Evolutionary Divergence of Human Muscle and Brain Metabolomes Parallels Human Cognitive and Physical Uniqueness." *PLOS Biology* 12, no. 5 (2014): e1001871.

Bradley, R. S. *Paleoclimatology: Reconstructing Climates of the Quaternary*. 3rd ed. San Diego: Academic Press, 2013.

Brain, C. K. *Swartkrans: A Cave's Chronicle of Early Man*. Pretoria: Transvaal Museum, 2004.

Brewer, M. B. "The Importance of Being We: Human Nature and Intergroup Relations." *American Psychologist* 62, no. 8 (2007): 728.

Brewer, M. B. and D. T. Campbell. *Ethnocentrism and Intergroup Attitudes: East African Evidence*. Beverly Hills: SAGE, 1976.

Bribiescas, R. G. *How Men Age: What Evolution Reveals about Male Health and Mortality*. Princeton, NJ: Princeton University Press, 2018.

Broecker, W. S. *The Glacial World According to Wally*. Palisades, NY: Eldigio Press, 1995.

Campo, M., D. M. Mackie, and X. Sanchez. "Emotions in Group Sports: A Narrative Review from a Social Identity Perspective." *Frontiers in Psychology* 10, no. 666 (2019).

Carey, S. *The Origin of Concepts*. New York: Oxford University Press, 2009.

Castro, L. and M. Toro. "The Evolution of Culture: From Primate Social Learning to Human Culture." *Proceedings of the National Academy of Sciences USA* 101, no. 27 (2004): 10235–10240.

Cheney, D. L. and R. M. Seyfarth. *How Monkeys See the World: Inside the Mind of Another Species*. Chicago: University of Chicago Press, 1990.

Choi, J. K. and S. Bowles. "The Coevolution of Parochial Altruism and War." *Science* 318, no. 5850 (2007): 636–640.

Chomsky, N. *Powers and Prospects: Reflections on Human Nature and the Social Order.* Boston: South End Press, 1996.

Chudek, M., P. E. Brosseau-Liard, S. Birch, and J. Henrich. "Culture-Gene Coevolutionary Theory and Children's Selective Social Learning." In *Navigating the Social World: What Infants, Children, and Other Species Can Teach Us*, edited by M. R. Banaji and S. A. Gelman, 181–185. Oxford: Oxford University Press, 2013.

Chudek, M. and J. Henrich. "Culture-Gene Coevolution, Norm-Psychology and the Emergence of Human Prosociality." *Trends in Cognitive Sciences* 15, no. 5 (2011): 218–226.

Churchill, S. E. *Thin on the Ground: Neandertal Biology, Archeology and Ecology.* Hoboken, NJ: John Wiley & Sons, 2014.

Cibot, M., T. Sabiiti, and M. R. MacLennan. "Two Cases of Chimpanzees Interacting with Dead Animals without Food Consumption at Bulindi, Hoima District Uganda." *Pan Africa News* 24, no. 1 (2017): 6–8.

Coale, A. J. "The Demographic Transition Reconsidered." In *International Population Conference, Liege.* Vol. 1, 53–72. Liège, Belgium: International Union for the Scientific Study of Population, 1973.

Coale, A. J. "The Decline of Fertility in Europe since the 18th Century as a Chapter in Demographic History." In *The Decline in Fertility in Europe*, edited by A. J. Coale and S. C. Watkins, 1–30. Princeton, NJ: Princeton University Press, 1986.

Coale, A. J. and S. C. Watkins, eds. *The Decline of Fertility in Europe.* Princeton, NJ: Princeton University Press, 1986.

Colleran, H. "The cultural evolution of fertility decline." *Philosophical Transactions of the Royal Society B: Biological Sciences* 371, no. 1692 (2016): 20150152.

Coontz, S. *Marriage, a History: How Love Conquered Marriage.* New York: Penguin, 2006.

Coqueugniot, H., J.-J. Hublin, F. Veillon, F. Houët, and T. Jacob. "Early Brain Growth in *Homo erectus* and Implications for Cognitive Ability." *Nature* 431, no. 7006 (2004): 299–302.

Cronin, H. *The Ant and the Peacock: Altruism and Sexual Selection from Darwin to Today.* Cambridge: Press Syndicate of the University of Cambridge, 1991.

Curry, O. S., D. A., Mullins, and H. Whitehouse. "Is It Good to Cooperate." *Current Anthropology* 60, no. 1 (2019): 47–69.

Dansgaard, W., S. J. Johnsen, H. B. Clausen, D. Dahl-Jensen, N. S. Gundestrup, C. U. Hammer, C. S. Hvidberg, et al. "Evidence for General Instability of Past Climate from a 250-Kyr Ice-Core Period." *Nature* 364 (1993): 218–220.

Darwin, C. *The Descent of Man and Selection in Relation to Sex* (second edition). New York: American Home Library, 1874.

Darwin, C. *The Expression of the Emotions in Man and Animals*. London: John Murray, 1872.

Dawkins, R. *The Selfish Gene*. Oxford: Oxford University Press, 1976.

Dean, L. G., R. L. Kendal, S. J. Schapiro, B. Thierry, and K. N. Laland. "Identification of the Social and Cognitive Processes Underlying Human Cumulative Culture." *Science* 335, no. 6072 (2012): 1114–1118.

Deblauwe, I. and G. P. J. Janssens. "New Insights in Insect Prey Choice by Chimpanzees and Gorillas in Southeast Cameroon: The Role of Nutritional Value." *American Journal of Physical Anthropology* 135, no. 1 (2008): 42–55.

Del Giudice, M., S. W. Gangestad, and H. S. Kaplan. "Life History Theory and Evolutionary Psychology." In *The Handbook of Evolutionary Psychology: Foundations*, edited by D. M. Buss, 88–114. New York: Wiley, 2016.

Delisle, R. G. "Cyclicity, Evolutionary Equilibrium, and Biological Progress." In *Charles Darwin's Incomplete Revolution: The Origin of Species and the Static Worldview*, 197–228. Cham, Switzerland: Springer International Publishing, 2019.

Derex, M., M. Beugin, B. Godelle, and M. Raymond. "Experimental Evidence for the Influence of Group Size on Cultural Complexity." *Nature* 503 (2013): 389–391.

Derex, M., C. Perreault, and R. Boyd. "Divide and Conquer: Intermediate Levels of Population Fragmentation Maximize Cultural Accumulation." *Philosophical Transactions of the Royal Society B: Biological Sciences* 373, no. 1743 (2018): 20170062.

d'Errico, F. and I. Colagè. "Cultural Exaptation and Cultural Neural Reuse: A Mechanism for the Emergence of Modern Culture and Behavior." *Journal of Biological Theory* 13, no. 4 (2018): 213–227.

Desmet, Klaus, Ignacio Ortuño-Ortín, and Romain Wacziarg. "Culture, Ethnicity, and Diversity." American Economic Review 107, no. 9 (2017): 2479-513.

Deutscher, G. The Unfolding of Language: An Evolutionary Tour of Mankind's Greatest Invention. New York: Henry Holt, 2005.

De Waal, F. *Are We Smart Enough to Know How Smart Animals Are?* New York: W. W. Norton, 2016.

Dhanani, S., V. Kumari, B. K. Puri, I. Treasaden, S. Young, and P. Sen. "A Systematic Review of the Heritability of Specific Psychopathic Traits Using Hare's Two-Factor Model of Psychopathy." *CNS Spectrums* 23, no. 1 (2017): 29–38.

Dickinson, E. R. *The World in the Long Twentieth Century: An Interpretive History*. Berkeley: University of California Press, 2018.

Ditlevsen, P. D., H. Svensmark, and S. Johnsen. "Contrasting Atmospheric and Climate Dynamics of the Last-Glacial and Holocene Periods." *Nature* 379 (1996): 810–812.

Donald, M. *Origins of the Modern Mind: Three Stages in the Evolution of Culture and Cognition*. Cambridge, MA: Harvard University Press, 1991.

Donnermeyer, J., C. Anderson, and E. Cooksey. "The Amish Population: County Estimates and Settlement Patterns." *Journal of Amish and Plain Anabaptist Studies* 1, no. 1 (2019): 4.

Du, J. and R. Mace. "Marriage Stability in a Pastoralist Society." *Behavioral Ecology* 30, no. 6 (2019): 1567–1574.

Duda, P. and J. Zrzavý. "Evolution of Life History and Behavior in Hominidae: Towards Phylogenetic Reconstruction of the Chimpanzee-Human Last Common Ancestor." *Journal of Human Evolution* 65, no. 4 (2013): 424–446.

Dunbar, R. "Religion, the Social Brain and the Mystical Stance." *Archive for Psychology of Religion* 42, no. 1 (2019): 46–62.

Durkheim, E. *The Elementary Forms of the Religious Life.* London: Allen & Unwin, 1915.

Edelman, G. M. *Neural Darwinism: The Theory of Neuronal Group Selection.* New York: Basic Books, 1987.

Efferson, C. and P. J. Richerson. "A Prolegomenon to Non-linear Empiricism in the Human Sciences." *Biology and Philosophy* 22, no. 1 (2007): 1–33.

Ellis, G. and M. Solms. *Beyond Evolutionary Psychology: How and Why Neurophysiological Modules Arise.* Cambridge: Cambridge University Press, 2017.

Evans, N. and S. C. Levinson. "The Myth of Language Universals: Language Diversity and Its Importance for Cognitive Science." *Behavior and Brain Sciences* 32 (2009): 429–492.

Everett, D. L. *Dark Matter of the Mind: The Culturally Articulated Unconscious.* Chicago: University of Chicago Press, 2016.

Fehr, E. and S. Gachter. "Altruistic Punishment in Humans." *Nature* 415, no. 6868 (2002): 137–140.

Feldblum, Joseph T., Emily E. Wroblewski, Rebecca S. Rudicell, Beatrice H. Hahn, T. Paiva, M. Cetinkaya-Rundel, Anne E. Pusey, and Ian C. Gilby. "Sexually Coercive Male Chimpanzees Sire More Offspring." *Current Biology* 24, no. 23 (2014): 2855–2860.

Feldman, R. "The Neurobiology of Human Attachments." *Trends in Cognitive Sciences* 21, no. 2 (2017): 80–99.

Finlay, B. L., R. B. Darlington, and N. Nicastro. "Developmental Structure in Brain Evolution." *Behavioral and Brain Sciences* 24, no. 2 (2001): 263–278.

Fisher, K. "Uncertain Aims and Tacit Negotiation: Birth Control Practices in Britain, 1925–50." *Population and Development Review* 26, no. 2 (2000): 295–317.

Frankenhuis, W. E. and A. Ploeger. "Evolutionary Psychology versus Fodor: Arguments for and against the Massive Modularity Hypothesis." *Philosophical Psychology* 20, no. 6 (2007): 687–710.

Frecska, E. and Z. Kulcsar. "Social Bonding in the Modulation of the Physiology of Ritual Trance." *Ethos* 17, no. 1 (1989): 70–87.

Frisch, R. E. *Female Fertility and the Body Fat Connection.* Chicago: University of Chicago Press, 2004.

Frost, K. and P. J. Richerson. "Is Science Postmodern: Cultural Evolution as an Example." In *Naturalistic Approaches to Culture*, edited by C. Pléh, G. Csbra, and P. Richerson, 29–44. Budapest: Académiai Kiadó, 2014.

Fuller, D. Q., T. Denham, M. Arroyo-Kalin, L. Lucas, C. J. Stevens, L. Qin, R. G. Allaby, and M. D. Purugganan. "Convergent Evolution and Parallelism in Plant Domestication Revealed by an Expanding Archaeological Record." *Proceedings of the National Academy of Sciences USA* 111, no. 17 (2014): 6147–6152.

Gangestad, S. W. and J. A. Simpson. "The Evolution of Human Mating: Trade-Offs and Strategic Pluralism." *Behavioral and Brain Sciences* 23, no. 4 (2000): 573–587.

Gat, A. *War in Human Civilization.* Oxford: Oxford University Press, 2008.

Gelfand, M. J., J. L. Raver, L. Nishii, L. M. Leslie, J. Lun, B. C. Lim, L. Duan, A. Almaliach, S. Ang, and J. J. S. Arnadottir. "Differences between Tight and Loose Cultures: A 33-Nation Study." *Science* 332, no. 6033 (2011): 1100–1104.

Gerbault, P., A. Liebert, Y. Itan, A. Powell, M. Currat, J. Burger, D. M. Swallow, and M. G. Thomas. "Evolution of Lactase Persistence: An Example of Human Niche Construction." *Philosophical Transactions of the Royal Society B: Biological Sciences* 366, no. 1566 (2011): 863–877.

Gibson, K. "Tools, Language and Intelligence: Evolutionary Implications." *Man* 26, no. 2 (1991): 255–264.

Gil-White, F. J. "Are Ethnic Groups Biological 'Species' to the Human Brain? Essentialism in Our Cognition of Some Social Categories." *Current Anthropology* 42, no. 4 (2001): 515–554.

Gil-White, F. J. "Is Ethnocentrism Adaptive." 2002, Unpublished.

Giuliano, P, and N. Nunn. "Understanding Cultural Persistence and Change". Review of Economic Studies (2020): forthcoming.

Gómez-Robles, A. "The Dawn of *Homo floresiensis.*" *Nature* 534, no. 7606 (2016): 188–189.

Goodman, A., I. Koupil, and D. W. Lawson. "Low Fertility Increases Descendant Socioeconomic Position but Reduces Long-Term Fitness in a Modern Postindustrial Society." *Proceedings of the Royal Society B: Biological Sciences* 279, no. 1746 (2012): 4342–4351.

Gosling, A. L., H. R. Buckley, E. Matisoo-Smith, and T. R. Merriman. "Pacific Populations, Metabolic Disease and 'Just-So Stories': A Critique of the 'Thrifty Genotype' Hypothesis in Oceania." *Annals of Human Genetics* 79, no. 6 (2015): 470–480.

Gray, P. B. and K. G. Anderson. *Fatherhood: Evolution and Human Parental Behavior.* Cambridge, MA: Harvard University Press, 2012.

Greenfield, P. M., H. Keller, A. Fuligni, and A. Maynard. "Cultural Pathways through Universal Development." *Annual Review of Psychology* 54, no. 1 (2003): 461–490.

Gurven, M., W. Allen-Arave, K. Hill, and M. Hurtado. "'It's a Wonderful Life': Signaling Generosity among the Ache of Paraguay." *Evolution and Human Behavior* 21, no. 4 (2000) 263–282.

Guthrie, R. D. *The Frozen Fauna of the Mammoth Steppe: The Story of Blue Babe.* Chicago: University of Chicago Press, 1990.

Guthrie, R. D. "Origin and Causes of the Mammoth Steppe: A Story of Cloud Cover, Woolly Mammoth Tooth Pits, Buckles, And Inside-Out Beringia." *Quaternary Science Reviews* 20 (2001): 549–574.

Guthrie, R. D. *The Nature of Paleolithic Art*. Chicago: University of Chicago Press, 2005.

Hakenbeck, S. E. "Genetics, Archaeology and the Far Right: An Unholy Trinity." *World Archaeology* 51, no. 4 (2019): 517–527.

Hakimi, O. and L. Cameron. "Effect of Exercise on Ovulation: A Systematic Review." *Sports Medicine* 47, no. 8 (2017): 1555–1567.

Harari, Y. N. *Sapiens: A Brief History of Humankind*. New York: Random House, 2014.

Hardy, A. "Was Man More Aquatic in the Past." *New Scientist* 7, no. 5 (1960): 642–645.

Hare, B. "Survival of the Friendliest: *Homo sapiens* Evolved via Selection for Prosociality." *Annual Review of Psychology* 68, no. 1 (2017): 155–186.

Hare, B., J. Call, and M. Tomasello. "Do Chimpanzees Know What Conspecifics Know?" *Animal Behaviour* 61, no. 1 (2001): 139–151.

Hare, B. and Vanessa Woods. *Survival of the Friendliest: Understanding Our Origins and Rediscovering Our Common Humanity*. New York: Random House, 2020.

Harpending, H. C., M. A. Batzer, M. Gurven, L. B. Jorde, A. R. Rogers, and S. T. Sherry. "Genetic Traces of Ancient Demography." *Proceedings of the National Academy of Sciences USA* 95, no. 4 (1998): 1961–1967.

Harrison, R. A. and A. Whiten. "Chimpanzees (*Pan troglodytes*) Display Limited Behavioural Flexibility When Faced with a Changing Foraging Task Requiring Tool Use." *PeerJ* 6 (2018): e4366.

Haslam, S. A. *Psychology in Organizations: The Social Identity Approach*. London: SAGE, 2004.

Hawkes, K., J. F. O'Connell, N. G. B. Jones, H. Alvarez, and E. L. Charnov. "Grandmothering, Menopause, and the Evolution of Human Life Histories." *Proceedings of the National Academy of Sciences USA* 95, no. 3 (1998): 1336–1339.

Hayashi, M. and T. Matsuzawa. "Mother-Infant Interactions in Captive and Wild Chimpanzees." *Infant Behavior and Development* 48 (2017): 20–29.

Hayden, B. "Richman, Poorman, Beggarman, Chief: The Dynamics of Social Inequality." In *Archaeology at the Millennium A Sourcebook*, edited by G. M. Feinman and T. D. Price, 231–272. Boston: Springer, 2001.

Hayes, C. *The Ape in Our House*. New York: Harper, 1951.

Hazlewood, N. *Savage: The Life and Times of Jemmy Button*. London: Macmillan, 2001.

Heilman, S. C. *Defenders of the Faith: Inside Ultra-Orthodox Jewry*. Berkeley: University of California Press, 2000.

Heldstab, S. A., C. P. van Schaik, and K. Isler. "Being Fat and Smart: A Comparative Analysis of the Fat-Brain Trade-Off in Mammals." *Journal of Human Evolution* 100 (2016): 25–34.

Henrich, J. "Demography and Cultural Evolution: Why Adaptive Cultural Processes Produced Maladaptive Losses in Tasmania." *American Antiquity* 69, no. 2 (2004): 197–214.

Henrich, J. *The Secret of Our Success: How Culture Is Driving Human Evolution, Domesticating Our Species, and Making Us Smarter.* Princeton, NJ: Princeton University Press, 2016.

Henrich, J. *The WEIRDest People in the World: How the West Became Psychologically Peculiar and Particularly Prosperous. Farrar,* New York: Straus and Giroux, 2020.

Henrich, J., R. Boyd, S. Bowles, C. Camerer, E. Fehr, H. Gintis, and R. McElreath. "In Search of *Homo economicus*: Behavioral Experiments in 15 Small-Scale Societies." *American Economic Review* 91, no. 2 (2001): 73–78.

Henrich, J., J. Ensminger, R. McElreath, A. Barr, C. Barrett, A. Bolyanatz, J. C. Cardenas, et al. "Market, Religion, Community Size and the Evolution of Fairness and Punishment." *Science* 327 (2010): 1480–1484.

Henrich, J. and F. J. Gil-White. "The Evolution of Prestige—Freely Conferred Deference as a Mechanism for Enhancing the Benefits of Cultural Transmission." *Evolution and Human Behavior* 22, no. 3 (2001): 165–196.

Henrich, J., S. Heine, and A. Norenzayan. "The Weirdest People in the World?" *Behavioral and Brain Sciences* 33 (2010): 61–135.

Henrich, J., R. McElreath, A. Barr, J. Ensminger, C. Barrett, A. Bolyanatz, J. C. Cardenas, et al. "Costly Punishment across Human Societies." *Science* 312 (2006): 1767–1770.

Henshilwood, C. S., F. d'Errico, C. W. Marean, R. G. Milo, R. D. Milo, and R. Yates. "An Early Bone Tool Industry from the Middle Stone Age at Blombos Cave, South Africa: Implications for the Origins of Modern Human Behaviour, Symbolism and Language." *Journal of Human Evolution* 41, no. 6 (2001): 631–678.

Henshilwood, C. S., F. d'Errico, R. Yates, Z. Jacobs, C. Tribolo, G. A. T. Duller, N. Mercier, et al. "Emergence of Modern Human Behavior: Middle Stone Age Engravings from South Africa." *Science* 295, no. 5558 (2002): 1278–1280.

Herrmann, B., C. Thöni, and S. Gächter. "Antisocial Punishment across Societies." *Science* 319 (2008): 1362–1367.

Hewlett, B. S. *Hunter-Gatherer Childhoods: Evolutionary, Developmental, and Cultural Perspectives.* London: Routledge, 2017.

Heyes, C. *Cognitive Gadgets: The Cultural Evolution of Thinking.* Cambridge, MA: Harvard University Press, 2018.

Hill, K. and A. M. Hurtado. "Cooperative Breeding in South American Hunter-Gatherers." *Proceedings of the Royal Society B* 276, no. 1674 (2009): 3863–3870.

Hinde, K. and L. A. Milligan. "Primate Milk: Proximate Mechanisms and Ultimate Perspectives." *Evolutionary Anthropology: Issues, News, and Reviews* 20, no. 1 (2011): 9–23.

Hitchcock, T. *English Sexualities, 1700–1800*. London: Macmillan International Higher Education, 1997.

Hitonaru, N. and M. Nakamura. "A Newborn Infant Chimpanzee Snatched and Cannibalized Immediately after Birth: Implications for 'Maternity Leave' in Wild Chimpanzee." *American Journal of Physical Anthropology* 165, no. 1 (2018): 194–199.

Hoffecker, J. F. and S. A. Elias. *Human Ecology of Beringia*. New York: Columbia University Press, 2007.

Hofstede, G. Culture's Consequences: International Differences in Work-Related Values. Newbury Park, CA: SAGE, 1984.

Hogg, M. A. "Social Identity Theory." In *Understanding Peace and Conflict through Social Identity Theory*, edited by S. McKeown, R. Haji, and N. Ferguson, 3–17. Cham, Switzerland: Springer, 2016.

Holden, C. J. and R. Mace. "The Cow is the Enemy of Matriliny: Using Phylogenetic Methods to Investigate Cultural Evolution in Africa." In *The Evolution of Cultural Diversity*, edited by R. Mace, C. J. Holden, and S. Shennan, 227–230. London: Routledge, 2016.

Holloway, R. L., D. C. Broadfield, and M. S. Yuan. "Morphology and Histology of Chimpanzee Primary Visual Striate Cortex Indicate That Brain Reorganization Predated Brain Expansion in Early Hominid Evolution." *The Anatomical Record Part A: Discoveries in Molecular, Cellular, and Evolutionary Biology* 273A, no. 1 (2003): 594–602.

Horns, J., R. Jung, and D. R. Carrier. "*In vitro* Strain in Human Metacarpal Bones during Striking: Testing the Pugilism Hypothesis of Hominin Hand Evolution." *Journal of Experimental Biology* 218, no. 20 (2015): 3215.

Hrdy, S. B. "Male-Male Competition and Infanticide among the Langurs (*Presbytis entellus*) of Abu, Rajasthan." *Folia primatologica* 22, no. 1 (1974): 19–58.

Hrdy, S. B. *The Woman That Never Evolved*. Cambridge, MA: Harvard University Press, 1981.

Hrdy, S. B. *Mothers and Others: The Evolutionary Origins of Mutual Understanding*. Cambridge MA: Harvard University Press, 2009.

Hublin, J. J., S. Neubauer, and P. Gunz. "Brain Ontogeny and Life History in Pleistocene Hominins." *Philosophical Transactions of the Royal Society B: Biological Sciences* 370, no. 1663 (2015): 20140062.

Hunt, L. *Inventing Human Rights: A History*. New York: W. W. Norton, 2007.

Hurst, C. E. and D. L. McConnell. *An Amish Paradox: Diversity and Change in the World's Largest Amish Community*. Baltimore: John Hopkins University Press, 2010.

Isler, K. and C. P. van Schaik. "The Expensive Brain: A Framework for Explaining Evolutionary Changes in Brain Size." *Journal of Human Evolution* 57, no. 4 (2009): 392–400.

Jablonka, E. "Epigenetic Inheritance and Plasticity: The Responsive Germline." *Progress in Biophysics and Molecular Biology* 111, no. 2–3 (2013): 99–107.

Jablonski, N. G. *Skin: A Natural History*. Berkeley: University of California Press, 2008.

Jacobs, Z., R. G. Roberts, R. F. Galbraith, H. J. Deacon, R. Grün, A. Mackay, P. Mitchell, R. Vogelsang, and L. Wadley. "Ages for the Middle Stone Age of Southern Africa: Implications for Human Behavior and Dispersal." *Science* 322 (2008): 733–735.

Janzen, R. and M. Stanton. *The Hutterites in North America*. Baltimore: John Hopkins University Press, 2010.

Jerison, H. J. *Evolution of the Brain and Intelligence*. New York: Academic Press, 1973.

Kaplan, H. S., K. Hill, J. Lancaster, and A. M. Hurtado. "A Theory of Human Life History Evolution: Diet, Intelligence, and Longevity." *Evolutionary Anthropology* 9, no. 4 (2000): 156–185.

Kaplan, H. S., J. B. Lancaster, S. E. Johnson, and J. A. Bock. "Does Observed Fertility Maximize Fitness among New Mexican Men?" *Human Nature* 6, no. 4 (1995): 325–360.

Karkanas, P., R. Shahack-Gross, A. Ayalon, M. Bar-Matthews, R. Barkai, A. Frumkin, A. Gopher, and M. C. Stiner. "Evidence for Habitual Use of Fire at the End of the Lower Paleolithic: Site-Formation Processes at Qesem Cave, Israel." *Journal of Human Evolution* 53, no. 2 (2007): 197–212.

Keeley, L. H. *War Before Civilization*. New York: Oxford University Press, 1996.

Keller, H. . *Cultures of Infancy*. Psychology Press, 2013.

Kelly, R. C. "The Evolution of Lethal Intergroup Violence." *Proceedings of the National Academy of Sciences* 102, no. 43 (2005): 15294–15298.

Kinzler, K. D., K. H. Corriveau, and P. L. Harris. "Children's Selective Trust in Native-Accented Speakers." *Developmental Science* 14, no. 1 (2011): 106–111.

Kinzler, K. D., K. Shutts, J. DeJesus, and E. S. Spelke. "Accent Trumps Race in Guiding Children's Social Preferences." *Social Cognition* 27, no. 4 (2009): 623–634.

Kirch, P. V. *On the Road of the Winds: An Archaeological History of the Pacific Islands before European Contact*. Berkeley: University of California Press, 2017.

Kirkpatrick, R. C. "The Evolution of Human Homosexual Behavior." *Current Anthropology* 41, no. 3 (2000): 385–413.

Kivell, T. L., P. Lemelin, B. G. Richmond, and D. Schmitt. *The Evolution of the Primate Hand*. New York: Springer, 2016.

K. Klein Goldewijk, A. Beusen, G. van Drecht, and M. de Vos, "The HYDE 3.1 Spatially Explicit Database of Human-Induced Global Land-Use Change Over the Past 12,000 Years," *Global Ecology and Biogeography* 20, no. 1 (2001): 73–86.

Klein, R. G. "Anatomy, Behavior, and Modern Human Origins." *Journal of World Prehistory* 9, no. 2 (1995): 167–98.

Klein, R. G. *The Human Career: Human Biological and Cultural Origins.* Chicago: University of Chicago Press, 2009.

Klein, R. G. and B. Edgar. *The Dawn of Human Culture: A Bold New Theory on What Sparked the "Big Bang" of Human Consciousness.* New York: John Wiley, 2002.

Klein Goldewijk, K., A. Beusen, G. van Drecht, and M. de Vos. "The HYDE 3.1 Spatially Explicit Database of Human-Induced Global Land-Use Change over the Past 12,000 Years." *Global Ecology and Biogeography* 20, no. 1 (2011): 73–86.

Kline, M. A. and R. Boyd. "Population Size Predicts Technological Complexity in Oceania." *Proceedings of the Royal Society B* 277 (2010): 2559–2564.

Kline, M. A., R. Shamsudheen, and T. Broesch. "Variation Is the Universal: Making Cultural Evolution Work in Developmental Psychology." *Philosophical Transactions of the Royal Society B: Biological Sciences* 373, no. 1743 (2018).

Knodel, J. "Starting, Stopping, and Spacing during the Early Stages of Fertility Transition: The Experience of German Village Populations in the 18th and 19th Centuries." *Demography* 24, no. 2 (1987): 143–162.

Konvalinka, I., D. Xygalatas, J. Bulbulia, U. Schjødt, E. M. Jegindø, S. Wallot, G. Van Orden, and A. Roepstorff. "Synchronized Arousal between Performers and Related Spectators in a Fire-Walking Ritual." *Proceedings of the National Academy of Sciences* 108, no. 20 (2011): 8514–8519.

Korotayev, A., A. Malkov, and D. Khalttourina. *Introduction to Social Macrodynamics: Compact Macromodels of the World System Growth.* Moscow: Editorial URSS, 2006.

Kramer, K. L. "Cooperative Breeding and Its Significance to the Demographic Success of Humans." *Annual Review of Anthropology* 39, no. 1 (2010): 417–436.

Kramer, K. L. and A. Veile. "Infant Allocare in Traditional Societies." *Physiology & Behavior* 193 (2018): 117–126.

Kraybill, D. B. *The Amish of Lancaster County.* Lanham, MD: Rowman & Littlefield, 2019.

Kraybill, D. B. and C. F. Bowman. *On the Backroad to Heaven: Old Order Hutterites, Mennonites, Amish, and Brethren.* Baltimore: Johns Hopkins University Press, 2001.

Kraybill, D. B., S. M. Nolt, and D. Weaver-Zercher. *The Amish Way: Patient Faith in a Perilous World.* San Francisco: Jossey-Bass, 2010.

Kropotkin, P. *Mutual Aid.* London, Heinemann, 1902.

Krubitzer, L. "The Organization of Neocortex in Mammals: Are Species Differences Really So Different?" *Trends in Neuroscience* 18, no. 9 (1995): 408–417.

Krubitzer, L. and D. S. Stolzenberg. "The Evolutionary Masquerade: Genetic and Epigenetic Contributions to the Neocortex." *Current Opinion in Neurobiology* 24 (2014): 157–165.

Kuzawa, C. W., H. T. Chugani, L. I. Grossman, L. Lipovich, O. Muzik, P. R. Hof, D. E. Wildman, C. C. Sherwood, W. R. Leonard, and N. Lange. "Metabolic Costs and Evolutionary Implications of Human Brain Development." *Proceedings of the National Academy of Sciences* 111, no. 36 (2014): 13010–13015.

Labov, William. Principles of Linguistic Change: Social Factors. Language in Society. Edited by Peter Trudgill. Vol. 29, Malden, MA: Blackwell, 2001.

Lahr, M. M., F. Rivera, R. K. Power, A. Mounier, B. Copsey, F. Crivellaro, J. E. Edung, et al. "Inter-Group Violence among Early Holocene Hunter-Gatherers of West Turkana, Kenya." *Nature* 529, no. 7586 (2016): 394–398.

Laland, K. N. *Darwin's Unfinished Symphony: How Culture Made the Human Mind*. Princeton, NJ: Princeton University Press, 2017.

Laland, K. N., J. Odling-Smee, and S. Myles. "How Culture Shaped the Human Genome: Bringing Genetics and the Human Sciences Together." *Nature Reviews Genetics* 11, no. 2 (2010): 137–148.

Laland, K. N., T. Uller, M. Feldman, K. Sterelny, G. B. Müller, A. Moczek, E. Jablonka, et al. "Does Evolutionary Theory Need a Rethink? Yes, Urgently versus No, All Is Well." *Nature* 514 (2014): 161–164.

Landers, J. "Stopping, Starting and Spacing: The Regulation of Fertility in Historical Populations." in *Human Reproductive Decisions: Biological and Social Perspectives*, edited by R. I. M. Dunbar, 180–206. London: Macmillan Education UK, 1995.

Larmuseau, M. H. D., P. van den Berg, S. Claerhout, F. Calafell, A. Boattini, L. Gruyters, M. Vandenbosch, K. Nivelle, R. Decorte, and T. Wenseleers. "A Historical-Genetic Reconstruction of Human Extra-Pair Paternity." *Current Biology* 29, no. 23 (2019): 4102–4107.e7.

Larson, G. and D. Q. Fuller. "The Evolution of Animal Domestication." *Annual Review of Ecology, Evolution, and Systematics* 45, no. 1 (2014): 115–136.

Leakey, M. D. and R. L. Hay. "Pliocene Footprints in the Laetolil Beds at Laetoli, Northern Tanzania." *Nature* 278, no. 5702 (1979): 317–323.

Lee, H. and Y. Kim. "Discovering a Gem: Development of the Group Emotions Model of Sport Fan Experience." *International Journal of Applied Sports Sciences* 25, no. 2 (2013): 130–149.

Lee, R. B. "Hunter-Gatherers and Human Evolution: New Light on Old Debates." *Annual Review of Anthropology* 47, no. 1 (2018): 513–531.

LeVine, R. and D. T. Campbell. *Ethnocentrism: Theories of Conflict, Ethnic Attitudes, and Group Behavior*. New York, Wiley, 1972.

Lewens, T. "Human Nature: The Very Idea." *Philosophy & Technology* 25, no. 4 (2012): 459–474.

Lew-Levy, S., S. M. Kissler, A. H. Boyette, A. N. Crittenden, I. A. Mabulla, and B. S. Hewlett. "Who Teaches Children to Forage? Exploring the Primacy of Child-to-Child Teaching among Hadza and BaYaka Hunter-Gatherers of Tanzania and Congo." *Evolution and Human Behavior* 41, no. 1 (2020): 12–22.

Li, F., S. L. Kuhn, O. Bar-Yosef, F.-y. Chen, F. Peng, and X. Gao. "History, Chronology and Techno-Typology of the Upper Paleolithic Sequence in the Shuidonggou Area, Northern China." *Journal of World Prehistory* 32, no. 2 (2019): 111–141.

Li, H. and R. Durbin. "Inference of Human Population History from Individual Whole-Genome Sequences." *Nature* 475, no. 7357 (2011): 493–496.

Lieberman, D. E. *The Story of the Human Body: Evolution, Health, and Disease.* New York: Pantheon, 2013.

Lindert, P. H. "English Population, Wages, and Prices:1541–1913." *Journal of Interdisciplinary History* 15 (1985): 609–634.

Linton, S. "Woman the Gatherer: Male Bias in Anthropology." In *Woman in Cross-Cultural Perspective: A Preliminary Sourcebook*, compiled by Sue-Ellen Jacobs, 9–21. Urbana: University of Illinois, Dept. of Urban and Regional Planning, 1971.

Liu, S., N. B. Brooks, and E. S. Spelke. "Origins of the Concepts Cause, Cost, and Goal in Prereaching Infants." *Proceedings of the National Academy of Sciences* 116, no. 36 (2019): 17747–17752.

Lonsdorf, E. V. "What Is the Role of Mothers in the Acquisition of Termite-Fishing Behaviors in Wild Chimpanzees (*Pan troglodytes schweinfurthii*)?" *Animal Cognition* 9, no. 1 (2006): 36–46.

Lonsdorf, E. V., M. A. Stanton, and C. M. Murray. "Sex Differences in Maternal Sibling-Infant Interactions in Wild Chimpanzees." *Behavioral Ecology and Sociobiology* 72, no. 7 (2018): 117.

Loulergue, L., A. Schilt, R. Spahni, V. Masson-Delmotte, T. Blunier, B. Lemieux, J.-M. Barnola, D. Raynaud, T. F. Stocker, and J. Chappellaz. "Orbital and Millennial-Scale Features of Atmospheric CH4 over the Past 800,000 Years." *Nature* 453, no. 7193 (2008): 383–386.

Lovejoy, C. O. "The Origin of Man." *Science* 211 (1981): 341–348.

Low, B. S. *Why Sex Matters: A Darwinian Look at Human Behavior.* Rev. ed. Princeton, NJ: Princeton University Press, 2015.

Lowe, A. E., C. Hobaiter, and N. E. Newton-Fisher. "Countering Infanticide: Chimpanzee Mothers Are Sensitive to the Relative Risks Posed by Males on Differing Rank Trajectories." *American Journal of Physical Anthropology* 168, no. 1 (2019): 3–9.

Lumsden, C. and E. O. Wilson. *Genes, Mind, and Culture: The Coevolutionary Process.* 25th Anniversary Edition. Hackensack, NJ: World Scientific, 2006.

Mace, R. "Human Behavioral Ecology and Its Evil Twin." *Behavioral Ecology* 25, no. 3 (2014): 443–449.

Marean, C. W. "Pinnacle Point Cave 13B (Western Cape Province, South Africa) in Context: The Cape Floral Kingdom, Shellfish, and Modern Human Origins." *Journal of Human Evolution* 59, no. 3–4 (2010): 425–443.

Marean, C. W. "An Evolutionary Anthropological Perspective on Modern Human Origins." *Annual Review of Anthropology* 44, no. 1 (2015): 533–556.

Marshack, A. *The Roots of Civilization: The Cognitive Beginnings of Man's First, Art, Symbol, and Notation.* New York: McGraw-Hill, 1971.

Martin, R. D. *How We Do It: The Evolution and Future of Human Reproduction.* New York: Basic Books, 2013.

Martrat, B., J. O. Grimalt, N. J. Shackleton, L. de Abreu, M. A. Hutterli, and T. F. Stocker. "Four Climate Cycles of Recurring Deep and Surface Water Destabilizations on the Iberian Margin." *Science* 317, no. 5837 (2007): 502–507.

Maryanski, A. "The Last Ancestor: An Ecological Network Model on the Origins of Human Sociality." *Advances in Human Ecology* 2 (1992): 1–32.

Maslin, M. A., S. Shultz, and M. H. Trauth. "A Synthesis of the Theories and Concepts of Early Human Evolution." *Philosophical Transactions of the Royal Society B: Biological Sciences* 370, no. 1663 (2015): 20140064.

Matt, S. J. *Homesickness: An American History.* Oxford: Oxford University Press, 2011.

Mattison, S. M., R. J. Quinlan, and D. Hare. "The Expendable Male Hypothesis." *Philosophical Transactions of the Royal Society B: Biological Sciences* 374, no. 1780 (2019): 20180080.

Maynard Smith, J. and E. Szathmáry. *The Major Transitions in Evolution.* Oxford: W.H. Freeman/Spectrum, 1995.

McBrearty, S. and A. S. Brooks. "The Revolution That Wasn't: A New Interpretation of the Origin of Modern Human Behavior." *Journal of Human Evolution* 39, no. 5 (2000): 453–563.

McElreath, R. *Statistical Rethinking: A Bayesian Course with Examples in R and Stan.* Chapman and Hall/CRC Texts in Statistical Science. Boca Raton, FL: CRC Press, 2015.

McElreath, R., R. Boyd, and P. Richerson. "Shared Norms Can Lead to the Evolution of Ethnic Markers." *Current Anthropology* 44, no. 1 (2003): 123–129.

McEvedy, C. and R. Jones. *Atlas of World Population History.* Harmondsworth, UK: Penguin, 1978.

McGrew, W. C., P. J. Baldwin, and C. E. G. Tutin. "Chimpanzees in a Hot, Dry and Open Habitat: Mt. Assirik, Senegal, West Africa." *Journal of Human Evolution* 10, no. 3 (1981): 227–244.

McGrew, W. C., P. J. Baldwin, and C. E. G. Tutin. "Diet of Wild Chimpanzees (*Pan troglodytes verus*) at Mt. Assirik, Senegal: I. Composition." *American Journal of Primatology* 16, no. 3 (1988): 213–226.

McHenry, H. M. "Sexual Dimorphism in Fossil Hominids and Its Socioecological Implications." In *The Archaeology of Human Ancestry*, edited by S. Shennan and J. Steele, 97–114. London: Routledge, 2005.

McNeill, W. H. *Keeping Together in Time: Dance and Drill in Human History.* Cambridge, MA: Harvard University Press, 1995.

Mesoudi, A. *Cultural Evolution: How Darwinian Theory Can Explain Human Culture and Synthesize the Social Sciences.* Chicago: University of Chicago Press, 2011.

Milam, E. L. *Creatures of Cain: The Hunt for Human Nature in Cold War America.* Princeton, NJ: Princeton University Press, 2019.

Miller, G. *The Mating Mind: How Sexual Choice Shaped the Evolution of Human Nature.* New York: Anchor, 2011.

Miller, J. A., M. A. Stanton, E. V. Lonsdorf, K. R. Wellens, A. C. Markham, and C. M. Murray. "Limited Evidence for Third-Party Affiliation during Development in Wild Chimpanzees (*Pan troglodytes schweinfurthii*)." *Royal Society Open Science* 4, no. 9 (2017): 170500.

Moerk, E. L. *The Mother of Eve—As a First Language Teacher.* Norwood, NJ: Ablex, 1983.

Moheau, J.-B. "Jean-Baptiste Moheau on the Moral Causes of Diminished Fertility." *Population and Development Review* 26, no. 4 (2000): 821–826.

Moheau, J.-B. and C. Behar. *Recherches et considérations sur la population de la France (1778).* Paris: INED, 1994.

Mollica, F. and S. T. Piantadosi. "Humans Store about 1.5 Megabytes of Information during Language Acquisition." *Royal Society Open Science* 6, no. 3 (2019): 181393.

Morgan, E. *The Aquatic Ape Hypothesis: The Most Credible Theory of Human Evolution.* London: Souvenir Press, 1982.

Muller, M. N., M. Emery Thompson, and R. W. Wrangham. "Male Chimpanzees Prefer Mating with Old Females." *Current Biology* 16, no. 22 (2006): 2234–2238.

Murray, C. M., M. A. Stanton, E. V. Lonsdorf, E. E. Wroblewski, and A. E. Pusey. "Chimpanzee Fathers Bias Their Behaviour towards Their Offspring." *Royal Society Open Science* 3, no. 11 (2016).

National Research Council (U.S.A), *Abrupt Climate Change: Inevitable Surprises.* Washington, DC: National Academy Press, 2002.

National Birth-Rate Commission. *The Declining Birth-rate: Its causes and effects.* London: Chapman and Hall, 1916.

Nave, G., C. Camerer, and M. McCullough. "Does Oxytocin Increase Trust in Humans? A Critical Review of Research." *Perspectives on Psychological Science* 10, no. 6 (2015): 772–789.

Nettle, D., C. Andrews, and M. Bateson. "Food Insecurity as a Driver of Obesity in Humans: The Insurance Hypothesis." *Behavioral and Brain Sciences* 40 (2017): e105.

Newson, L., T. Postmes, S. E. G. Lea, and P. Webley. 2005. "Why are modern families small? Toward an evolutionary and cultural explanation for the demographic transition." *Personality and Social Psychology Review*, no. 9 (2005): 360-75.

Newson, L. and P. J. Richerson. "Why Do People Become Modern: A Darwinian Mechanism." *Population and Development Review* 35, no. 1 (2009): 117–158.

Nikolskiy, P. and V. Pitulko. "Evidence from the Yana Palaeolithic Site, Arctic Siberia, Yields Clues to the Riddle of Mammoth Hunting." *Journal of Archaeological Science* 40, no. 12 (2013): 4189–4197.

Nisbett, R. E. *The Geography of Thought: How Asians and Westerners Think Differently—And Why.* New York: Free Press, 2003.

Nolt, S. M. . *A History of the Amish.* New York: Good Books, 2015.

O'Neill, M. C., B. R. Umberger, N. B. Holowka, S. G. Larson, and P. J. Reiser. "Chimpanzee Super Strength and Human Skeletal Muscle Evolution." *Proceedings of the National Academy of Sciences* 114, no. 28 (2017): 7343–7348.

Oelze, V. M., B. T. Fuller, M. P. Richards, B. Fruth, M. Surbeck, J.-J. Hublin, and G. Hohmann. "Exploring the Contribution and Significance of Animal Protein in the Diet of Bonobos by Stable Isotope Ratio Analysis of Hair." *Proceedings of the National Academy of Sciences* 108, no. 24 (2011): 9792–9797.

Opel, N., R. Redlich, K. Dohm, D. Zaremba, J. Goltermann, J. Repple, C. Kaehler, et al. "Mediation of the Influence of Childhood Maltreatment on Depression Relapse by Cortical Structure: A 2-Year Longitudinal Observational Study." *Lancet Psychiatry* 6, no. 4 (2019): 318–326.

Ostrom, E., R. Gardner, and J. Walker. *Rules, Games, and Common-Pool Resources.* Ann Arbor: University of Michigan Press, 1994.

Ozcelik, T., N. Akarsu, E. Uz, S. Caglayan, S. Gulsuner, O. E. Onat, M. Tan, and U. Tan. "Mutations in the Very Low-Density Lipoprotein Receptor VLDLR Cause Cerebellar Hypoplasia and Quadrupedal Locomotion in Humans." *Proceedings of the National Academy of Sciences* 105, no. 11 (2008): 4232–4236.

Pääbo, Svante. *Neanderthal Man: In Search of Lost Genomes.* New York: Basic Books, 2014.

Palmqvist, P., B. Martínez-Navarro, J. A. Pérez-Claros, V. Torregrosa, B. Figueirido, J. M. Jiménez-Arenas, M. Patrocinio Espigares, S. Ros-Montoya, and M. De Renzi. "The Giant Hyena *Pachycrocuta brevirostris*: Modelling the Bone-Cracking Behavior of an Extinct Carnivore." *Quaternary International* 243, no. 1 (2011): 61–79.

Panchanathan, K. and W. M. Baum. "Editorial: The Many Faces of Behavioral Evolution." *Behavioural Processes* 161 (2019): 1–2.

Panksepp, J. and L. Biven. *The Archaeology of Mind: Neuroevolutionary Origins of Human Emotions.* New York: W. W. Norton, 2012.

Park, M. S., A. D. Nguyen, H. E. Aryan, H. S. U, M. L. Levy, and K. Semendeferi. "Evolution of the Human Brain: Changing Brain Size and the Fossil Record." *Neurosurgery* 60, no. 3 (2007): 555–562.

Parker, L. A. *Writing the Revolution: A French Woman's History in Letters.* Oxford: Oxford University Press, 2013.

Paul, R. A. *Mixed Messages: Cultural and Genetic Inheritance in the Constitution of Human Society.* Chicago: University of Chicago Press, 2015.

Perreault, C. "The Pace of Cultural Evolution." *PLOS ONE* 7, no. 9 (2012): e45150.

Pinker, S. *The Language Instinct: How the Mind Creates Language.* New York: W. Morrow, 1994.

Pinker, S. *The Blank Slate: The Modern Denial of Human Nature.* London: Allen Lane, 2002.

Pinker, S. *The Better Angels of Our Nature: Why Violence Has Declined*. New York: Penguin, 2012.

Pinker, S. *Enlightenment Now: The Case for Reason, Science, Humanism, and Progress*. New York: Penguin, 2018.

Polanyi, M. "The Logic of Tacit Inference." *Philosophy* 41, no. 155 (1966): 1–18.

Pollak, R. A. and S. C. Watkins. "Cultural and Economic Approaches to Fertility: Proper Marriage or Mesalliance?" *Population and Development Review* 19, no. 3 (1993): 467–496.

Pontzer, H., M. H. Brown, D. A. Raichlen, H. Dunsworth, B. Hare, K. Walker, A. Luke, et al. "Metabolic Acceleration and the Evolution of Human Brain Size and Life History." *Nature* 533, no. 7603 (2016): 390–392.

Potts, Richard. "Evolution and Climate Variability." *Science* 273 (1996): 922–923.

Poulin-Dubois, D. and P. Brosseau-Liard. "The Developmental Origins of Selective Social Learning." *Current Directions in Psychological Science* 25, no. 1 (2016): 60–64.

Powell, A., S. Shennan, and M. G. Thomas. "Late Pleistocene Demography and the Appearance of Modern Human Behavior." *Science* 324 (2009): 1298–1301.

Price, T. D. and O. Bar-Yosef. "The Origins of Agriculture: New Data, New Ideas: An Introduction to Supplement 4." *Current Anthropology* 52, no. S4 (2011): S163–S174.

Pringle, H. "Ice Age Communities May Be Earliest Known Net Hunters." *Science* 277, no. 5330 (1997): 1203–1204.

Prohaska, A., F. Racimo, A. J. Schork, M. Sikora, A. J. Stern, M. Ilardo, M. E. Allentoft, et al. "Human Disease Variation in the Light of Population Genomics." *Cell* 177, no. 1 (2019): 115–131.

Prüfer, K., C. de Filippo, S. Grote, F. Mafessoni, P. Korlević, M. Hajdinjak, B. Vernot, L. Skov, P. Hsieh, and S. Peyrégne. "A High-Coverage Neandertal Genome from Vindija Cave in Croatia." *Science* 358, no. 6363 (2017): 655–658.

Prüfer, K., K. Munch, I. Hellmann, K. Akagi, J. R. Miller, B. Walenz, S. Koren, et al. "The Bonobo Genome Compared with the Chimpanzee and Human Genomes." *Nature* 486 (2012): 527.

Prufer, K., F. Racimo, N. Patterson, F. Jay, S. Sankararaman, S. Sawyer, A. Heinze, et al. "The Complete Genome Sequence of a Neanderthal from the Altai Mountains." *Nature* 505, no. 7481 (2014): 43–49.

Pusey, A., C. Murray, W. Wallauer, M. Wilson, E. Wroblewski, and J. Goodall. "Severe Aggression among Female *Pan Troglodytes schweinfurthii* at Gombe National Park, Tanzania." *International Journal of Primatology* 29, no. 4 (2008): 949.

Pusey, A. E. and K. Schroepfer-Walker. "Female Competition in Chimpanzees." *Philosophical Transactions of the Royal Society B: Biological Sciences* 368, no. 1631 (2013): 20130077.

Rakoczy, H., F. Warneken, and M. Tomasello. "Young Children's Selective Learning of Rule Games from Reliable and Unreliable Models." *Cognitive Development* 24, no. 1 (2009): 61–69.

Rasmussen, S., M. E. Allentoft, K. Nielsen, L. Orlando, M. Sikora, K.-G. Sjögren, A. G. Pedersen, et al. "Early Divergent Strains of *Yersinia pestis* in Eurasia 5,000 Years Ago." *Cell* 163, no. 3 (2015): 571–582.

Raubenheimer, D. and J. M. Rothman. "Nutritional Ecology of Entomophagy in Humans and Other Primates." *Annual Review of Entomology* 58, no. 1 (2013): 141–160.

Reader, S. M., Y. Hager, and N. Laland Kevin. "The Evolution of Primate General and Cultural Intelligence." *Philosophical Transactions of the Royal Society B: Biological Sciences* 366, no. 1567 (2011): 1017–1027.

Reich, D. *Who We Are and How We Got Here: Ancient DNA and the New Science of the Human Past*. Oxford: Oxford University Press, 2018.

Reno, P. L., R. S. Meindl, M. A. McCollum, and C. O. Lovejoy. "Sexual Dimorphism in *Australopithecus afarensis* Was Similar to That of Modern Humans." *Proceedings of the National Academy of Sciences* 100, no. 16 (2003): 9404–9409.

Richards, R. J. *Darwin and the Emergence of Evolutionary Theories of Mind and Behavior*. Chicago: University of Chicago Press, 1987.

Richerson, P. J. "The Use and Non-Use of the Human Nature Concept by Evolutionary Biologists." In *Why We Disagree about Human Nature*, edited by E. Hannon and T. Lewens, 145–169. Oxford: Oxford University Press, 2018.

Richerson, P. J., R. Baldini, A. Bell, K. Demps, K. Frost, V. Hillis, S. Mathew, et al. "Cultural Group Selection Plays an Essential Role in Explaining Human Cooperation: A Sketch of the Evidence, Together with Commentaries and Authors' Response." *Behavioral and Brain Sciences* 39, no. e30 (2016): 1–68.

Richerson, P. J. and R. Boyd. "The Evolution of Human Ultrasociality." In *Indoctrinability, Ideology, and Warfare; Evolutionary Perspectives*, edited by I. Eibl-Eibesfeldt and F. K. Salter, 71–95. New York: Berghahn Books1998.

Richerson, P. J. and R. Boyd. *Not by Genes Alone: How Culture Transformed Human Evolution*. Chicago: University of Chicago Press, 2005.

Richerson, P. J. and R. Boyd. "The Darwinian Theory of Cultural Evolution and Gene-Culture Coevolution." In *Evolution since Darwin: The First 150 Years*, edited by M. A. Bell, D. J. Futuyma, W. F. Eanes, and J. S. Levinton. Sunderland MA: Sinauer, 2010.

Richerson, P. J. and R. Boyd. "Why Possibly Language Evolved." *Biolinguistics* 4, no. 2–3 (2010): 289–306.

Richerson, P. J. and R. Boyd. "Rethinking Paleoanthropology: A World Queerer Than We Supposed." In *Evolution of Mind, Brain, and Culture*, edited by G. Hatfield and H. Pittman, 263–302. Philadelphia: University of Pennsylvania Museum of Archaeology and Anthropology, 2013.

Richerson, P. J. and R. Boyd. "The Human Life History Is Adapted to Exploit the Adaptive Advantages of Culture." *Philosophical Transactions of the Royal Society B* 375, no. 1803 (2020): 20190498.

Richerson, P. J., R. Boyd, and R. L. Bettinger. "Was Agriculture Impossible during the Pleistocene but Mandatory during the Holocene? A Climate Change Hypothesis." *American Antiquity* 66, no. 3 (2001): 387–411.

Richerson, P. J., R. Boyd, and R. L. Bettinger. "Cultural Innovations and Demographic Change." *Human Biology* 81, no. 2–3 (2009): 211–235.

Richmond, B. and S. Peterson. *An Introduction to Systems Thinking*. Hanover, NH: High Performance Systems., Inc., 2001.

Ridley, M. *The Rational Optimist: How Prosperity Evolves*. New York: Harper Perennial, 2010.

Rightmire, G. P., D. Lordkipanidze, and A. Vekua. "Anatomical Descriptions, Comparative Studies and Evolutionary Significance of the Hominin Skulls from Dmanisi, Republic of Georgia." *Journal of Human Evolution* 50, no. 2 (2006): 115–141.

Roebroeks, W. "Art on the Move." *Nature* 514, no. 7521 (2014): 170–171.

Roebroeks, W. and P. Villa. "On the Earliest Evidence for Habitual Use of Fire in Europe." *Proceedings of the National Academy of Sciences* 108, no. 13 (2011): 5209–5214.

Rogers, A. R. "Does Biology Constrain Culture?" *American Anthropologist* 90 (1989): 819–831.

Rogers, A. R. "Genetic Evidence for a Pleistocene Population Explosion." *Evolution* 49 (1995): 608–615.

Rogers, A. R., N. S. Harris, and A. A. Achenbach. "Neanderthal-Denisovan Ancestors Interbred with a Distantly Related Hominin." *Science Advances* 6, no. 8 (2020): eaay5483.

Rogers, E. M. *Diffusion of Innovations*. New York: Free Press, 1995.

Rogoff, B. *Apprenticeship in Thinking: Cognitive Development in Social Context*. New York: Oxford University Press, 1990.

Romer, Paul M. "Increasing Returns and Long-Run Growth. *Journal of Political Economy* 94, no. 5 (1986): 1002–1037.

Ross, C. T. and P. J. Richerson. "New Frontiers in the Study of Cultural and Genetic Evolution." *Current Opinion in Genetics and Development* 29 (2014): 103–109.

Rush, J. A. *Spiritual Tattoo: A Cultural History of Tattooing, Piercing, Scarification, Branding, and Implants*. Berkeley, CA: Frog Books, 2005.

Sahle, Y., H. Reyes-Centeno, and C. Bentz. *Modern Human Origins and Dispersal*. Tübingen, Germany: Kerns Verlag, 2019.

Sánchez Goñi, M. F., S. Desprat, A. L. Daniau, F. C. Bassinot, J. M. Polanco-Martínez, S. P. Harrison, J. R. M. Allen, et al. "The ACER Pollen and Charcoal Database: A Global Resource to Document Vegetation and Fire Response to Abrupt Climate Changes during the Last Glacial Period." *Earth System Science Data* 9, no. 2 (2017): 679–695.

Sánchez Goñi, M. F., S. Desprat, W. J. Fletcher, C. Morales-Molino, F. Naughton, D. Oliveira, D. H. Urrego, and C. Zorzi. "Pollen from the Deep-Sea: A Breakthrough in the Mystery of the Ice Ages." *Frontiers in Plant Science* 9, no. 38 (2018).

Sankararaman, S., S. Mallick, N. Patterson, and D. Reich. "The Combined Landscape of Denisovan and Neanderthal Ancestry in Present-Day Humans." *Current Biology* 26, no. 9 (2016): 1241–1247.

Sapolsky, R. M. *Behave: The Biology of Humans at Our Best and Worst*. New York: Penguin, 2017.

Scelza, B., S. Prall, N. Swinford, S. Gopalan, E. Atkinson, R. McElreath, J. Sheehama, and B. Henn. "High Rate of Extrapair Paternity in a Human Population Demonstrates Diversity in Human Reproductive Strategies." *Science Advances* 6, no. 8 (2020): eaay6195.

Scerri, E. M. L., M. G. Thomas, A. Manica, P. Gunz, J. T. Stock, C. Stringer, M. Grove, et al. "Did Our Species Evolve in Subdivided Populations across Africa, and Why Does It Matter?" *Trends in Ecology & Evolution* 33, no. 8 (2018): 582–594.

Schiefenhövel, W. "Preferential Female Infanticide and Other Mechanisms Regulating Population Size among the Eipo." In *Population and Biology*, edited by N. Keyfitz, 169–192. Liège, Belgium: Ordina Editions, 1984.

Schiffels, S. and R. Durbin. "Inferring Human Population Size and Separation History from Multiple Genome Sequences." *Nature Genetics* 46, no. 8 (2014): 919–925.

Schoch, W. H., G. Bigga, U. Böhner, P. Richter, and T. Terberger. "New Insights on the Wooden Weapons from the Paleolithic Site of Schöningen." *Journal of Human Evolution* 89 (2015): 214–225.

Sear, R. and R. Mace. "Who Keeps Children Alive? A Review of the Effects of Kin on Child Survival." *Evolution and Human Behavior* 29, no. 1 (2008): 1–18.

Secord, J. A. *Victorian Sensation: The Extraordinary Publication, Reception, and Secret Authorship of Vestiges of the Natural History of Creation*. Chicago: University of Chicago Press, 2003.

Semendeferi, K., A. Lu, N. Schenker, and H. Damasio. "Humans and Great Apes Share a Large Frontal Cortex." *Nature Neuroscience* 5, no. 3 (2002): 272–276.

Seyfarth, R. M. and D. L. Cheney. "The Evolutionary Origins of Friendship." *Annual Review of Psychology* 63, no. 1 (2012): 153–177.

Seymour, R. S., V. Bosiocic, and E. P. Snelling. "Fossil Skulls Reveal That Blood Flow Rate to the Brain Increased Faster Than Brain Volume during Human Evolution." *Royal Society Open Science* 3, no. 8 (2016): 160305.

Shaffer, L. J. "An Anthropological Perspective on the Climate Change and Violence Relationship." *Current Climate Change Reports* 3, no. 4 (2017): 222–232.

Shahack-Gross, R., F. Berna, P. Karkanas, C. Lemorini, A. Gopher, and R. Barkai. "Evidence for the Repeated Use of a Central Hearth at Middle Pleistocene (300 Ky Ago) Qesem Cave, Israel." *Journal of Archaeological Science* 44 (2014): 12–21.

Shennan, S. *The First Farmers of Europe: An Evolutionary Perspective*. Cambridge: Cambridge University Press, 2018.

Shennan, S. *Genes, Memes, and Human History: Darwinian Archaeology and Cultural Evolution*. London: Thames and Hudson, 2002.

Shermer, M. *The Moral Arc: How Science and Reason Lead Humanity toward Truth, Justice, and Freedom*. London: Macmillan, 2015.

Sherwood, C. C., F. Subiaul, and T. W. Zawidzki. "A Natural History of the Human Mind: Tracing Evolutionary Changes in Brain and Cognition." *Journal of Anatomy* 212, no. 4 (2008): 426–454.

Sievert, L. L. "Human Senescence." In *Basics in Human Evolution*, edited by M. P. Muehlenbein, 309–320. Amsterdam: Elsevier, 2015.

Sikora, M., A. Seguin-Orlando, V. C. Sousa, A. Albrechtsen, T. Korneliussen, A. Ko, S. Rasmussen, et al. "Ancient Genomes Show Social and Reproductive Behavior of Early Upper Paleolithic Foragers." *Science* 358, no. 6363 (2017): 659–662.

Simon, H. "A Mechanism for Social Selection And Successful Altruism." *Science* 250, no. 4988 (1990): 1665–1668.

Skoglund, P., S. Mallick, M. C. Bortolini, N. Chennagiri, T. Hünemeier, M. L. Petzl-Erler, F. M. Salzano, N. Patterson, and D. Reich. "Genetic Evidence for Two Founding Populations of the Americas." *Nature* 525, no. 7567 (2015): 104–108.

Smaldino, P. E., J. C. Schank, and R. McElreath. "Increased Costs of Cooperation Help Cooperators in the Long Run." *American Naturalist* 181, no. 4 (2013): 451–463.

Soffer, O., J. M. Adovasio, and D. C. Hyland. "The 'Venus' Figurines: Textiles, Basketry, Gender, and Status in the Upper Paleolithic." *Current Anthropology* 41, no. 4 (2000): 511–537.

Sosis, R. and C. Alcorta. "Signaling, Solidarity, and the Sacred: The Evolution of Religious Behavior." *Evolutionary Anthropology* 12 (2003): 264–274.

Stadler, N. *Yeshiva Fundamentalism: Piety, Gender, and Resistance in the Ultra-Orthodox World*. New York: NYU Press, 2009.

Starkweather, K. E. and R. Hames. "A Survey of Non-Classical Polyandry." *Human Nature* 23, no. 2 (2012): 149–172.

Stearns, S. C., N. Allal, and R. Mace. "Life History Theory and Human Development." In *Foundations of Evolutionary Psychology*, edited by C. Crawford and D. Krebs, 47–69. Hoboken, NJ: Lawrence Erlbaum, 2008.

Stetzik, L., D. Ganshevsky, M. N. Lende, L. E. Roache, S. Musatov, and B. S. Cushing. "Inhibiting ERα Expression in the Medial Amygdala Increases Prosocial Behavior in Male Meadow Voles (*Microtus pennsylvanicus*)." *Behavioural Brain Research* 351 (2018): 42–48.

Szreter, S. and K. Fisher. *Sex before the Sexual Revolution: Intimate Life in England 1918–1963*. Cambridge: Cambridge University Press, 2010.

Tabbaa, M., B. Paedae, Y. Liu, and Z. Wang. "Neuropeptide Regulation of Social Attachment: The Prairie Vole Model." *Comprehensive Physiology* 7 (2017): 81–104.

Teicher, M. H. and J. A. Samson. "Annual Research Review: Enduring Neurobiological Effects of Childhood Abuse and Neglect." *Journal of Child Psychology and Psychiatry* 57, no. 3 (2016): 241–266.

Thompson, M. E. "Reproductive Ecology of Female Chimpanzees." *American Journal of Primatology* 75, no. 3 (2013): 222–237.

Thompson, M. E. and R. W. Wrangham. "Diet and Reproductive Function in Wild Female Chimpanzees (*Pan troglodytes schweinfurthii*) at Kibale National Park, Uganda." *American Journal of Physical Anthropology* 135, no. 2 (2008): 171–181.

Tobias, P. V. "The Brain of *Homo habilis*: A New Level of Organization in Cerebral Evolution." *Journal of Human Evolution* 16, no. 7 (1987): 741–761.

Tomasello, M. "Do Apes Ape?" In *Social Learning in Animals: The Roots of Culture*, edited by C. M. Heyes and B. G. Galef Jr., 319–346. New York: Academic Press, 1996.

Tomasello, M. *Becoming Human: A Theory of Ontogeny*. Cambridge, MA: Belknap Press, 2019.

Tooby, J. and L. Cosmides. "The Psychological Foundations of Culture." In *The Adapted Mind: Evolutionary Psychology and the Generation of Culture*, edited by J. Barkow, L. Cosmides, and J. Tooby, 19–136. New York: Oxford University Press, 1992.

Toth, N. and K. Schick. "Early Stone Industries and Inferences Regarding Language and Cognition." In *Tools, Language and Cognition in Human Evolution*, edited by K. R. Gibson and T. Ingold, 346–362. Cambridge: Cambridge University Press, 1993.

Trevathan, W. *Ancient Bodies, Modern Lives: How Evolution Has Shaped Women's Health*. Oxford: Oxford University Press, 2010.

Trhlin, M. and J. Rajchard. "Chemical Communication in the Honeybee (*Apis mellifera L.*): A Review." *Veterinární Medicína* 56, no. 6 (2011): 265–273.

Trinkaus, E. and A. P. Buzhilova. "Diversity and Differential Disposal of the Dead at Sunghir." *Antiquity* 92, no. 361 (2018): 7–21.

Trinkaus, E., A. P. Buzhilova, M. B. Mednikova, and M. V. Dobrovolskaya. *The People of Sunghir: Burials, Bodies, and Behavior in the Earlier Upper Paleolithic*. Oxford: Oxford University Press, 2014.

Turchin, P. *Historical Dynamics*. Princeton, NJ: Princeton University Press, 2003.

Turchin, P. *Ultrasociety: How 10,000 Years of War Made Humans the Greatest Cooperators on Earth*. Chaplin, CT: Beresta Books, 2015.

Turner, J. C., M. A. Hogg, P. J. Oakes, S. D. Reicher, and M. S. Wetherell. *Rediscovering the Social Group: A Self-Categorization Theory*. Oxford: Basil Blackwell, 1987.

Uno, K. T., P. J. Polissar, K. E. Jackson, and P. B. deMenocal. "Neogene Biomarker Record of Vegetation Change in Eastern Africa." *Proceedings of the National Academy of Sciences* 113, no. 23 (2016): 6355–6363.

Valentine, J. W. and E. M. Moores. "Plate-Tectonic Regulation of Faunal Diversity and Sea Level: A Model." *Nature* 228, no. 5272 (1970): 657–659.

Van Baal, J. and J. Verschueren. *Dema: Description and Analysis of Marind-Anim Culture (South New Guinea)*. The Hague: Martinus Nijhoff, 1966.

van de Walle, E. "Fertility Transition, Conscious Choice, and Numeracy." *Demography* 29, no. 4 (1992): 487–502.

Vidali, Valerio, Jennifer Uman, and Alix Barzelay, *Jemmy Button* (Dorking, UK: Templar, 2014).

Vogt, S., N. A. Mohmmed Zaid, H. El Fadil Ahmed, E. Fehr, and C. Efferson. "Changing Cultural Attitudes towards Female Genital Cutting." *Nature* 538, no. 7626 (2016): 506–509.

Volk, A. A. and J. A. Atkinson. "Infant and Child Death in the Human Environment of Evolutionary Adaptation." *Evolution and Human Behavior* 34, no. 3 (2013): 182–192.

Volk, T. and J. Atkinson. "Is Child Death the Crucible of Human Evolution?" *Journal of Social, Evolutionary, and Cultural Psychology* 2, no. 4 (2008): 247–260.

Vygotsky, L. *Mind in Society*. Cambridge, MA: Harvard University Press, 1978.

Walker, C. S., K. K. Walker, G. Paulo, and A. E. Pusey. "Morphological Identification of Hair Recovered from Feces for Detection of Cannibalism in Eastern Chimpanzees." *Folia Primatologica* 89, no. 3–4 (2018): 240–250.

Walker, R. S., M. V. Flinn, and K. R. Hill. "Evolutionary History of Partible Paternity in Lowland South America." *Proceedings of the National Academy of Sciences* 107, no. 45 (2010): 19195–19200.

Walker, R. S., K. R. Hill, M. V. Flinn, and R. M. Ellsworth. "Evolutionary History of Hunter-Gatherer Marriage Practices." *PloS ONE* 6, no. 4 (2011).

Walker, R. S., S. Wichmann, T. Mailund, and C. J. Atkisson. "Cultural Phylogenetics of the Tupi Language Family in Lowland South America." *PLoS ONE* 7, no. 4 (2012): e35025.

Wallace, A. R. "The Origin of Human Races and the Antiquity of Man Deduced from the Theory of 'Natural Selection.'" *Journal of the Anthropological Society of London* 2 (1864): clviii–clxxxvii.

Ward, P. D. and J. Kirschvink. *A New History of Life: The Radical New Discoveries about the Origins and Evolution of Life on Earth*. New York: Bloomsbury, 2015.

Warneken, F. "How Children Solve the Two Challenges of Cooperation." *Annual Review of Psychology* 69, no. 1 (2018): 205–229.

Warneken, F. and M. Tomasello. "Varieties of Altruism in Children and Chimpanzees." *Trends in Cognitive Sciences* 13, no. 9 (2009): 397–402.

Wells, J. C. K. "The Evolution of Human Fatness and Susceptibility to Obesity: An Ethological Approach." *Biological Reviews* 81, no. 2 (2006): 183–205.

Werner, E. E. *Cross Cultural Child Development: A View from the Planet Earth.* Monterey, CA: Brooks/Cole,1979.

White, T. D., B. Asfaw, Y. Beyene, Y. Haile-Selassie, C. O. Lovejoy, G. Suwa, and G. WoldeGabriel. "*Ardipithecus ramidus* and the Paleobiology of Early Hominids." *Science* 326, no. 5949 (2009): 64–86.

Whitehead, H. *Sperm Whales: Social Evolution in the Ocean.* Chicago: University of Chicago Press, 2003.

Whitehouse, H. and J. A. Lanman. "The Ties That Bind Us: Ritual, Fusion, and Identification." *Current Anthropology* 55, no. 6 (2014): 674–695.

Whitehead, H. and L. Rendell. *The Cultural Lives of Whales and Dolphins.* Chicago: University of Chicago Press, 2015.

Whitehead, H. and P. J. Richerson. "The Evolution of Conformist Social Learning Can Cause Population Collapse in Realistically Variable Environments." *Evolution and Human Behavior* 30 (2009): 261–273.

Whiten, A. "Culture Extends the Scope of Evolutionary Biology in the Great Apes." *Proceedings of the National Academy of Sciences* 114, no. 30 (2017): 7790–7797.

Whiten, A. "Social Learning and Culture in Child and Chimpanzee." *Annual Review of Psychology* 68, no. 1 (2017): 129–154.

Whiten, A., J. Goodall, W. C. McGrew, T. Nishida, V. Reynolds, Y. Sugiyama, C. E. G. Tutin, R. W. Wrangham, and C. Boesch. "Cultures in Chimpanzees." *Nature* 399 (1999): 682–685.

Whiten, A., N. McGuigan, S. Marshall-Pescini, and L. M. Hopper. "Emulation, Imitation, Over-Imitation and the Scope of Culture for Child and Chimpanzee." *Philosophical Transactions of the Royal Society B* 364 (2009): 2417–2428.

Whiting, Beatrice, and John W. M. Whiting. *Children of Six Cultures, A Psycho-Cultural Analysis.* Cambridge, MA: Harvard University Press, 1975.

Wiessner, P. and A. Tumu. *Historical Vines: Enga Networks of Exchange, Ritual, and Warfare in Papua New Guinea.* Washington, DC, Smithsonian Institution Press, 1998.

Wilkins, A. S., R. W. Wrangham, and W. T. Fitch. "The 'Domestication Syndrome' in Mammals: A Unified Explanation Based on Neural Crest Cell Behavior and Genetics." *Genetics* 197, no. 3 (2014): 795–808.

Wilkinson, G. S., G. G. Carter, K. M. Bohn, and D. M. Adams. "Non-kin Cooperation in Bats." *Philosophical Transactions of the Royal Society B: Biological Sciences* 371, no. 1687 (2016): 20150095.

Will, M., N. J. Conard, and C. A. Tyron. "Timing and Trajectory of Cultural Evolution on the African Continent 200,000–30,000 Years Ago." In *Modern Human Origins and Dispersal,* edited by Y. Sahle, H. Reyes-Centeno, and C. Bentz, 25–72. Tübingen, Germany, Kerns Verlag, 2019.

Williams-Hatala, E. M., K. G. Hatala, M. Gordon, A. Key, M. Kasper, and T. L. Kivell. "The Manual Pressures of Stone Tool Behaviors and Their Implications for the Evolution of the Human Hand." *Journal of Human Evolution* 119 (2018): 14–26.

Wilson, E. O. *On Human Nature*. Cambridge, MA: Harvard University Press, 1978.

Wilson, E. O. *The Social Conquest of Earth*. New York: Liveright, 2012.

Wilson, M. "Six Views of Embodied Cognition." *Psychonomic Bulletin and Review* 9, no. 4 (2002): 625–636.

Wilson, M. L., C. Boesch, B. Fruth, T. Furuichi, I. C. Gilby, C. Hashimoto, C. L. Hobaiter, et al. "Lethal Aggression in *Pan* is Better Explained by Adaptive Strategies Than Human Impacts." *Nature* 513 (2014): 414–417.

Winterhalder, B. and E. A. Smith. "Analyzing Adaptive Strategies: Human Behavioral Ecology at Twenty-Five." *Evolutionary Anthropology: Issues, News, and Reviews* 9, no. 2 (2000): 51–72.

Wolf, M., S. Naftali, R. C. Schroter, and D. Elad. "Air-Conditioning Characteristics of the Human Nose." *Journal of Laryngology & Otology* 118, no. 2 (2004): 87–92.

Woodroffe, R. and A. Vincent. "Mother's Little Helpers: Patterns of Male Care in Mammals." *Trends in Ecology & Evolution* 9, no. 8 (1994): 294–297.

Wrangham, R. *Catching Fire: How Cooking Made Us Human*. New York: Basic Books, 2009.

Wrangham, R. *The Goodness Paradox: The Strange Relationship between Virtue and Violence in Human Evolution*. New York: Pantheon, 2019.

Wrangham, R. W. and D. Peterson. *Demonic Males: Apes and the Origins of Human Violence*. Boston: Houghton Mifflin, 1996.

Yang, M. A. and Q. Fu. "Insights into Modern Human Prehistory Using Ancient Genomes." *Trends in Genetics* 34, no. 3 (2018): 184–196.

Young, R. W. "Evolution of the Human Hand: The Role of Throwing and Clubbing." *Journal of Anatomy* 202, no. 1 (2003): 165–174.

Zachos, J., M. Pagani, L. Sloan, E. Thomas, and K. Billups. "Trends, Rhythms, and Aberrations in Global Climate 65 Ma to Present." *Science* 292, no. 5517 (2001): 686–693.

Zelizer, V. A. *Pricing the Priceless Child: The Changing Social Value of Children*. Princeton, NJ: Princeton University Press, 1994.

Zhu, Z., R. Dennell, W. Huang, Y. Wu, S. Qiu, S. Yang, Z. Rao, et al. "Hominin Occupation of the Chinese Loess Plateau Since about 2.1 Million Years Ago." *Nature* 559, no. 7715 (2018): 608–612.

Zupancich, A., N. Solodenko, T. Rosenberg-Yefet, and R. Barkai. "On the Function of Late Acheulean Stone Tools: New Data from Three Specific Archaeological Contexts at the Lower Palaeolithic Site of Revadim, Israel." *Lithic Technology* 43, no. 4 (2018): 255–268.

Index

For the benefit of digital users, indexed terms that span two pages (e.g., 52–53) may, on occasion, appear on only one of those pages.

Figures are indicated by *f* following the page number